IEE MANAGEMENT OF TECHNOLOGY SERIES 11

Series Editors: G. A. Montgomerie
B. C. Twiss

INTEGRATION AND MANAGEMENT OF TECHNOLOGY FOR MANUFACTURING

Other volumes in this series:

Volume 1 **Technologies and markets** J. J. Verschuur
Volume 2 **The business of electronic product development** F. Monds
Volume 3 **The marketing of technology** C. G. Ryan
Volume 4 **Marketing for engineers** J. S. Bayliss
Volume 5 **Microcomputers and marketing decisions** L. A. Williams
Volume 6 **Management for engineers** D. L. Johnston
Volume 7 **Perspectives on project management** R. N. G. Burbridge
Volume 8 **Business for Engineers** B. Twiss
Volume 9 **Exploitable UK research for manufacturing industry** G. Bryan (Editor)
Volume 10 **Design in wealth creation** K. K. Schwarz
Volume 11 **Integration and management of technology for manufacturing** E. H. Robson, H. M. Ryan and D. Wilcock (Editors)
Volume 12 **Winning through retreat** E. W. S. Ashton
Volume 13 **Interconnected manufacturing systems—an overview** H. Nicholson
Volume 14 **Software engineering for technical managers** R. J. Mitchell (Editor)

INTEGRATION AND MANAGEMENT OF TECHNOLOGY FOR MANUFACTURING

Edited by

E H Robson, H M Ryan and D Wilcock

Peter Peregrinus Ltd. on behalf of the Institution of Electrical Engineers

Published by: Peter Peregrinus Ltd., London, United Kingdom

Peter Peregrinus Ltd.,
Michael Faraday House,
Six Hills Way, Stevenage,
Herts. SG1 2AY, United Kingdom

British Library Cataloguing in Publication Data
Integration and management of manufacturing.
1. Manufacturing industries. Technological development.
I. Robson, E. H. (Edward Heron), 1942-
II. Ryan, H. M. (Hugh McLaren)
III. Wilcock, D. 670.42'7

ISBN 0 86341 206 8

Printed in England by Short Run Press Ltd., Exeter

Contents

Preface

To survive in the increasingly competitive economic conditions of the contemporary world, manufacturing organisations are continually seeking ways of improving their market shares by becoming more effective and efficient and by producing goods of higher quality. It appears that advanced manufacturing technology (AMT) can provide at least some of the means of achieving these improvements. In March 1989 an international conference on this subject (SAMT'89) was held at Sunderland Polytechnic in England. This brought together not only academics who are undertaking research in the fields of manufacturing technology and its management, but also personnel who are involved in the implementation, development and on-going management of new technology within manufacturing organisations in many regions inside and outside the United Kingdom. Over seventy papers were presented on theories and concepts of AMT and its management and on the practical experiences of those who are, or have been, concerned with its implementation. This book includes forty of those papers which were selected because they concentrated on the management and strategic issues involved in the integration of AMT into a manufacturing organisation wishing to achieve optimum improvements in performance and quality.

The book is divided into four parts with distinct although overlapping themes. Each part begins with a short introduction by one of the Editors which is intended to identify a common context within which the reader can place the chapters in the part. Taken as a whole, the book provides a wide coverage of important ideas and experiences for those who wish to study the effects of AMT or who wish to effectively embrace its potential benefits. It is hoped, therefore, that it will make a contribution to an improvement in the understanding of modern manufacturing techniques.

The Editors wish to record their thanks to all of those who contributed to the success of the conference, and especially to those who are authors of the chapters contained in this book. They would also like to acknowledge the financial support given to the conference by Sunderland Polytechnic, the Borough Council of Sunderland and the Great Britain – Sasakawa Foundation. Finally the Editors are pleased to record their gratitude to Mrs. Avril Silk, who not only acted as secretary to the conference, but also patiently typed the manuscript.

E.H. Robson
H.M. Ryan
D. Wilcock

Sunderland Polytechnic, July 1989

List of Contributors

Mr A. P. Akimov
Research Computer Centre of Akademy of Science
40 Vavilov Street
117967 Moscow
USSR

Mr R. W. Baines
North Staffordshire Polytechnic
Dept of Mech & Computer-Aided Engineering
Beaconside
Stafford SR18 0AD

L. F. Baxter
Dept of Management Studies
Glasgow Business School
53 – 59 Southpark Avenue
Glasgow G12 8LF

Dr D. Bennett
Aston Business School
Aston Triangle
Birmingham

U. S. Bititci
University of Strathclyde
Dept of Design
Manufacture & Engineering Management
James Weir Building
75 Montrose Street
Glasgow G1 1XJ

Professor P. R. Bunn
Paisley College of Technology
High Street
Paisley
Renfrewshire PA1 2BE

J. L. Catchpole
University of Durham
School of Engineering & Applied Science
Science Laboratories
South Road
Durham DH1 3LE

Mr C. Chapman
Wellworthy Ltd
Production Control Manager
Netherhampton Road
West Harnham
Salisbury
Wiltshire SP2 8NN

Mr S. J. Childe
Plymouth Polytechnic
Dept of Computing
CIM Institute
Drake Circus
Plymouth
Devon P14 8AA

Dr D. B. Chuah
Teesside Polytechnic
Dept of Mechanical Engineering & Metallogy
Borough Road
Middlesbrough TS1 3BA

Mr G. H. Colquhoun
Liverpool Polytechnic
School of Engineering & Technology Management
Byrom Street
Liverpool

Mr A. J. Coppen
Bailey's Farm
Sutton-cum-Lound
Retford
Nottingham DN22 8PN

Mr C. Davis
Marketing Manager
John Brown Automation Ltd
Torrington Avenue
Coventry CV4 9XQ

Professor A. de Pennington
Dept of Mechanical Engineering
University of Leeds
Leeds LS2 9JT

S. Downey
University of Sheffield
Dept of Mech & Process Engineering
Mappin Street
Sheffield S1 3JD

M. P. Elliott
School of Manufacturing Technology
Sunderland Polytechnic
Sunderland SR2 7EE

Mr S. N. Farimani-Toroghi
Leeds Polytechnic
Faculty of Information & Engineering Systems
The Grange
Beckett Park
Leeds LS6 3QS

Mr N. Ferguson
Glasgow Business School,
Buyer/Supplier Relationships Project
Dept of Management Studies
53 – 59 Southpark Avenue
Glasgow G12 8LF

Dr W. Gerrard
Glasgow College
School of Engineering
70 Cowcaddens Road
Glasgow G4 0BA

R. Gill
Dept Mechanical Engineering
Middlesex Polytechnic
Bounds Green Road
London N11 2NQ

Mr N. Godwin
Coventry Polytechnic
Dept of Computer Science
Priory Street
Coventry CV1 5FB

Mr F. Gohari
University of Strathclyde
Dept of Design Manufacture & Engineering Management
James Weir Building
75 Montrose Street
Glasgow G1 1XJ

M. Gore
Coventry Polytechnic
Priory Street
Coventry CV1 5FB

Mr H. Hensser
Mycalex Motors Ltd,
Wilkinson Road
Love Lane Industrial Estate
Cirencester
Glos. G17 1YT

P. Hewitt
School of Manufacturing Technology
Sunderland Polytechnic
Sunderland SR2 7EE

Mr T. M. Hogg
Nissan Motor Manufacturing (UK) Ltd
Washington Road
Sunderland
Tyne & Wear

C. Holly
Dept of Mechanical Engineering
Middlesex Polytechnic
Bounds Green Road
London N11 2NQ

Dr P. Johnson
Engineering Manager
Lucas Engineering & Systems Ltd
PO Box 52
Shirley
Solihull
West Midlands B90 4JJ

Dr C. Jones
University College of Swansea
Dept of Management Science & Statistics
Singleton Park
Swansea SA2 8PP

Mrs C. M. Jones
University of Warwick
S I B S
Coventry CV4 7AL

A. H. Juri
Dept of Mechanical Engineering
University of Leeds
Leeds

C. Kimble
Newcastle Business School
Newcastle Polytechnic
Ellison Place
Newcastle NE1 8ST

Mr J. Kuvik
USIP
Hrachova 30
Bratislava 82711
Czechoslovakia

Mr R. J. Laird
University of Ulster
Jordanstown
Shore Road
Newtownabbey
Co Antrim
Northern Ireland BT37 5FB

Mr D. J. Leech
University College of Swansea
Dept of Management Science & Statistics
Singleton Park
Swansea SA2 8PP

D. M. Love
Mech & Production Engineering
Aston University
Aston Triangle
Birmingham B4 7ET

D. K. MacBeth
Dept of Management Studies
Glasgow Business School
53–59 Southpark Avenue
Glasgow G12 8LF

Mr H. Mather
Hal Mather Inc.
Management Counceling & Education
4412 Paces Battel
NW Atlanta
GA. 30327
USA

Mr K. Maughan
School of Computer Studies and Mathematics
Sunderland Polytechnic
Priestman Building
Green Terrace
Sunderland SR1 3SD

Mr C. Morris
Artix Ltd
North West Industrial Estate
Peterlee
Co. Durham SR8 2HX

Dr R. Murray-Shelley
Dept of Electrical & Electronic Engineering
Wales Polytechnic
Pontypridd
Mid Glamorgan CF37 1DL

G. C. Neil
Dept of Management Studies
Glasgow Business School
53–59 Southpark Avenue
Glasgow G12 8LF

Prof C. C. New
Cranfield School of Management
Cranfield Institute of Technology
Cranfield
Bedford MK43 0AL

Dr M. H. Oakley
Aston University Business School
Aston Triangle
Birmingham

Mr K. V. Pandya
Teesside Polytechnic
Division of Computer Studies & Electrical Engineering
School of Information Engineering
Middlesbrough
Cleveland TS1 3BA

S. N. Peck
Leeds Polytechnic
Faculty of Information Systems
The Grange
Beckett Park
Leeds LS6 3QS

Dr D. T. S. Perera
Sheffield City Polytechnic
Dept of Mechanical & Production Engineering
Pond Street
Sheffield S1 1WB

A. Pollock
Artix Ltd
North West Industrial Estate
Peterlee
Co Durham SR8 2HX

Dr V. B. Prabhu
Newcastle Polytechnic
Dept of Management & Administration
Ellison Place
Newcastle NE1 8ST

Dr S. K. Rajput
Aston University
Innovation, Design & Operations Management
Research Unit
Aston Triangle
Birmingham

C. W. Rapley
School of Manufacturing Technology
Sunderland Polytechnic
Sunderland SR2 7EE

Mr D. C. Richardson
Sunderland Polytechnic
Dept of Mechanical Engineering
Edinburgh Building
Chester Road
Sunderland

M. Richter
Fraunhofer Institute of Industrial Engineering
Nobelstrasse 12c
D-7000 Stuttgart 80
West Germany

Dr K. Ridgway
The University of Sheffield
Dept of Mech & Process Eng.,
Mappin Street
Sheffield S1 3JD

Mr A. Saia
University of Leeds
Dept of Mechanical Engineering
Geometric Modelling Project
Leeds LS2 9JT

Dr M. Sarhadi
University of Durham
School of Engineering & Applied Science
Science Laboratories
South Road
Durham DH1 3LE

Mr U. A. Seidel
Fraunhofer Institute of Industrial Engineering
Nobelstrasse 12c
D-7000 Stuttgart 80
West Germany

Professor V. V. Shafransky
Computer Centre Akademy of Science
40 Vavilov Street
117967 Moscow
USSR

Dr R. Sotudeh
Teesside Polytechnic
Division of Computer Studies & Electrical Engineering
School of Information Engineering
Middlesbrough
Cleveland TS1 3BA

Mr E. T. Sweeney
University of Warwick
Manufacturing Systems Engineering Group
Dept of Engineering
Coventry CV4 7AL

Mr K. Thaler
Fraunhofer Institute for Industrial Engineering
Nobelstrasse 12,
D-7000 Stuttgart 80
West Germany

Mr R. A. G. Twose
Lucas Aerospace Ltd
Shaftmoor Lane
Hall Green
Birmingham

Mr W. Walker
School of Computer Studies and Mathematics
Sunderland Polytechnic
Priestman Building
Green Terrace
Sunderland SR1 3SD

Dr A. Watkins
University College of Swansea
Dept of Management Science & Statistics
Singleton Park
Swansea SA2 8PP

J. Whitehurst
Westland Helicopters Ltd
Westland Works
Yeovil
Somerset BA20 2YB

Mr R. G. Williams
University College of Swansea
Dept of Management Science & Statistics
Singleton Park
Swansea SA2 8PP

Mr R. J. Wilson
Director of Operations
Westland Helicopters Ltd
Westland Works
Yeovil
Somerset BA20 2YB

Mr H. S. Woodgate
International Computers Ltd
Kings House
33 Kings Road
Reading
Berks RG1 3PX

Editors

Dr E. H. Robson
Teaching Company
Hillside House
79 London Street
Faringdon
Oxon SN7 8AA

Prof H. M. Ryan and Dr D. Wilcock
Sunderland Polytechnic
Edinburgh Building
Chester Road
Sunderland SR1 35D

Part 1:

Strategic issues

Chapter 1

Introduction

E. H. Robson

Teaching Company Directorate, UK

The advance of technology is having enormous impact on manufacturing. Not only does it continually create possibilities of developing new artefacts and of modifying existing ones, but it also provides improved methods and processes for making things. Inevitably these opportunities require a high level of decision making by those individuals and organisations endeavouring to survive in the manufacturing sector. However, progress, especially in information technology, is providing exciting opportunities for improving the ways in which manufacturing activities are monitored, managed and integrated into the other associated functions required to support a business enterprise. For example, there are the computer's capacity to accurately capture, record, manipulate and display data, the capability for high-speed transference of information between computer systems, and the ability, using microprocessor engineering, to consistently and precisely control machinery and materials flows. These give significant means to improve the overall efficiency of a complex organisation and to maintain up-to-date knowledge of the states of its constituent parts. Moreover, the rate of change of technological development implies that companies must operate in a highly dynamic environment. The important challenge for all companies involved in manufacturing is to extract maximum benefit from these opportunities. This requires them to address, in addition to more detailed and specific matters related to the implementation of AMT, a range of strategic issues, many of which are raised within the chapters of this section.

The introduction of new technology into one area of an organisation will almost certainly have an impact on other areas; indeed it may well impose considerable alterations in the ways those other areas carry out their functions. A good example of this is the implementation of a computer-aided design facility into the draughting area which usually affects the areas of design, engineering and production, as illustrated in Chapter 5.

Thus to introduce automation in a significant way into one area of a manufacturing organisation requires a careful, strategic examination of the routines and practices of that and related areas of the organisation so that the full benefits of its integration can be realised. In so many cases such examinations reveal inefficiencies, and even redundancies, resulting from complex, outmoded or badly structured operations. Thus the need for changes in working routines and practices are invariably identified. In Chapter 2, Mather is adamant that simplification of these practices is the key to success. Coppen in Chapter 3 believes that any change has to reflect a careful balance between quality and cost of the manufacturing function. In Chapter 4, Laird outlines a programme for coping with the inconveniences that such changes impose, and a technique for planning the implementation of advanced technology is discussed in Chapter 8. The other chapters in this part introduce new approaches to examining some of the more relevant strategic issues concerning the introduction of new technology, including its financial consequences. A continual high-level review of the operation of a manufacturing company in the light of contemporary developments in technology will always be necessary and, indeed, technology itself is providing some of the tools to assist with that reviewing (see, for example, the discussion in Chapter 7).

Chapter 2

Simplification before automation

Hal Mather
Hal Mather, Inc., USA

2.1 Introduction

There is much interest worldwide today in advanced manufacturing technology (AMT). The lights-out factory of the future fires the imagination of almost everyone. No question, we are moving towards factories automated far more than traditional plants. Technology, either here today or currently under development, will support this trend. And it will give a competitive edge to those successful in applying AMT well. The missing first step, though, in many automation programmes is simplification of the business process. Without this step, automated chaos is the inevitable end result. The resulting investment without payback will destroy every company except those with very deep pockets. More importantly, it will sour senior managers on any other AMT project, no matter how worthwhile. The key is to visualise a utopian business first. Simplification of all business areas will be obvious as the first step towards utopia. The second step, simple automation of a simple process, will put you ahead of the pack, earning a fabulous return from your total investment.

2.2 Where is AMT heading?

A futuristic view of a manufacturer has been suggested by Richard P. Rumelt. 'A design engineer draws a part on a video screen and enters the specifications into a clearing-house for bid. Seconds later, an acceptable bid is made by one or more local facilities, and an electronic contract is formed. Two hundred units of the specified part are delivered the next week having been made at a facility with just the right machines and job schedule to minimise their cost. This may sound like science fiction today.

But there is no doubt that we are moving towards this scenario. An order will be received electronically, designed uniquely for that customer on a computer-aided design (CAD) station, the process created on a computer-aided process planning (CAPP) system, bearing in mind the resource capacities, current work loads, and optimal schedule, instructions given to the automated storage and retrieval system (ASRS) to pick materials and tools, an automated guided vehicle (AGV) will move these to the correct processes, robots and direct numerical control (DNC) machines will produce the products and deliver them to the shipping dock. Artificial intelligence (AI) and local area networks (LANS) will tie all these elements together and make adjustments for any conflicts.

Data will move from customer to supplier electronically and flow through the supplier smoothly even to his supplier, not with the stop/start motion using people interfaces as today. We do not have all the technology to perform this routinely yet. Much of it is available but some critical elements are still missing. Current development activities will fill the gaps in the next few years.

A number of companies have tried to get ahead of their toughest competition using technology. The best American example is General Motors, using technology to leapfrog Toyota and the other Japanese competitors. Their Saturn project, a completely new company set up to produce small cars in a radically new way, is just such an example. To their misfortune, they found they were trying to leapfrog a unicorn. When you play this game, you had better be sure of yourself before you jump. The stakes are too high to risk not clearing the objective. Even though General Motors has invested billions of dollars, they have been forced to reduce the size of the Saturn plant to produce half as many cars and have pushed back the start-up date by several years. The technical difficulties were much more than anyone at the company expected. Contrast this with the General Motors/Toyota joint venture in California. It is the most efficient of all General Motors plants even though it is the least automated. The simplification strategy used at this plant is really paying off. And so does the limited automation, only used where it gives a provable competitive edge.

2.3 Visualise utopia

A utopian firm can be visualised as a series of pipes. Products flow smoothly and continuously from vendors into the plant, through the plant and out to customers. Information flows smoothly in the reverse direction. It directs all the product flows to stay in balance and for each product to stay in the correct sequence with other products.

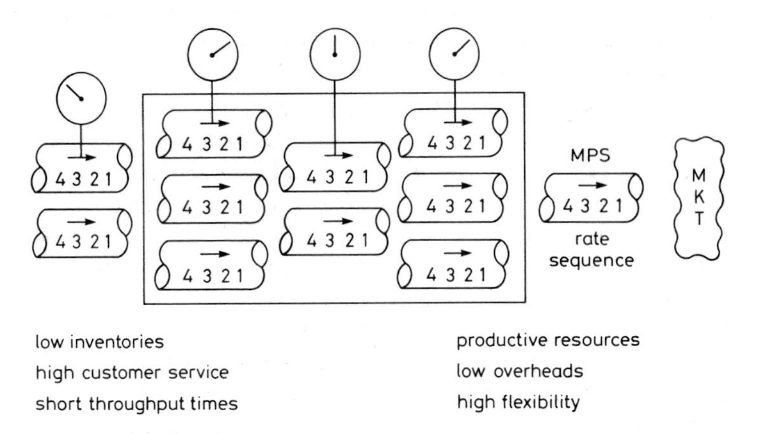

Fig. 2.1 The oil refinery analogy

Fig. 2.1 shows such a picture. I have labelled it the 'oil refinery analogy' not because I am holding up oil refineries as the epitome of good manufacturing but to reinforce the liquids flowing concept. On the right is the marketplace, drawn to suggest a dynamic entity, something like an amoeba that forms, breaks apart and reforms again. The rectangle is the business, the rectangle suggesting an inherently more rigid entity than the marketplace. The inherent rigidity comes about because changes in staffing, machinery and space take time to implement. Decisions to change these resources are not entered into lightly. The linkage between the rectangle and the amoeba is an output plan for the business. This is shown as a pipe of product leaving the plant. The pipe's diameter represents the needed rate of output, the numbers 4, 3, 2, 1 signify the sequence products that are needed in the marketplace. The needed rate of output, (which could be expressed as dollars/month, units per week, tons or hours/day etc.) is used to evaluate the upstream resources. These are represented as pipes either within the rectangle, hence work centres, or pipes flowing into the rectangle, hence vendors. The evaluation is to check whether all these upstream resources are capable of flowing product or information at the rate needed to support the output pipe. If discrepancies are found, the only two choices are to change the flow-rate capabilities of upstream resources somehow or change the output pipe's rate to suit the feeding pipes. Product and information flow can be affected by many factors; for example, scrap, rework, absenteeism, machine breakdowns etc. Hence planning adequate flow rates through all pipes is a necessary but not sufficient activity. Measuring actual flow rates and reacting to the inevitable discrepancies is mandatory to get a balanced flow of product through all pipes. This measuring and reaction activity is represented by the gauges linked to the pipes.

Now that the flow rates are kept in balance throughout the system it is time to ensure that products are being made in the correct sequence. Flowing the right amount of the wrong things is of no value. The numbers 4, 3, 2, 1 in each pipe signify that all resources are making the right things at the right time to support the output pipe. A utopian factory is one where the flow rate and sequence of flow from the vendors is exactly balanced and synchronised to the flow rate and sequence in the primary work centres, the primary work centres' flow rate and sequence is balanced and synchronised to the flow rate and sequence in the secondary work centres, and so on through the plant, the last work centre's flow rate and sequence being balanced and synchronised to the output plan that supports the marketplace.

A balanced flow and synchronised sequence through all resources will give extremely low inventories, equal to the processing time of the product. This is less than a day for most products, so three-digit inventory turns should be easily possible. High customer satisfaction will result from the perfect execution of the output plan. Throughput times will be short because of the low inventories. High flexibility to marketplace changes will be a by-product of the short throughput times. Low overheads will be because the process is under control. Only value-added activities will be performed. And the resources will be truly productive because they will only be making exactly the right amounts of exactly the right things customers need in the short term.

2.4 Roles of environment and technology

The utopian picture just described is what industry has been moving towards for many years. All management technologies and philosophies, for example, MRP II, JIT and CIM, have this scenario as their end objective. How do you get there? The answer is a combination of simplifying the environment and applying technology. The right combination is critical. As an example, a computer controlling stop lights in a congested city centre does not do much for the traffic flow. The environment prevents technology being the solution. Contrast this with an interstate highway away from any large cities. The environment of divided highways and limited access controls the traffic flow without technology being needed. The same condition exists when applying AMT. Plants without adequate environmental conditioning need help. But AMT will *not* solve their problems. AMT will be as impotent as computer controlled stop lights downtown. Simplifying the environment first then applying AMT will be a winner.

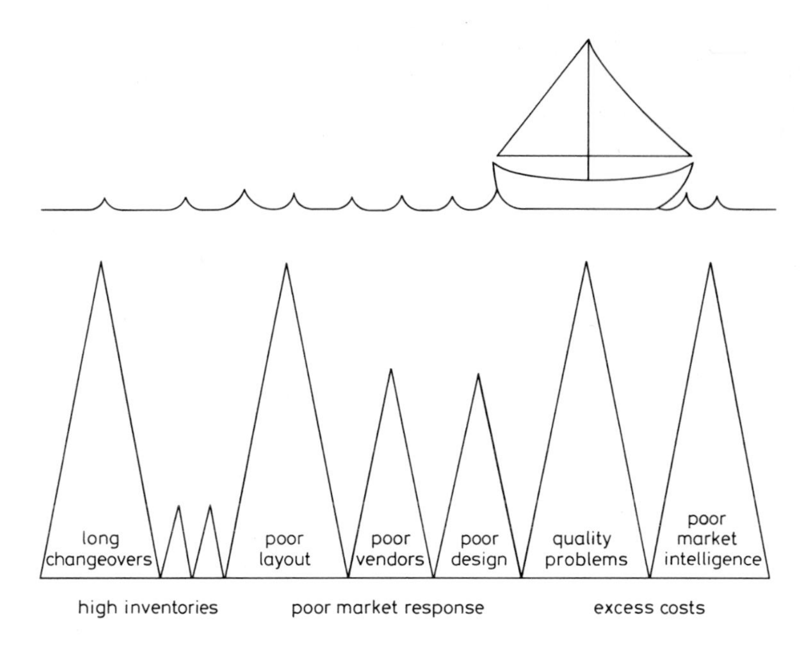

Fig. 2.2 The lake and rocks analogy

The lake and rocks analogy from Fig. 2.2 is a good one to show the correct approach. A boat can sail smoothly on a lake full of rocks as long as the water level is high. The depth of water is analogous to inventory. The rocks cause poor market response and excess costs. Only a few rocks are shown here but there are literally hundreds of them. And they exist in every department. Sales policies and practices; accounting rules, budgets, reports, and costing methods; product variety and design architecture; machine breakdowns, restrictive work rules, inaccurate data, other manufacturing problems these are only a few examples of the problem that push a company away from utopia. Some companies' attempts at reducing inventories, improving customer service and reducing costs have been to implement bigger and more complex systems. But these have not reduced the rocks, they have let the company 'sail around' them. The risk of running aground is high, as is also the cost of operating this complexity. And the realised benefits fall far short of the potential. The key is SIM before CIM as shown in Fig. 2.3. SIM is an abbreviation for simplification. The complexity of CIM is a direct function of the simplicity of the operation. And the complexity of CIM is indirectly proportional to its chance of success. A JIT factory is one where simplification of the environment has progressed well in all areas. Referring back to Fig. 2.2,

JIT crushes the rocks and eliminates them. Costs now decrease, complexity decreases, inventories drop and customer service improves. Simple automation can now give incremental benefits with high payback. You are on your way to utopia.

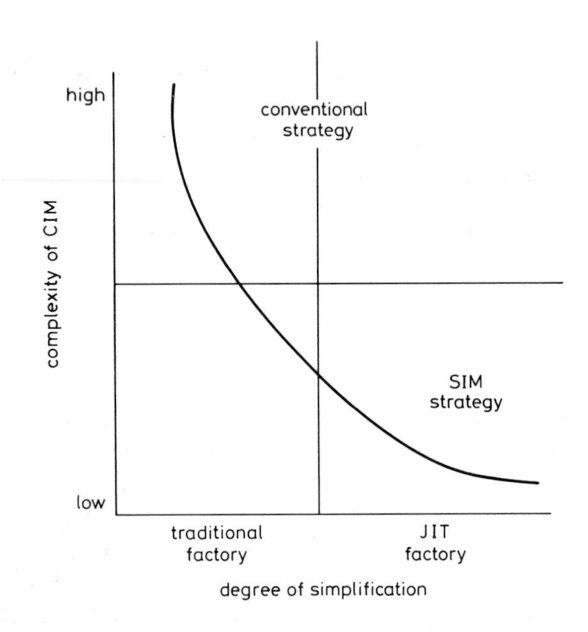

Fig 2.3 SIM before CIM: the benefits of simplification

2.5 Summary

Automation of manufacturing has been in progress since the advent of steam power. The pace of automation has accelerated in recent years due to computer technology and the silicon revolution. Existing technology plus some under active development will expand automation potential to far more areas. Computer-integrated enterprise is fast approaching feasibility. The pace of technology is outstripping the environmental preparation to successfully apply the new technology. The manufacturing process, product design, strategy to use AMT well, sales practice and many more all need critical examination and revision. Simplification of the process *must* precede automation of these processes. Once simplified, a company can determine the best mix of technologies to maximise profits, whether human intensive, capital intensive or automation intensive, and where to apply them. Simplification before automation have to be the watchwords. And they are the only words to guarantee a competitive advantage from automation.

Chapter 3

Towards a Manufacturing Manifesto

A. J. Coppen
Teaching Company Directorate, UK

3.1 Introduction

Until the past few years the performance of a manufacturing activity and management of many companies were judged mainly in terms of cost and conformance to budget. There is, now, much greater emphasis on the contribution that the manufacturing function can make to the company's ability to win orders. The function is expected to be effective as well as being efficient. Pressure from the market is demanding ever higher levels of quality and delivery performance together with greater flexibility and shorter lead times. The acceptable levels of inventory in many companies are being reduced at the same time. These objectives are not independent of each other and trade-offs between them are often necessary. The optimum balance between these objectives will be different for each company. It is suggested that a clear statement of the requisite balance between these objectives for the company — the Manufacturing Manifesto — would provide guidance to manufacturing managers who have now, inevitably, to make judgments between conflicting objectives.

3.2 The changed role of manufacturing management

In the past many companies regarded their manufacturing operations as an unavoidable but necessary cost. However, it is now becoming recognised that the performance of this activity can be, and often is, a very important factor in the company's ability to attract orders. Customers are, increasingly, needing to be convinced that their suppliers will deliver products on time without any deviations from their specification before even considering that company as a potential supplier and negotiating a price (or so they say). In many companies the performance of the manufacturing activities has, until the past few years, been assessed by the cost of producing the product and by their conformance to the budget. Quality and delivery have tended to be afforded lip service and records of the company's performance in these areas are, surprisingly, often not available, even now. The investment in inventory and plant under the control of the manufacturing function, which typically amounts to some 80% of the company's capital, has also not, until the recent past, been a factor that received much attention from the manufacturing manager.

It is not surprising that manufacturing managers, whose performance was measured almost entirely by their conformance to a cost budget, should have given priority to cost and have concentrated on being efficient. They have made less effort on contributing to the company's effectiveness in the marketplace by ensuring a high level of product quality and a reliable delivery performance. Neither have they been too concerned with controlling the capital tied up in the activity. The situation now is very different. Pressures from the market are forcing manufacturing managers to find ways of improving their quality and delivery standards and, in addition, many are expected to reduce the amount of capital tied up in inventory. Many companies are cutting their inventories and are expecting more frequent deliveries of smaller batches of product from their suppliers. The lower level of stock now held by many companies gives them a much smaller buffer from which to draw when their customer changes his order schedule. This, in turn, usually initiates a change of order schedule back to his suppliers. Thus the manufacturing manager has yet another factor to consider — he must be able to respond effectively to changes in his order pattern; changes in specification, batch quantity and delivery date. Product life cycles are now much shorter and the ability to get new products to the market quickly is essential. This is yet another additional problem for a manufacturing manager.

Thus the manufacturing activity now has a major contribution to make to the *effectiveness* of the company in the market by the standards of quality and delivery it maintains relative to the competition, the flexibility with which it can respond to customer demands and the lead time it can offer. However, there is still a need for the manufacturing activity to be *efficient* by controlling and reducing both operating costs and the capital tied up in manufacture despite the smaller batch sizes that are now necessary. These objectives are not mutually independent nor can all of theses be achieved completely by a company — there has to be a trade-off between them. For example, the achievement of high levels of effectiveness usually adversely affects the efficiency of the manufacturing activities.

The relative importance of each objective to a company depends on the corporate aims of a company, the market(s) that it serves, the availability of funds etc. For example, the balance of importance between these factors for a company acting as a sub-contractor to the motor industry is likely to be very different from that of a supplier of machine tools. This balance between these objectives will alter when changes in the market and policies of the company occur. However, many companies do not recognise nor address this need to strike a balance between their objectives for their manufacturing function and do not set a clear policy on this balance. It is often left to the executives of the company to make *ad hoc* decisions when balancing the importance between the measures of effectiveness and efficiency without having a coherent company policy to guide them. Such a policy would be the Manufacturing Manifesto.

3.3 Measures of company performance

When setting the Manifesto for a company's manufacturing activities it would be very helpful if there was a single measure of performance for a company that is objective and stable. Probably the best that is available is the price/earnings (P/E) ratio used by financial institutions. This is based on many subjective elements such as fashion, sentiment, judgment of potential growth, and the effectiveness of the company's financial performance, as well as the more objective measures such as the history of growth of the company's turnover, the record of the return on the assets employed and the yield.

The manufacturing function can make only very minor contributions to the subjective elements of the judgment of a company's P/E ratio but can have a large influence on both the return on investment (ROI) and also the growth in the turnover of the company. Growth and ROI can also be used as measures of performance for divisions of a company where a P/E ratio is not available.

Thus the manufacturing function of a company can improve the performance of the company or a division by:

- *(a)* helping to increase sales by contributing to the company's effectiveness
- *(b)* reducing the amount of investment in the manufacturing operations, and by its traditional activity of
- *(c)* reducing the operating costs and becoming more efficient.

The balance of importance between these factors provides the basis for developing the Manufacturing Manifesto.

3.4 Manufacturing Manifesto

Like the Marketing Mission, with which it overlaps, the Manufacturing Manifesto is a definition of policy for the function. It is not a methodology or a technology nor is it a solution to manufacturing problems, such as CIM, JIT, MRP or OPT. These might, however, provide alternative approaches to the achievement of the Manufacturing Manifesto. Companies develop their manufacturing capacity, capability and competence to suit their current product range. However, the boundaries of these characteristics often are not defined clearly. Changes in the range of products to be manufactured resulting from a corporate planning exercise can move production into areas of lower manufacturing competence and capability without this being obvious to those in the company — apart from those in the manufacturing departments. The definition of a new and different Manufacturing Manifesto for the revised corporate plan will reveal to the planners that changes to manufacturing technology and methodology are likely to be necessary and the resource implications can easily be recognised.

It is clear that there should be congruence between parts of the Marketing Mission statements and the Manufacturing Manifesto, both forming components of the company's Corporate Plan. The misunderstandings that can occur between these two functions as to each other's tasks and operating limits should be reduced. Similarly, the Manifesto should help communication between the design and manufacturing departments as to the range of products and their features that manufacturing has to be able to make. Each of the factors constituting the Manifesto should, wherever possible, be quantified although it may be difficult to quantify some of the factors such as the range of product features to be offered to customers.

3.5 Illustrative example

To illustrate this approach, a comparison of the Manufacturing Manifestos of two hypothetical companies — one a supplier of motor car components and the other a manufacturer of large capital plant — is given in Tables 3.1 — 3.3

Table 3.1 Manufacturing Manifesto: Market driven factors

Company product	Car components	Large capital plant
Capacity	$10^2 \rightarrow 10^4$ Components/week	Size Product range Manhours
Delivery		
1 Period	Weekly schedule	Several months
2 Conformance	Vital to deliver on time	Some latitude often acceptable

Table 3.1 *Continued*

Company product	Car components	Large capital plant
Quality conformance	S Q C necessary and conformance to customer quality system specification	BS 5750, customer code and performance to specification
Lead time	12 months or so usually	Included in delivery time
Flexibility 1 Delivery	Significant and frequent schedule changes	Design/specification changes during manufacture
2 Product	Narrow range	Wide range

Table 3.2 Manufacturing Manifesto: Internal factors

Company product	Car components	Large capital plant
Cost	Very tight	Performance can influence
Capital/cash flow	Control of WIP	Long lead times→High WIP

Table 3.3 Manufacturing Manifesto: Main discriminating factors

Company product	Car components	Large capital plant
	1 Quality	1 Price/performance
	2 Price	2 Delivery/lead-time
	3 Delivery/ performance	3 Product flexibility

Chapter 4

Seven steps to technology change

R. J. Laird
University of Ulster, UK

'Change is not made without inconvenience, even from worse to better'

Richard Hooker, 1554 – 1600

4.1 Introduction

In this modern technological age it is frequently asserted that everything is changing and that the rate of change is increasing. Some fear it, some tolerate it while others welcome it, but whatever the approach, change occurs. Perhaps the attitude to change is related to Hooker's observation in the sixteenth century about inconvenience, an aspect of life which few relish. To minimise inconvenience during the process of change must be a management objective. Change is going to take place for a variety of reasons. Increased output, reduction in unit costs, improved efficiency or better quality are only a few of them. The introduction of new technology into a complex manufacturing environment calls for skills of planning and leadership. There will be financial consequences, effects on human and physical resources, continuity of supply and quality of the product or service. To minimise inconvenience and to maintain continuity requires management to take a considered approach to the change to new technology. This chapter examines the steps in this approach as advanced by managers themselves.

4.2 Source of information

The views of top executives and senior managers are presented in this work. They were interviewed in their companies during 1987 as part of a skills survey undertaken for the Northern Ireland Training Authority (Doherty *et al*, 1987).

One hundred companies which had recently introduced new technology were selected for the survey. They ranged in size from 40 to over 4000 employees and came from the industry sectors of food and drink, engineering, textiles, clothing, general manufacture and service. The companies were located in all areas of Northern Ireland. Their involvement with new technology varied. Some had introduced a single computer numerical control (CNC) machine while others had invested heavily in automated materials handling equipment or computerised workshop data collection. The companies had introduced new technology in some form either as production and process equipment, material and equipment handling or computer systems in management. The essential aspect was that they had recently experienced a change from traditional methods to using new technology in some part of their operation.

4.3 The steps

These steps detail the features which the interviewees considered to be important to their successful change to new technology. Companies differed on the relative weighting placed on each step but they believed that the steps must be considered as part of this process.

4.3.1 Step 1: Keep-up-to-date with developments

To keep abreast of technology developments in one's own field of interest was seen as an essential requisite for the selection of equipment. While it was recognised that the use of specialist consultants could help with this task, there was a basic need to be aware of developments in the market-place. Managers in small and medium sized enterprises saw this process as being very difficult owing to them fulfilling several functions. Nevertheless, it was essential to keep up to date. Some large companies had assigned personnel to continuously monitor and investigate appropriate new technology, thereby relieving line managers of this responsibility. A variety of means of keeping up to date were recommended. Trade fairs, shows, exhibitions, industrial visits and short specialist courses were all rated highly. Professional journals were also recognised sources of information.

4.3.2 Step 2: Obtain good advice

Keeping up-to-date is a necessary background but it does not select equipment or systems. It had to be supplemented by good advice. This was recognised to come from other users, consultants, suppliers and from within their own organisation. Aspects which should be investigated included:

(*a*) the capabilities and operation of the equipment to ensure that it is suitable for the task
(*b*) technical and maintenance aspects with a costing
(*c*) provision of spares and maintenance from the supplier
(*d*) the service available from the local agent including range of services, expertise, stability, sense of urgency and cost
(*e*) siting of the plant for maximum benefit involving support services.

It was recognised that there were many sources of advice from within organisations which had not been fully utilised in drawing up the equipment specification. Some suggested that two independent consultants should be used for large investments.

4.3.3 Step 3: Inform personnel

Informing people and obtaining their support for new technology was rated highly by many managers. It is frequently the case that industrialists can cite examples of new equipment lying unused because the human aspects were not managed effectively. Two different approaches were distinguished for the exchange of information:

(*a*) Consult employees to obtain their suggestions and involvement in the change.
(*b*) Inform employees before equipment is installed.

Both methods were agreed to be effective. Consultation was claimed to yield useful input and foster good industrial relations.

Techniques of information transfer varied. Three distinct methods were used:

(*a*) Use the existing management organisation to pass on information and to obtain view points. It was argued that this method reinforced the manager/subordinate link especially when much information transfer occurs through the Trade Union system.
(*b*) Special briefing sessions at which all the personnel involved in planning and implementing the change are present. This allows all questions to be taken and answered on the spot by those who know the answers. It was believed to be an efficient means of briefing and of dispelling rumours.
(*c*) Use consultants to brief employees about future changes. Since the consultants have investigated the situation fully they are best able to take technical queries. Management representatives should also be on hand to answer domestic questions.

It was in this step that qualities of leadership, honesty and frankness were called for to give confidence to the employees and enlist their support.

4.3.4 Step 4: Purchase well known equipment

A prerequisite for success was argued by some managers to depend on obtaining well known equipment. By this it was inferred that well known equipment will be reliable in service, will have been fully developed and will be easily maintained with a comprehensive support service. Some emphasised that it was essential for new technology to be reliable and not to jeopardise the earnings of employees by giving persistent faults. Frequent stoppages would hinder the acceptance of new technology into traditional manufacturing environments and would be counter-productive to successful change. A number of companies which had bought less well known brands of equipment or which had developed their own equipment using production staff strongly advocated this step.

4.3.5 Step 5: Select the team

To ensure that changes in new technology were successful it was considered essential that the people who were involved with its introduction and operation were carefully selected. There was a need to demonstrate that new technology improved the conditions of the employees either by increased earnings potential or safeguards against redundancy. It was necessary to get off to a good start by the effective use of the equipment by people who were interested in it and well motivated. The process of building a suitable team required management to take a long term view at their recruiting policy in addition to the immediate manning of equipment. Key factors are:

(*a*) Recruit personnel of higher educational ability and train them to accept new technology appropriate to the business.
(*b*) Involve in the team personnel who are well motivated, keen to learn and who are not afraid of new technology.
(*c*) Involve younger people as they are more receptive to the concepts of new technology.
(*d*) Recruit more and better trained engineers into the organisation to improve the understanding and application of the new equipment.
(*e*) Select operators who are capable of increased output. They can be used to demonstrate that higher earnings levels can be achieved. This will encourage future support for change to modern facilities and equipment as it will be seen that earnings can be improved thereby.

4.3.6 Step 6: Train personnel

This step was the most important of all; successful change required that personnel should be trained to an adequate level. This requirement applied to supervisors, technicians, staff and particularly operative and maintenance groups.

The training of operators was approached in several ways:

(*a*) Train before installation by the supplier, manufacturer or consultant.
(*b*) Train a company instructor at the supplier's plant before installation. Use this instructor to train employees after equipment installation at own premises.
(*c*) Train employees after installation on one's own premises by using the supplier or consultant.

Each approach had its merits but the emphasis must be on obtaining a proficient employee to operate the equipment as soon as it is operational. The other key training area was for maintenance staff. Four different approaches were distinguished:

(*a*) *Before installation:* Train a number of maintenance staff at the manufacturer's plant. This has the advantage of having resident maintenance ability from installation. On the other hand the training would be unlikely to cover the variety of faults which could occur during the early period of operation.
(*b*) *After installation:* Train at the manufacturer's plant at some time between six and twelve months after installation. This approach would allow the maintenance operator time to become familiar with the equipment and then to obtain more benefit from a course by asking the relevant questions. It has the disadvantage in that there may be no resident maintenance expertise immediately following installation.
(*c*) *Assist with the installation* and train at six to twelve months thereafter at the manufacturer's plant. This approach may overcome some of the disadvantages of the previous method.
(*d*) *No training:* Small companies with limited resources opted for this tactic, particularly for computing equipment which was supported by a local service centre. The nature of the equipment and the quality of local service support are essential considerations when deciding to follow this course of action.

4.3.7 Step: Introduce at a suitable pace

The rate of change to new technology was considered to be an important factor contributing to the success of the exercise. Emphasis was placed on changing at a pace which allowed each phase to be fully consolidated before advancing further. No particular duration was suggested as this was related to many factors which were different for each situation. Among these were the complexity of the equipment, the attitude of the employees, the extent of training at all levels and the degree of planning to integrate the new equipment into the plant.

4.4 Effects of change to new technology on employment groups

During and after the change the behaviour of different employment groups was noted in most companies. This behaviour gave management valuable indicators to the way in which they had managed the change and to aspects which should be considered in future changes.

4.4.1 Production employees

Three main areas of interest were identified for this group:
Job security: It was necessary for management to make an early declaration on the job security issue. Production employees preferred to know if redundancies would have to occur or if continued full employment would result. Lack of a clear policy statement on this issue allowed rumour to spread, discontent to set in and a negative attitude towards change to new technology; all of these being counter-productive to the fundamental reasons for change. Given this clear lead employees and Trade Unions were willing to accept new technology, even though accompanied by minor redundancies, as their own longer-term job security was enhanced.
Earnings: In traditional manufacturing methods the quantity of output depended largely on the industry of the operator. Consequently, their earnings could be boosted by extra effort in production — incentive work arrangements. Changes to new equipment altered this situation to one in which output was more dependent on the machine performance. To ensure that it was seen that new equipment could boost earnings management selected operators of the right aptitude and potential to demonstrate this effect. Favourable responses usually followed. Linked with this question of earnings was the question of machine serviceability. It was unacceptable from the operator's viewpoint to have a new machine which was unreliable, difficult to maintain and produced faulty output; earnings were adversely affected. It was therefore necessary for management to pay particular attention to Steps 4 and 6 so that operators would welcome new equipment.
Retraining: To avoid redundancies many employees had to accept retraining to be able to operate the new equipment. Traditional skills were replaced by skills involving machine setting, material presentation and machine minding. Consequently, training periods were short and specific to the equipment coming into service. The importance of this aspect to successful change has been emphasised by Step 6.

4.4.2 Supervisors

Too often it has been the case that management have concerned themselves rightly with the needs of the labour force only to overlook the importance of supervision. At the more junior level of management, supervisors have to be involved in the changes, and to have the opportunity to express their views and receive training appropriate to their function. The behaviour of management, in particular Steps 2, 3 and 6, generally produced the deserved reaction from this group. Ignore suggestions, uninform and forget to train can produce a negative reaction from a group which is very much involved in the exercise. It was observed that there were several important characteristics for a supervisor of new technology. These were:

(*a*) Have a good knowledge of the capabilities of the equipment.
(*b*) Understand the nature of the operation being performed.
(*c*) Understand the relative effects of operator and equipment on the quantity and quality of output.
(*d*) Respond more quickly to problems which arise in relation to the equipment; this was necessary owing to higher production rates.
(*e*) Organise and arrange faster material input and component output flows for the equipment; this was related to the observed performance of the operator, earning levels and increasing efficiency.
(*f*) Possess higher academic and technical qualifications to be able to understand the effects of new technology on output, the manufacturing environment and human behaviour.

These characteristics provide pointers for the selection and training of future supervisors.

4.4.3 Technicians

This group was heavily involved in the change to new technology. To be effective, especially in small companies, the technician required a variety of abilities such as:

(*a*) a good education in relevant subjects
(*b*) multi-skilled ability in mechanics, hydraulics, pneumatics and electronics; the latter assumed greater importance owing to the sophistication of control systems
(*c*) a knowledge of the operation of the equipment.

With this type of background a technician in the maintenance function was required to diagnose faults quickly and effect repair. The value of a well qualified and trained technician was appreciated, especially in the non-engineering sectors of industry. Frequently, he was the only individual with a detailed electrical and mechanical knowledge to maintain equipment in service.

4.4.4 Staff

The effects of change to new technology on this group refers mostly to the introduction of computerisation. Included within it are professional and clerical grades from all support and administrative functions in the organisation. Management recognised the importance of Steps 3 and 6 to obtaining the involvement and support from this group. Although some older members were initially hesitant it was reported that favourable reactions were obtained. In fact some companies observed a degree of elitism among employees who worked with the new technology and a strong desire in others to become involved also.

4.5 Consequences for maintenance

The maintenance activity has been referred to in several parts of this chapter. It was generally agreed among interviewees that this function now assumed greater importance. There were several reasons for this:

(a) *The cost effect of down time* Owing to higher production rates and considerable reliance on computerisation it was essential for equipment to continue in service. Stoppages were expensive in terms of lost production or delay in information processing.

(b) *The cost effect of incorrect adjustment:* To have production equipment wrongly set would soon result in large quantities of reject output. When companies only required production operators to be machine minders the responsibility for setting rested with the maintenance function. On other occasions the process of adjustment may have been outside the ability of the operator and may have required special tools and involved methods.

(c) *Sophistication of new equipment:* Previously production equipment would have been predominantly mechanical or electrical, but new equipment also requires a knowledge of electronics and sometimes other advanced disciplines. This effectively removes from the operator all maintenance activities other than the simplest and places them with the maintenance technician.

Companies solved their maintenance problems by a combination of:

- adequate in-house spares cover
- necessary in-house skills
- reliance on the equipment supplier.

It was recognised that new equipment was improving in reliability but that fast and efficient repair were essential to maintain output.

4.6 The effect on quality

Change in some cases was initiated to obtain better quality in the product or service. This was assessed as providing a wider range of services or producing a product to a tighter specification more consistently. Companies were able to compete in new markets as a result of their modernisation. New equipment also produced other benefits within the organisation. Some of these were:

(*a*) reduced inspection levels on components, which were produced to a more consistent size
(*b*) more careful planning to prevent errors in the information to the equipment as this could result in considerable loss
(*c*) supply of more consistent material or information which forced other support activities to provide a better quality service
(*d*) improved methods of material handling and presentation. In addition to higher throughputs new equipment demanded consistency in input, thereby imposing a discipline on support services
(*e*) better business management resulting from timely and detailed information.

4.7 Conclusion

No sensible management will embark on change to make conditions worse. Yet how often do conditions deteriorate before they get better. In a modern competitive economy managements can no longer afford set backs which may have been tolerated in past times. The expectations of customers, the demands of employees and the sophistication of technology itself requires the process of change to be carefully considered, well planned and expertly led. Managers themselves have given their views based on their experiences. It is relevant to note that the vast majority of them (89%) considered that their changes were successful and characterised as smooth and uneventful. They succeeded in minimising inconvenience by knowing their operation and possible solutions, involving their people and following the seven steps to technology change.

Chapter 5

CIM — experiences on the road to integration

C. Morris and A. Pollock
Artix Ltd., Peterlee, UK

5.1 Introduction

In 1973 a new company, DJB Engineering, was formed at Peterlee in north-east England. Its aim was to design and build an articulated dump truck for the mining and construction industries. It rapidly grew from a handful of original employees to become the world's largest manufacturer of such products. In 1985 the rights to the product range were bought by Caterpillar Inc. It was realised early in the life of DJB Engineering that some degree of computer assistance was needed. Whilst manual methods may be sufficient to control the design and manufacture of low volumes, they are unsuitable as volumes rise. The management of the start-up company was sufficiently aware to realise that only on-line, real-time systems would satisfy their needs — a difficult requirement to satisfy in the mid 1970s. At that time the only such systems available were mainframe based, at costs which were much too high to be considered. For instance an IBM system with 16 terminals was quoted at a price of £800K. The alternative was to use a minicomputer based system which was targetted to on-line usage but for which all the application software had to be generated.

The major minicomputer supplier then, as now, was DEC supplying PDP 11 based systems programmed in BASIC. This approach was rejected on two grounds. The hardware growth path could have been limited as the PDP 11 was a 16 bit machine and the first of the 32 bit machines was appearing. Secondly the programming language of BASIC was rejected in favour of COBOL.

The supplier chosen was Interdata, having since changed its name to Perkin Elmer Data Systems and then to Concurrent Computer Corporation. However, the original choice has proved to be valid as all hardware and software have been transparently upgradable. Software written in the early days will still execute without re-compilation or linking in spite of many upgrades to operating system and hardware. The system software also needed considerable enhancement. It was realised at inception that transaction processing systems were communication systems rather than computer systems. Hence a system software environment was created which used the concept of server and client programs whether for a single processor or a network. Thus processors for single or multiple functions can be added as required to boost the computer power available.

Another early decision was that data processing support should be provided as a central cost. No department or user has been directly charged for the software development or operating services required. Overall control has been maintained by a management committee which allocates the use of resources. This has resulted in the optimisation of software development usage in the most efficient manner for overall company operation. The result has been the development of a system of application software which is fast and efficient in its assistance to users. One of the earliest aims was to provide a response to a terminal query in less than a second. This has been found to be readily and consistently achievable with all but the most complex of transactions. This has furthered use of computer-based systems as it enhances the objective of making the systems the easiest and fastest way to record or obtain information.

5.2 Application system development

The basis of all production systems is the engineering database. This is created and maintained by the Engineering Department. Currently a single database for both Engineering and Production needs is used although the higher levels of the product structure have to be duplicated by Production because of conflicting requirements. The debate on whether structures should be shared or duplicated has been proceeding almost from the start of the systems development. Additional information is added for the benefit of the Production function. This ranges from the controlled introduction of changes and new products to the quantity and location of parts held in stock. Also information is held to control the inventory, issue and usage of these parts. This information gave the basis for a Material Requirements Planning (MRP) system which was introduced in 1980.

Standard routing information has since been added representing the sequence of manufacturing procedures required by each part together with the time, manpower and machine resources required. This obviously allows production capacity management calculations to be performed. It is also used as the basis for a production monitoring system, loading and tracking production batches through the plant. Supporting systems cover the purchasing function, an integrated sales order system for service and spare parts plus all the necessary financial systems. The whole is based on a database system requiring information to be entered once to be available for all functions needing access. The result has allowed movement to a closed loop material resource planning system (MRP II). This optimises use of resources within Artix, allowing for higher volumes of production together with higher levels of product variation without proportional increases in clerical staffing.

5.3 Hardware systems development

As mentioned earlier, computer systems are based on hardware from Concurrent Computer Corporation. The following network is now established:

Peterlee site

CCC 3230 (4Mb) Database 56 users 36 parts
CCC 3210 (4Mb) User 72 users
CCC 3230 (3Mb) Development 24 users
CCC 3205 (2Mb) Communication SNA Gateway

Stockton site

CCC 3220 (1Mb) Database
CCC 3205 (1Mb) User 30 users 12 parts

Luton site

CCC 3210 (4Mb) User + Dbase 16 users 12 parts

Network

Ethernet V1 + IEE 802.3 using fibre + copper + 4 bridges to V29
The topology is shown in Fig. 5.1

5.4 CAM then CAD?

Until 1987, the company was not convinced that CAD was cost effective. However in 1983 a CAM system for sheet metal profiling was in use. Steel is bought in flat plates and must be cut, or profiled, to the necessary shape, processed further and then welded to make the required part. For reasons of economy many varied parts may be made from a single plate and the process of laying out these profiles is difficult and time consuming. In addition there were three different types of machine tool, all requiring different methods of programming. With no existing single programming system available, suitable graphics based software was commissioned on DEC PDP ll computers and has proved to be extremely effective.

When it was decided to move to CAD systems the availability of an integrated CAM facility was required. It was thought that this would ease

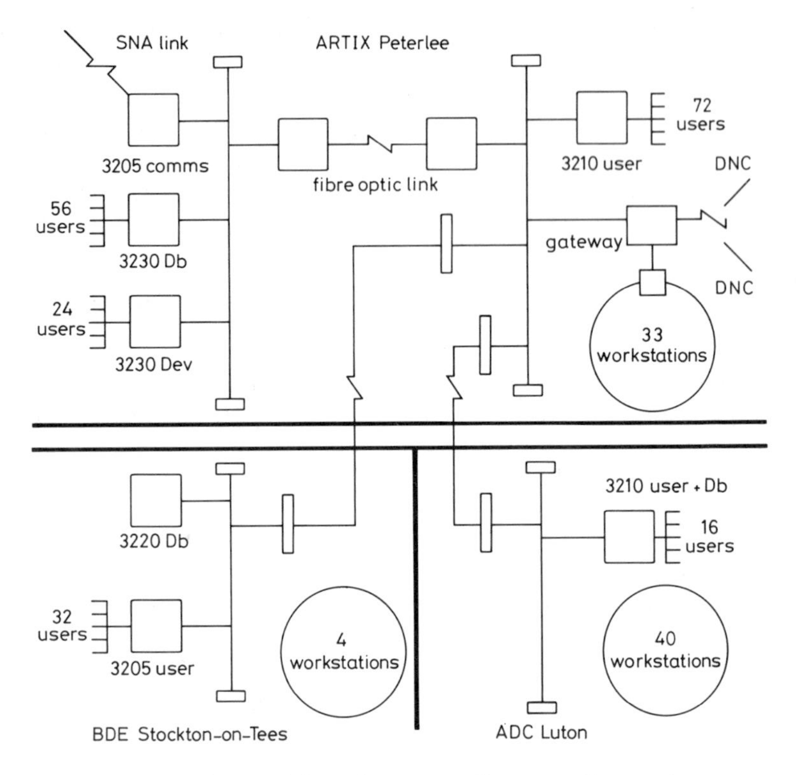

Fig. 5.1 Topology of network from Concurrent Computer Corporation

the transfer of drawings to production and speed the introduction of new parts. Many software products were reviewed and the MCS product, Anvil 5000, was chosen. The choice of hardware platform was investigated in parallel with the software selection. A workstation network was preferred as being more cost effective, flexible and easily expansible than a mini or mainframe based configuration. A network of workstations is now in use.

5.5 The justification for CAD/CAM at Artix

The introduction of CAD at Artix was seen as an opportunity to automate some of the design functions of the Engineering Department. The benefits of CAD as an electronic drawing board and pencil were well recognised. However, it was expected that a major proportion of its future benefit would stem from the ability to transfer design data (created at the draughting stage) directly into the production control system. In this way the efficiency of the Engineering Department would be increased in three areas:

(a) Data would be entered once, by an individual designer, and subsequently manipulated by computer, thus reducing repetitive input and the risk of error.
(b) The amount of administrative work carried out by Engineering in support of the design process would be reduced.
(c) The quality and consistency of data would improve, owing to stricter enforcement of standards by software and more incentive to use the routines provided.

In order to implement the above, it was recognised that some form of automatic control of computer-based drawing files would be required. Database access software would also be required so that designers could display drawing information on their CAD workstations as accessory functions to the draughting.

5.6 The paper-based design and draughting system

Artix currently produces all ADT design work exclusively for Caterpillar. Both design and draughting aspects are hence strongly influenced by existing Caterpillar engineering standards. Drawings are produced within Caterpillar borders on sheet sizes ranging from A to E (approximately A4 – A0). The part number system consists of (usually) a six digit alphanumeric identifier, followed by a change level indicating the version of the part. For example:

7U1234 – 00 indicates the first version of a standard production part
7U9867 – 03 indicates the fourth version of a standard production part
7U4569 – A indicates the first version of a development part

When making a change to an existing part, if the type and application of the part are unchanged (e.g. the enlarging of a hole to make fitting the part easier), then the same part number is used, but its change level is incremented. This is carried out by retrieving the master paper copy from the drawing tanks, and changing it with pencil and rubber. If the change would affect the application of the part (e.g. a smaller bolt is required to fit the part), then a new part number (at '00' or 'A' change level) must be used. A new part number is, of course, used for all new designs. A simple similar parts index is kept to reduce duplicate designs. Assembly drawings are produced on B, C, D or E sized borders and the geometry of the components are identified by item numbers and quantities displayed in parts callout balloons. Each item is listed with quantity, part number, unit of measure and description in the parts list, which is displayed in the drawing area. The parts list is displayed under headings showing the component type. The component type may be one of thirteen, reflecting the use of the part when applied to the assembly — self explanatory examples of which include Manufacturing Purposes Only or Parts Service Only. An example to illustrate these points is shown in Fig. 5.2

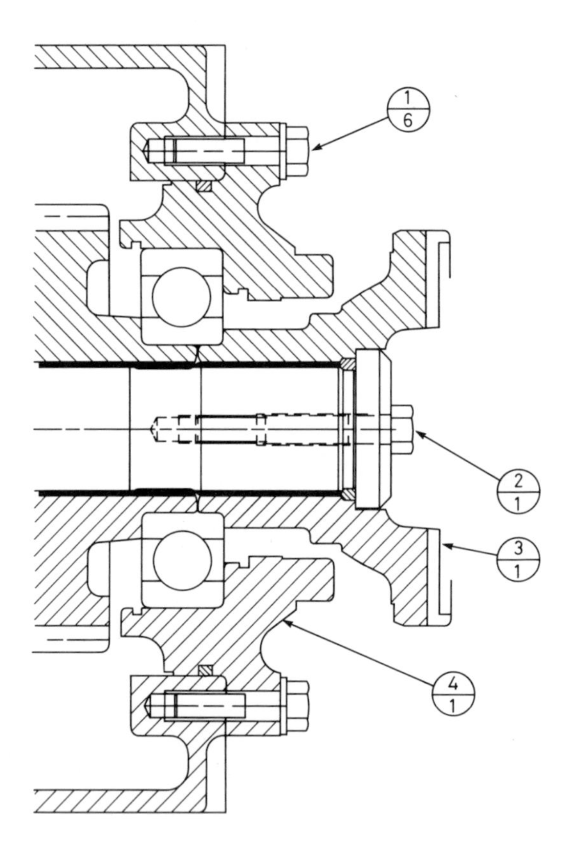

Fig. 5.2 Example of paper-based designs and draughting system

5.7 The paper-based part-release process

As described earlier, the company has long used production control software. With the paper-based system, designs were implemented in the production system in two steps:

(*a*) As a part is created its description and other base data are entered on the item master file (IMF) residing in the main database. This data is mainly created by the Engineering Department. The IMF provides the main catalogue for all part data during the production process.

(*b*) The parts list for all new parts is then provisionally entered onto the product structure file (PSF). Parts on the list must have already been created on the IMF, but the final parts list structure is not accessible by the production control software until Engineering is satisfied with its integrity. Once a parts list is released onto the PSF, it is considered as a live, bona-fide part.

When a drawing is considered as complete by a designer, it is checked for design integrity by the Chief Designer or senior design personnel. At the same time, the administrative paperwork required for part release is put together by Engineering. If problems in either of the above areas come to light, then the part is returned to the designer for changes to be made. Once the part has been checked and passed it can be released, along with its parts list, into the production control systems.

5.8 The plan for computerisation of the paper-based system

A system has been planned where designers can take out a part number and create a parts list whilst seated at their CAD workstation. This involves a hardware and software link between the Apollo token ring network and the Concurrent network. Together with these functions, an automatic release procedure is included to integrate with the drawing-check procedures described above. Application software is required on the relevant computer for these operations. The main database of parts initially resided on the Apollo network, owing to the large amount of collective memory space available (about 155 Mb/machine). As the database expands, data will be archived or transferred elsewhere.

5.9 First steps towards a computerised design department

Within three months of CAD operation, the familiarity of the system began to increase, as did the number of production drawings. It was realised that, as the number of drawing movements began to increase, restricted access areas should be implemented to house drawings in different states of completion. The directories constructed for this purpose were set up as described below:

User's working directory: All drawings undergoing work, be it construction or modification.
Pre-released directory: A 'holding tank' for drawings moved from user's working directories during design and admin. integrity checking. No designer access.
Released directory: All most recent change-level drawings with full released status. Available for copy to a user's working directory, but otherwise no access.
Superseded directory: All drawings that have a more recent version currently available in the released directory. Available for copy to a user's working directory, but otherwise no access.

Initially designers were reluctant to move drawings to the pre-release directory when finished, without keeping a copy in their own workspace. This was perhaps due to the ease with which a personal copy could be made, and a justified lack of confidence in the new system.

The programming tools available with Apollo/MCS hardware/software combination were as follows:

The MCS Anvil CAD package provides a programming system called GRAPL (Graphic Application Programming Language. This language allows parametic graphics programs to be written and also provides a facility to interface between Anvil and its external environment.

The Apollo workstations provide a high-level system programming language called Shellscript. All operating system commands can be incorporated to provide automatic sequences, with limited program structure and file access.

Several standard language compilers are available for the Apollo, and much use has been made of 'C' programming for more complicated tasks.

The first part management routines were written in Apollo Shellscript to give designers simple methods of moving and copying drawings.

5.10 A menu of CAD accessory software

As the number and diversity of application programs grew, it was realised that a collective area should be created for all part management and draughting-aid functions. GRAPL was the obvious solution to provide a menu of accessory functions. A menu system was constructed that could be displayed by pressing function key F1 on the keyboard. This displayed a list of sub-menus which were divided into topical areas, explained below:

Additional draughting: GRAPL macro programs to aid with draughting tasks such as drawing bolts to Caterpillar standards.
Drawing management: Functions aiding the management of such items as drawings borders, title blocks and headers.

Data output: Functions concerned with any form of data output from the CAD system, including drawing plots and printing of ASCII files.
Workspace management: Functions concerned with managing a user's workspace, such as listing/renaming/deleting files.
Part management: Functions concerned with managing parts, such as releasing/copying/renaming.
Auxiliary functions: Other functions that can be run to aid in broader aspects of design, such as PC emulator window.

The current limit to the number of topical sub-menus that can be displayed by Anvil is 19, and the menus may be nested up to 10 deep. This provides ample scope for any future development.

The menu system allows Anvil external routines to be invoked (written in C, Shellscript, Fortran etc.) while a drawing is displayed. They are invoked by a single switch program written in C. Some of the current menu software and much of the software under development communicates data from the Anvil part to an external process and back again. Development in this area has been hampered by the restriction of GRAPL as a real Anvil-interface language, and this is discussed later.

The existence of the menu system presented a further application opportunity to control the application of accessory software. All the functions that use the 'C' switch program can be invoked in either 'wait' or 'background' mode. In this way, designers can be forced to complete an accessory function before returning to an Anvil drawing or otherwise. An obvious example of a function that should be invoked in 'background' mode is the plotting routine, where users initiate a plot (larger plots may take more than half an hour to complete) and may then return to Anvil while the plotting is being carried out.

5.11 CAD administration routines

As the complexity of the system increased, it became necessary to develop routines that could be run by CAD administration only, and not by the designers themselves. A selection of the routines required are described below:

(*a*) A drawing database switch program to bar access to the database when it was under alteration, or parts of it became temporarily unavailable (e.g. a workstation is taken off the network).
(*b*) Functions to reset various routines after unexpected problems or crashes. These apply to the plotting, release, PC emulator routines etc.
(*c*) A comprehensive function for listing the drawing database. Listing

may be performed in one of many ways, in order to find a part or groups of parts.

(*d*) Drawing movement routines. These include functions to move drawings back to a designer from any part of the database, and functions to move drawings into the released or superseded sections.

(*e*) A program to carry out operations on selected directories where Anvil can be run (currently around 45). This has proven one of the most time-saving software routines yet written for administrative purposes.

Currently, some ten administrative programs exist, many displaying further menus or functions.

5.12 Reorganisation of development effort

At the start of the project to computerise the part number and parts list creation, the amount of preliminary development work was severely underestimated. No automatic transfer of parts lists could be created in a reliable and accessible form on the computer drawing. In turn, a parts list could not be produced inside a drawing until an automatic and universal system of border manipulation was devised. This is turn could not be done until all the future requirements of the computerised border were recognised. If this was not done, large numbers of drawings would soon exist on the system that would all have to be changed to be compatible with new software.

Documentation was soon required for the accessory functions that were being written, as the number of new CAD users was increasing. Initially user guides were written for individual functions, but it became obvious that a complete CAD system document was required to cover all aspects of operation, from responsibilities and procedures to standards and user guides. This document is being continually updated and extended.

It was soon found that the development of systems and routines must be carried out starting at the absolute source of data flow. It was therefore nearly a year after the start of the project before enough of the basic work had been carried out to allow the start of Bill of Materials development.

As more accessory part management functions were developed, it was realised that some form of Engineering Part Management database (and access software) would have to be developed. Several commercially available databases were examined, such as Oracle, Ingress, Informix and Empress, but were eventually rejected mainly on the grounds of cost for a 17 – 32 end user system. Instead the plans for the hardware link between the Apollo network and Concurrent machine was expanded to include a general Engineering database that would contain part files as well as the master files required for all drawing, part number and parts list transactions.

5.13 Engineering standards

At the outset of CAD development, it was agreed that the ability to transfer Bill of Materials data from its source, the assembly drawing, into the production control system should be an automatic process. Apart from a reduction in effort required by Engineering administration, this move was seen as a golden opportunity to increase the quality and standard of data emanating from the Engineering department. As described earlier, much of this operation is carried out in conjunction with Caterpillar engineering standards. With the paper system, there was less inherent incentive to adhere to these standards, but computerisation of many routines has had a twofold effect:

(*a*) The engineering standards are easier to use, as only the current relevant data for an operation is displayed. This means that in many cases it is easier and more attractive to use the routines provided, than to try and do it any other way.

(*b*) Where previously with the paper system, the adhesion to some standards was not easy to enforce, often the computerised functions will either check the data entered by the user, or provide a menu of options where only a limited choice is valid.

A good example is the entry of part descriptions. The Caterpillar standard says that the first word (noun) of the description must be one of a standard library of about 500 engineering terms. This section of the part number creation program that prompts for part descriptions now displays (in an information window on the screen) a list of available first words. When one is entered, it is validated against this library — there is no other way of giving a description. When the part number and description is itemised in the parts list, the first word of the description (together with AS for assembly and GP for group etc.) is extracted and displayed. These operations should produce 100% correct nomenclature for parts.

5.14 The Apollo-Concurrent part-management interface routines

Before a parts list can be drawn up and added to an assembly drawing, each of the component parts must have at least had their base data entered onto the IMF. So before a Bill of Materials system could be designed, a function to create the base data for a new part (on the drawing and on the IMF) had to be written. Because the GRAPL programming language is not capable of communicating directly with external system files, most of the data processing and validation had to be carried out by a 'C' program. It was at this point in development that a major problem was encountered with GRAPL. A subroutine called GPASS is provided which is the only method of passing data to or from a drawing and an external process. The GPASS routine is only capable of passing numeric data. A routine for passing character data into GRAPL has been developed, by encoding it into its ASCII code, and decoding it back into characters in the GRAPL program!

5.15 The benefit of hindsight

After two years of CAD operation and development, there are now more than three times the original number of workstations. With the benefit of hindsight, some of the area that appeared slightly grey in original specification for the system can now be laid down with ironic clarity! On the whole, the pure draughting facilities have met or exceeded expectations in terms of performance.

Where expectations have not been met is in the general area of interfacing Anvil to external systems. As mentioned above the only interface method currently available is the GRAPL language with its severe limitations in terms of capability and reliability.

For those considering a CAD system for use as more than an electronic drawing board, it is hoped that this chapter provides a helpful insight into the experiences of one company. In terms of software interfacing, several points should be considered in the choice of a CAD system:

(*a*) What accessory software is provided to link the CAD drawing to the external system? Does this software tool have sufficient programming capability and scope?

(*b*) Can the basic design nomenclature (such as part numbers, descriptions and Bill of Materials data) be passed to and from the drawing easily?

(*c*) Can data from a displayed drawing be accessed easily by automatic functions? (or will your part management functions have to prompt the user to type in information that is already displayed in front of him?)

(*d*) Does the CAD package provide any inbuilt Bill of Materials or parts list handling facilities? Are they relevant, too general, or too specific?

(*e*) Does the CAD package provide any drawing management functions? (searching for drawings, or listing drawings etc).

Do not expect too much of CAD; the concept of it providing a complete design environment is not yet with us. The hope that a part need only be input once and then called off where required to maintain designs and Bills of Material etc. is mistaken. We must wait for high performance 3-dimensional CAD with significant entity manipulation properties and possibly the inclusion of artificial intelligence.

With the advent of CAD the logical progression is for further automation of existing manual tasks in the drawing office. With this in mind the construction is planned of a dedicated Engineering database. All existing part management routines will be co-ordinated by this database, giving a level of drawing office efficiency which could not have been satisfied by the purchase of a CAD system alone.

Chapter 6

Statistical cost estimating — the construction of an estimating relationship

D. J. Leech and R. G. Williams
University College of Swansea, UK

6.1 Introduction

Estimating the costs of capital investments or of research and development programmes is a major concern of many organisations. A number of approaches can be applied depending on the data available, the type of project and the degree of detail to which the project has been specified. The main approaches are the synthetic method, the analogy method and statistical methods.

The synthetic method relies on breaking down the work involved in the project into its most elemental parts, and then applying historical data about their costs or duration to these elements and aggregating these to provide the total cost or duration. So that the work elements can be identified a project needs to be well defined, and so this method is not appropriate at the early stages of project definition. In addition there may not always be such detailed historical information available on each work element, and in any case such information can be expensive whether provided in-house or bought in from work study consultants.

In the analogy method the estimator makes a subjective selection of previously undertaken analogous projects and extrapolates from these projects, taking into account perceived differences, to provide the required estimate. It does not need such extensive information about the current or previous projects as the synthetic method and it can therefore be applied at an early stage of project definition. Its disadvantage is that it relies on the subjective perception of the estimator and is prone to error as a result of psychological and behavioural factors.

Statistical cost estimating reduces the subjectivity required by formalising the analysis mathematically. There are two main approaches; both rely on historical data, but need not be as detailed as that for the synthetic method. The first approach fits a relationship to the empirical data by regression methods to derive a model which represents the relationship between certain characteristics or a project and its cost and duration. The second approach is similar but uses a predetermined general relationship which is thought to be a reasonable representation of the way the characteristics affect cost or duration. This relationship is then modified using the historical data available so that it becomes specific to the organisation or to the type of project. This process is sometimes known as calibration. The literature on estimating methods uses a variety of terms for the different types of statistical estimating methods. These include: step counting, exponential methods, factorial methods and parametric methods.

6.2 Statistical cost-estimating methods

The most basic statistical method is the ordinary least squares regression model which involves fitting a straight line to sets of data concerning projects undertaken in the past. The data includes information of the characteristic being estimated, such as the project's cost, duration or effort in man-months, and other characteristics which are thought to have an important influence on the estimate. The analysis of the data requires a deal of skill in firstly recognising the relevant important characteristics and then in the application of the linear regression techniques. A number of assumptions are made by the classical linear regression model about the way the data observed behaves. Violations of these assumptions occur when: relevant characteristics are omitted, irrelevant characteristics are included, the relationship between the dependent variable and an independent variable is not linear, the parameters of the underlying estimating relationship do not remain constant over the period the data was collected, there are errors in the data on the variables, when two or more of the independent variables are correlated in some way, and when the number of observations is less than or equal to the number of independent variables.

Practical difficulties can also arise in the collection of data on past projects: the data may not have been recorded in the first instance, or it may be dispersed at distant locations, or the types of records kept may be different for different projects. There is merit therefore in generalised cost estimating relationships which have forms which are adaptable enough to be valid for a range of types of activities undertaken in different organisations. Because of their generality they can lack accuracy, and for this reason then tend to be used to establish order of magnitude estimates at the early stages of a project rather than to provide the more accurate estimates which may be required and can be provided by other methods at later stages.

The most common forms of estimating relationships are linear and logarithmic or exponential. The linear model is of the form;

$$Y = A + BX_1 + CX_2 + \ldots \qquad (6.1)$$

and the exponential model of the form

$$Y = A\,X_1{}^B\,X_2{}^C \ldots \qquad (6.2)$$

or in linear form,

$$\log(Y) = A + B\log(X_1) + C\log(X_2) \ldots \qquad (6.3)$$

Linear estimating relationships can be fitted to empirical data using commercially available computer packages, which will also perform some of the analysis necessary to test whether the model is a reasonable representation of the data. Nevertheless, some skill is needed to interpret the analysis and to be able to suggest modifications to the model if it is not suitable.

Table 6.1 displays the weight and prices of the basic models of a range of cars (Ford, Rover and Vauxhall). Fig. 6.1 shows the data and a line of best fit of a linear regression and Fig. 6.2 shows the data together with a line of best fit of a log linear regression. There is little difference found between the two relationships; the logarithmic relationship tends to make lower estimates at lower weights than the straight line relationship and vice versa for the heavier cars. It cannot be said that one of the lines provides better estimates than the other, although it may be said that they both provide inaccurate estimates for certain cars of certain weights.

Table 6.1 Kerbside weights and costs of basic models of Ford, Rover and Vauxhall cars in 1988.

Cars	Ford		Rover		Vauxhall	
	Weight (lb)	Cost (£)	Weight (lb)	Cost (£)	Weight (lb)	Cost (£)
	1653	5004	1378	4299	1631	4993
	1851	6121	1687	4998	1852	6310
	1940	7730	1973	8057	1951	6927
	2293	8720	1984	6514	2165	7889
	2612	11995	2194	8347	2570	11373
			2866	12680	3098	16079

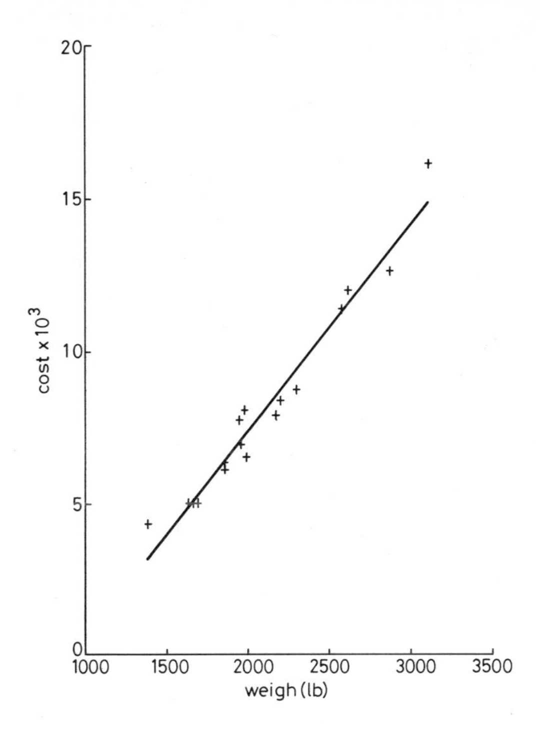

Fig. 6.1 Linear cost curve

6.2.1 Exponential models

Exponential models can be justified in the following way. Suppose that a project is to be undertaken which is similar to one previously undertaken but of a different size. Then the costs of the two will be related by

$$C_1 = C_2 \, (S_1/S_2)^n \qquad (6.4)$$

where C_1 is the cost at scale S_1

C_2 is the cost at scale S_2

n is a scale exponent

When $n = 1$ the costs are linearly related, but when $n < 1$ the marginal cost decreases with magnitude, i.e. there are economies of scale, and conversely when $n > 1$ there are diseconomies of scale.

If the costs and magnitude at scale 1 in eqn. 6.4 are replaced by a notional unit cost k and by unity, respectively, eqn. 6.4 becomes

$$C = k \, S^n \qquad (6.5)$$

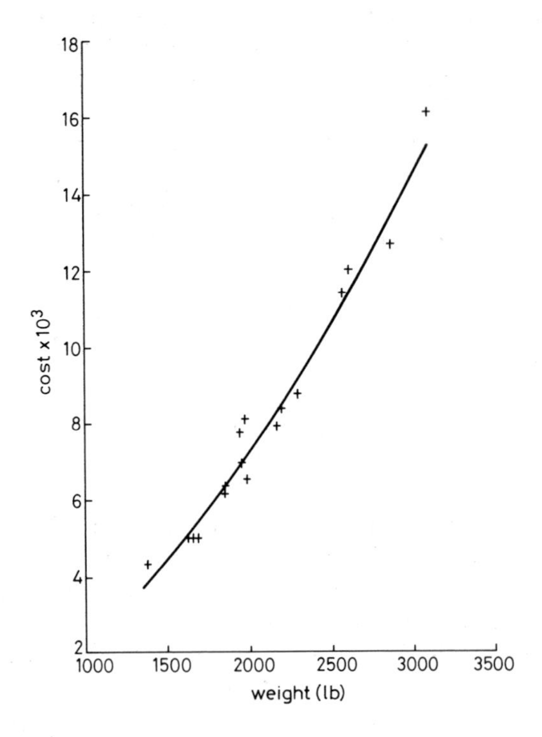

Fig. 6.2 Logarithmic cost curve

If a project similar to the one under consideration has already been undertaken then the index n is often found to be in the range 0.4 to 0.8 and most typically 0.6 to 0.7. (Institution of Chemical Engineers, 1988). This relationship is sometimes known as the 'six-tenths power law' or the 'six-tenths factor' (Krieg, 1979) as the index is often assumed to be equal to 0.6.

It should be noted that this relationship applies when comparing projects which differ in scale of magnitude only and may not apply when the project under consideration differs from others in terms of, for example, technology, manufacturing techniques, construction methods, changed tolerances of machining, and general quality.

A development of the relationship so that it can take into account other differences between projects as well as just physical scale can be made by adjusting the notional unit cost. It is thought that costs will not increase linearly with the degree of difficulty of the project but exponentially, and so the notional unit cost is raised to the power of an index which represents the difference between the difficulty of a basic notional unit scale project and the current project being considered. The cost estimating relationship then becomes

$$C = k^i S^n \tag{6.6}$$

The notional unit cost k will increase as the difficulty of the project increases, and it also is likely that economies of scale will become less as the degree of difficulty increases. There is therefore an inverse relationship between the indices i and n. The value of i will be greater than one, as k represents the unit cost of a project with basic difficulty. Also, if it is assumed that there are economies of scale the value of the index i will be in the range $0 < i < 1$. A simple relationship, which satisfies these requirements of i and n, is

$$i = 1 / (1 - n) \tag{6.7}$$

so that (6.6) becomes

$$C = k^{1/(1 - n)} S^{n} \tag{6.8}$$

It is thought that it is this type of estimating relationship which is used by some commercial computerised 'black box' cost estimating packages. A major difficulty of using this form of cost estimating relationship lies in the difficulty of estimating an appropriate value for the index n. Although the index is based originally on the marginal cost of the item being considered, when the relationship is developed to take into account more than just differences in scale, the value of n looses its relevance in this respect. There may be merit to reduce misinterpretation therefore in rewriting (6.8), using the relationship (6.7), so that it becomes

$$C = k^{i} S^{(1 - 1/i)} \tag{6.9}$$

The index i thus becomes a measure of the degree of difficulty of a project compared to some notional base project. The value of k depends on the values of n chosen to represent the calibrating projects and can be found by regression methods. A difficulty here is that values for the index of difficulty have to be assumed in order to find k. It is reasonable to assume that i will have a value such that $1 - 1/i$ will be in the region of 0.6 — 0.75.

6.2.2 The effect of learning

To arrive at an estimate of the unit cost of a manufactured item, the effect of learning should be accounted for. Learning implies that the manufacturing costs of an item produced later in a production run will be less than that produced earlier in the run. One established method uses the relationship

$$C_n = C_1 \times n^{(\log L \log 2)} \tag{6.10}$$

where C_n is the cost of the nth item

C_1 is the cost of the first item

L is the learning factor

The total cost of the first N items is

$$C_1 \; [N^{(1+b)} - 1] \; / \; [1+b] \qquad (6.11)$$

where $b = \log L / \log 2$
and the average cost of N items is

$$C_1 \; [N^{(1+b)} - 1] \; / \; N \, [1+b] \qquad (6.12)$$

Combining (6.12) and (6.9) the estimated cost becomes

$$C = k^i \, S^{(1 - 1/i)} \, [N^{(1+b)} - 1] - 1] \; / \; N \, [1+b] \qquad (6.13)$$

A regression of cost against weight in (6.13) for the data in Table 6.1 indicates that, when $N = 200000$, $i = 4$ and $L = 0.9$, k is in the region of 5 for estimating costs in 1988 pounds.

6.2.3 Sensitivity of the cost estimate to values of i

Fig. 6.3 shows the relationship between the degree of difficulty i and the estimated cost of an item weighing 2000 lb. It can be seen that the gradient of the curve increases as the scale of difficulty increases, which indicates that the estimated cost gets more sensitive to changes in i as the value of i increases. The scale of difficulty for a typical car weighing 1973 lb is 3.34 and its estimated cost £8108, but if its scale of difficulty is estimated to be 3.36 then its cost is estimated to be £8488. A difference of 0.6% in the estimate of i will result in a difference of 4.6% in the cost estimate. This example is at a fairly low scale of difficulty, but for projects with a high degree of difficulty, such as equipment for manned space flight, the scale of difficulty may have a value over 5 and the cost estimate is very sensitive to very small changes in the estimated value of i.

Fig. 6.4 shows the actual weights and costs of cars together with the estimating envelope of the suggested relationship when $i = 3.50$ (the upper continuous line) and when $i = 3.17$ (the lower continuous line). It can be seen that the width of the envelope increases with weight, as a consequence of an increasing sensitivity of the estimate to differences in weight as i increases. It appears therefore that not only is the estimate more sensitive to variations in i as i increases but is also prone to more variation as a result of differences in the weight as i increases.

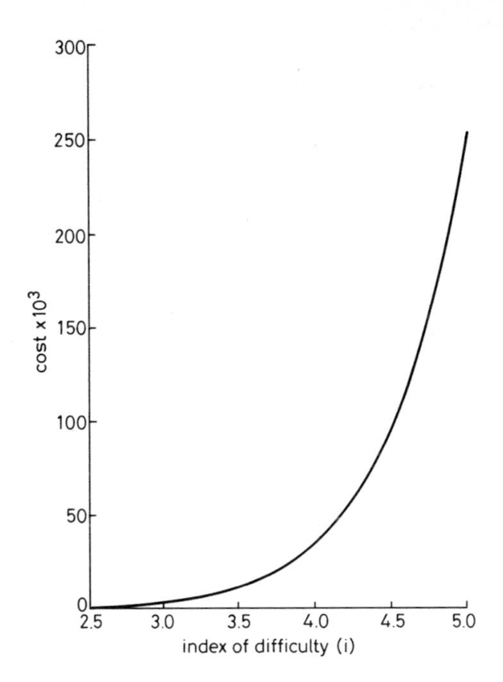

Fig. 6.3 Sensitivites of estimates to the index of difficulty

6.2.4 Estimating the index of difficulty

The index of difficulty can be estimated subjectively, based on the empirical experience of an estimator on previous projects. There may be difficulties in this method as the index is notional and dimensionless. It may be possible to calibrate an estimator in the same sense as work study personnel are calibrated to estimate the working rates of individuals. An alternative approach is to derive a mathematical procedure to provide a value of i. The input parameters chosen to be characteristic of the type of project under consideration would constrain the model only to those types of projects for which these characteristics would be appropriate. These may include the difficulty of assembly, the number of parts, the workability of materials and the precision of fabrication. Even if these types of characteristics are incorporated into a mathematical expression for providing a value of i there is still a need for estimating subjectively some of these characteristics.

6.3 Conclusion

A cost estimating model of the form derived can provide accurate estimates of costs, provided that the appropriate value of the index of

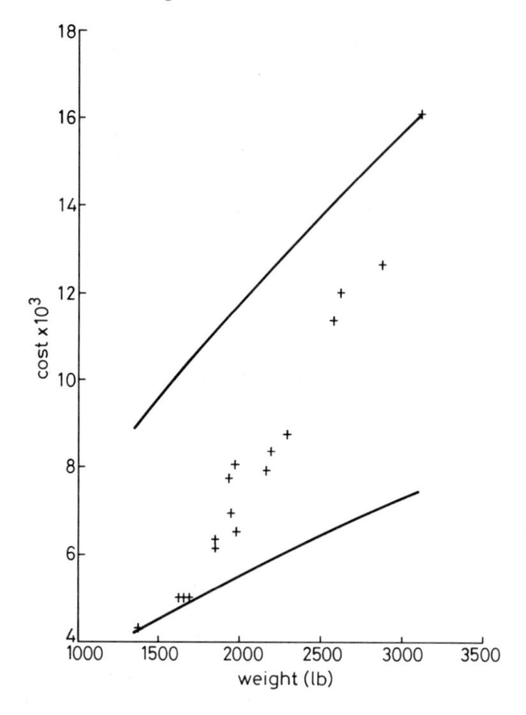

Fig. 6.4 Envelope of estimates for two values of the index of difficulty

difficulty is chosen. At early stages of a project there may be little knowledge of the physical nature of the item being estimated and so the estimating relationship requires only one specific input characteristic regarding the project. Skill and experience in estimating the scale of difficulty can be developed so that improved quality of estimates can be made at the early stages of a project.

Chapter 7

The use of expert systems in demand forecasting

E. T. Sweeney
University of Warwick, UK

7.1 Introduction

Accurate forecasting of product demand is essential if the management of a manufacturing system is to be as effective as possible. In virtually every production planning decision some kind of forecast needs to be considered. In recent years the greater uncertainty in economic and financial affairs, caused in part by the rapid rate of technological development, has sharpened the focus on the need for improved forecasting. An expert system is a computer system which possesses a set of facts, heuristics or knowledge about a specific domain of human expertise, and by manipulating these facts intelligently it is able to make useful inferences for the user of the system. These systems make use of rules of inference to draw conclusions or make decisions within defined areas. Their power comes from the presence of facts and procedures which have been identified by human experts as the key components in the problem solving process. In recent years, developments in computer hardware and software have facilitated the use of expert systems in industry. The aim is to explore the potential usefulness of such systems in forecasting.

7.2 Expert systems in demand forecasting

A large number of both qualitative and quantitative forecasting techniques are currently available. By using computers, quantitative techniques (time series and causal models) which involve much calculation and processing and incorporate a wide range of variables can be run off accurately and in a short time. Access to relevant databases has also been simplified, and the effect of variations in data considered sensitive to change can easily be explored. However, each forecasting technique has its special use and care must be taken to select the correct technique for a particular application. A large number of complex factors must be considered in the selection of an appropriate technique. The application of knowledge-based systems could, therefore, be of great benefit. Such a system would involve the embodiment within a computer of the knowledge needed in deciding which technique to use.

7.3 Knowledge acquisition

7.3.1 Introduction

The knowledge acquisition phase involved the gathering of facts and heuristics concerning the factors which influence the decision concerning which forecasting technique to use. This knowledge came mainly from the author's experience of implementing demand forecasting strategies, but a number of colleagues and published works (primarily Chambers *et al.*, 1971; Makrifakis *et al.*, 1979; Levenbach *et al.*, 1982) were consulted on some of the more subtle and difficult points. So, for the most part, the author performed the role of both the forecasting expert and the knowledge engineer. The assuming of both functions did not in any way detract from the knowledge acquisition process; indeed it reduced the time needed and helped ensure that problems which are frequently encountered at this stage (Kidd, 1987) were largely avoided.

7.3.2 Techniques considered

Sixteen commonly used forecasting techniques were considered (see Table 7.1). It was seen as important that knowledge about a range of both qualitative and quantitative methods be embodied in the expert system in order to ensure widespread applicability. With regard to quantitative techniques both time series and causal models were investigated. It can be seen from Table 7.1 that the methods considered varied from the very simple (e.g. using the demand figure for the previous period) to the quite complex (e.g. econometric modelling).

7.3.3 Factors influencing choice of technique

Many complex and often inter-related factors need to be considered in choosing an appropriate forecasting strategy. Table 7.2 outlines the general classification of those factors considered in this study, as well as describing in more detail what is meant by each one.

Table 7.1 Forecasting techniques considered in the study

Type of technique	Technique
Qualitative	The Delphi method Market research Panel consensus Visionary forecast Historical analogy
Time series	Past average Figure for last period Moving average Exponential smoothing Box – jenkins X – 11 Trend projection
Causal	Regression Econometric model Input – output model Life cycle analysis

7.3.4 Choice of technique

Once the factors influencing the choice of technique have been identified, the situations in which a particular technique is appropriate need to be established. The nature of any situation can be assessed by examining the status of the relevent influencing factors (Table 7.2).

To achieve this, each of the 16 techniques under consideration (Table 7.1) was examined in turn. Each was examined in terms of the influencing factors with a view to establishing its appropriateness in a range of situations. For example, for the visionary forecasting technique the knowledge acquisition process would arrive at a statement such as the following:

> 'A visionary forecast is usually poor in terms of its accuracy and turning point identification, is easy to understand, takes only days to arrive at, gives a long term forecast (over two years), is inexpensive to implement and requires little, if any tabulated data.'

A statement in similar format was arrived at for each of the 16 techniques. These statements must then be transformed into a format which allows for their embodiment in a computer.

Table 7.2 Factors influencing choice of forecasting techniques

Factor	Explanation
Accuracy of forecast	Is accuracy considered to be of importance?
Turning point identification	Is the identification of turning points considered to be of importance?
Ease of understanding	Is ease of understanding of the forecasting methodology and interpretation of the forecast considered to be of importance?
Time span	What is the time span of the forecast? Immediate term (hours) Short term (days) Medium term (weeks) Long term (months)
Time horizon	What is the time horizon of the forecasting? Short range (0—3 months) Medium range (3 months—2 years) Long range (over 2 years)
Cost of forecast	What financial resources are available in making the forecast?
Availability of data	What relevant data are available? Demand patterns for previous periods Market research reports Factors influencing demand patterns Similar product information In-company product and service flows
Data characteristics	What characteristics are apparent in existing data? Secular trends Seasonal variations Cyclical variations Irregularities

7.4 Knowledge representation

A number of knowledge representation schemes were evaluated in view of the nature and format of the acquired knowledge. These included production rules, frame-based representation and semantic nets (Forsyth, 1984). It was decided to use production rules primarily because the knowledge could be easily and accurately represented in this form. The knowledge was, therefore, rewritten in terms of antecedent – consequent (IF..THEN) rules. The rule for the visionary forecasting technique presented above would then be expressed as follows:

IF accuracy is not of importance
  AND turning point identification is not of importance
  AND ease of understanding is not of importance
  AND time span is short term
  AND time horizon is long term
  AND available resources are very limited
  THEN the use of visionary forecast appears to be appropriate.

No reference is made in the above statement to availability or characteristics of data. This is to ensure that, although in the making of a visionary forecast tabulated data is not generally used, the existence of such data should not preclude its use. Care was taken to ensure that points of this nature were recognised in forming the production rules from the acquired knowledge for all the techniques under consideration.

7.5 Software development tools

7.5.1 Expert system shells

An idea which has appealed to many as an aid to developing working expert systems is to make use of a standard framework or shell, and a large number of commercially available shells have become widely used in the last few years (Bramer, 1985). Such a shell can be thought of as a full expert system with its knowledge base taken away.

i.e. EXPERT SYSTEM – KNOWLEDGE BASE = SHELL

thus providing the developer with an ideal framework for the construction of a knowledge base in a given field. Shells provide a carefully co-ordinated collection of tools supporting a variety of knowledge representation and reasoning schemes. A typical shell would include the following (Ranlefs, 1984):

Interpreters and/or compliers of knowledge representation techniques; inference support; knowledge base management system tools for explaining rationales or justifications of courses of action; tools to support acquisition and modification of knowledge; tools for external interfaces (e.g. displays).

Their provision of facilities such as those outlined above can greatly reduce the development time of an expert system. Thus, the increasing availability of such shells is likely to result in more widespread acceptance and use of expert systems. The way in which the forecasting knowledge described above is represented lends itself readily to embodiment using a shell which employs antecedent-consequent rules.

7.5.2 Crystal

Crystal is a general purpose expert system shell which can be used on personal computers. It uses production rules in the building of a knowledge base. Starting with a single line description of what the application is seeking to achieve (e.g. '*an* appropriate forecasting technique can be recommended'), Crystal allows indefinite expansion of that thought. Each step is a sequence of logical alternatives. The forecasting production rules outlined above were implemented using Crystal. The system (FOREX) is described below.

7.6 System description (FOREX)

7.6.1 Recommendation of techniques

The decision concerning which forecasting technique to use is based on the contents of the FOREX knowledge base and the responses given by the user to system prompts concerning the status of the relevant factors outlined in Table 7.2. At the end of consultation FOREX will recommend one or more techniques and, if required, give an explanation as to how it has reached its decision. Information concerning typical application areas is also displayed. This, and the explanation facility, help improve the credibility of the expert system's decision making process in the user's mind. Alternatively if no technique appears to be suitable an appropriate message is displayed. Expert system approaches are increasingly being used in the development of computer assisting learning (CAL) systems (Lewis *et al.*, 1987). Expert system shells often provide an ideal framework for the development of CAL coursewares. As well as recommending a forecasting technique FOREX can explain what implementing the technique involves, state the underlying theoretical basis and give references if additional information is required. This information is presented both as pages of text and, where appropriate, graphically.

7.7 Further work

The decision about which forecasting technique(s) to use can be considered to be the first stage in arriving at an actual forecast (see Fig. 7.1). Stage 1 of this process can be carried out using FOREX, as already described. Stage 2 can be achieved, albeit to a somewhat limited extent, using the CAL aspect of FOREX, described above. Work is currently being undertaken at improving this aspect of FOREX and involves a study of various computer based pedagogical approaches as well as the utilisation of advanced features of CRYSTAL such as interfacing with specially written routines (in, for example, C).

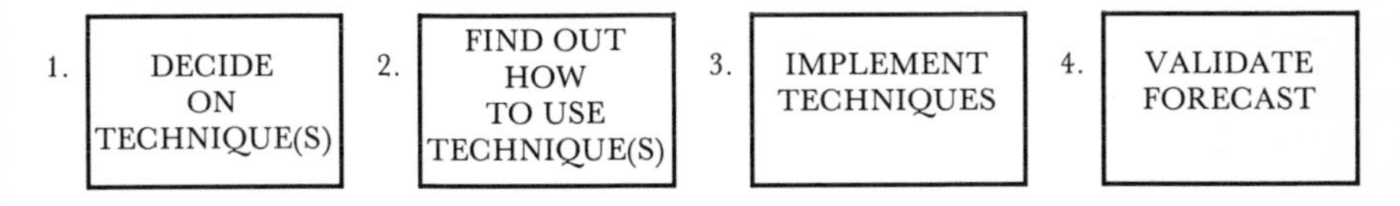

Fig. 7.1 Stages in the forecasting process (modified from (3))

Integration of FOREX with existing commercially available forecasting software would provide a more complete and comprehensive framework in which to develop an accurate forecast. Work being carried out at present aims to integrate FOREX with a number of commercially available software packages as well as with appropriate databases.

7.8 Conclusion

Traditionally, computerised forecasting has involved the development of a forecast using one of the quantitative techniques discussed in this chapter. By using computers, calculations and processing which often incorporate a wide range of variables can be run off accurately and in a short time. The effects of any amendments can also be explored by changing variable data. The application of expert systems provides a framework in which the knowledge required to select an appropriate technique can be embodied. In addition, it can provide a facility for the production of 'intelligent' teaching systems. By providing access to the knowledge needed in selecting the forecasting technique most appropriate to a particular set of circumstances, the use of the expert system strategies outlined in this chapter could have a major impact on the nature of the forecasting process, in relation to product demand as well as other factors, within manufacturing industry.

Chapter 8

A graphical technique for planning the introduction of AMT: the GRAI method

K. Ridgway and S. Downey
University of Sheffield, UK

8.1 Introduction

This chapter discusses the use of the GRAI method in planning developments in production management systems and the planning and implementation of AMT. The GRAI method was developed at the University of Bordeaux 1 in 1980. It is a structured technique which utilises a system of grids and nets to analyse management information systems. Unlike the IDEFO method which views complex systems as a combination of functions which are decomposed into a set of interacting sub functions (Ming Wang *et al.*, 1988), the GRAI method concentrates on the identification of decision centres and associated information requirements. This method has been used on a number of occasions to analyse the production management systems in French companies. Maisonneuve (1987) used the method to analyse the production management systems in a small company manufacturing gas bottles and developed a specification to assist the company to implement a production management software package. Bourely (1987) used the method to examine inconsistencies in the management and information systems at the Bordeaux Blood Transfusion Centre and develop a computerised information system to control the storage and distribution of blood supplies. From these examples it can be seen that the GRAI method is particularly useful in the planning and specification of computer aided production management (CAPM) systems. However, as Pun *et al.* (1986) point out the method is also particularly useful in the design of flexible manufacturing systems (FMS). The objective of a FMS is high productivity and efficiency. This can only be achieved if attention is paid to supervisory functions such as scheduling and routing. The GRAI method can be used to analyse the supervisory functions and thus can be an invaluable aid when planning the introduction of AMT.

The GRAI method is discussed below and one particular application in a small British company is described. In this application the method was used to analyse the company's production management systems and assist in the implementation of a CAPM package.

8.2 The GRAI method

The GRAI method is based on the premise that a manufacturing company contains three sub-systems as shown in Fig. 8.1

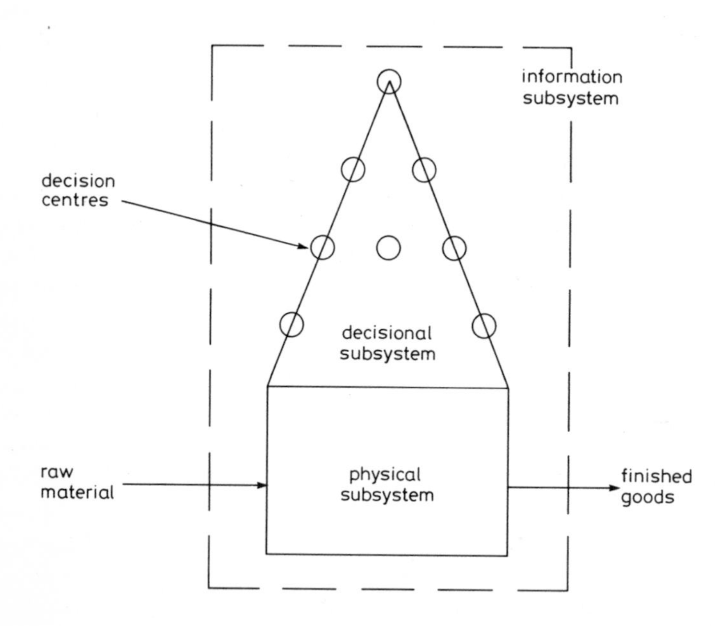

Fig. 8.1 Physical, decisional and informational sub-systems

The three sub-systems are:

- The physical sub-system which transforms raw material into end product.
- The decisional sub-system which controls the physical sub-system and has a hierarchical structure.
- The information sub-system which provides the decisional sub-system with relevant data.

In the ideal production management system the information provided at a decision centre is consistent with the decision made.

The GRAI method makes the following assumptions:

(i) All decisions are made at decision centres.
(ii) Decision centres are hierarchical.
(iii) The position of a decision centre is defined by two characteristics:
  (*a*) the production function.
  (*b*) the horizon H (the length of time over which the decision extends) and the period P (the length of time before the decision is reviewed) (For example a four year budget plan reviewed annually has a horizon of 4 years and a period of 1 year).

Using the above description of a decision centre, the technique uses two graphical tools to analyse the system and identify any inconsistencies.

8.2.1 The graphical tools

(*a*) *GRAI grid*: The GRAI grid (Fig. 8.2) is used to identify the relative positions of decision centres within the production management hierarchy. It comprises columns representing the various functions of the company. These are typically planning, purchasing, manufacturing, quality control and delivery. Rows represent the hierarchical position of the decision centre defined by the horizon and period previously discussed. The relationship between decision centres is indicated using two symbols:

→ denotes transmission of a decision

⇒ denotes transmission of information (only used for the transfer of important information).

(*b*) *GRAI Nets*: The GRAI nets (Figs. 8.4 and 8.5) are used to examine all the physical and management activities and information which are required to make a decision. A decisional activity is represented by a vertical arrow between two circles. Information and physical activities are represented by horizontal arrows.

8.2.2 Application

In using the GRAI method it is necessary to identify the management team involved in decision making. This is termed the synthesis group and their involvement and commitment are essential to the success of the analysis. The synthesis group is interviewed by an analyst who uses the graphical tools to analyse the structure of the company. The analyst may need to be assisted/advised by a GRAI specialist who is more experienced in the use of the technique. The analysis is carried out in two distinct

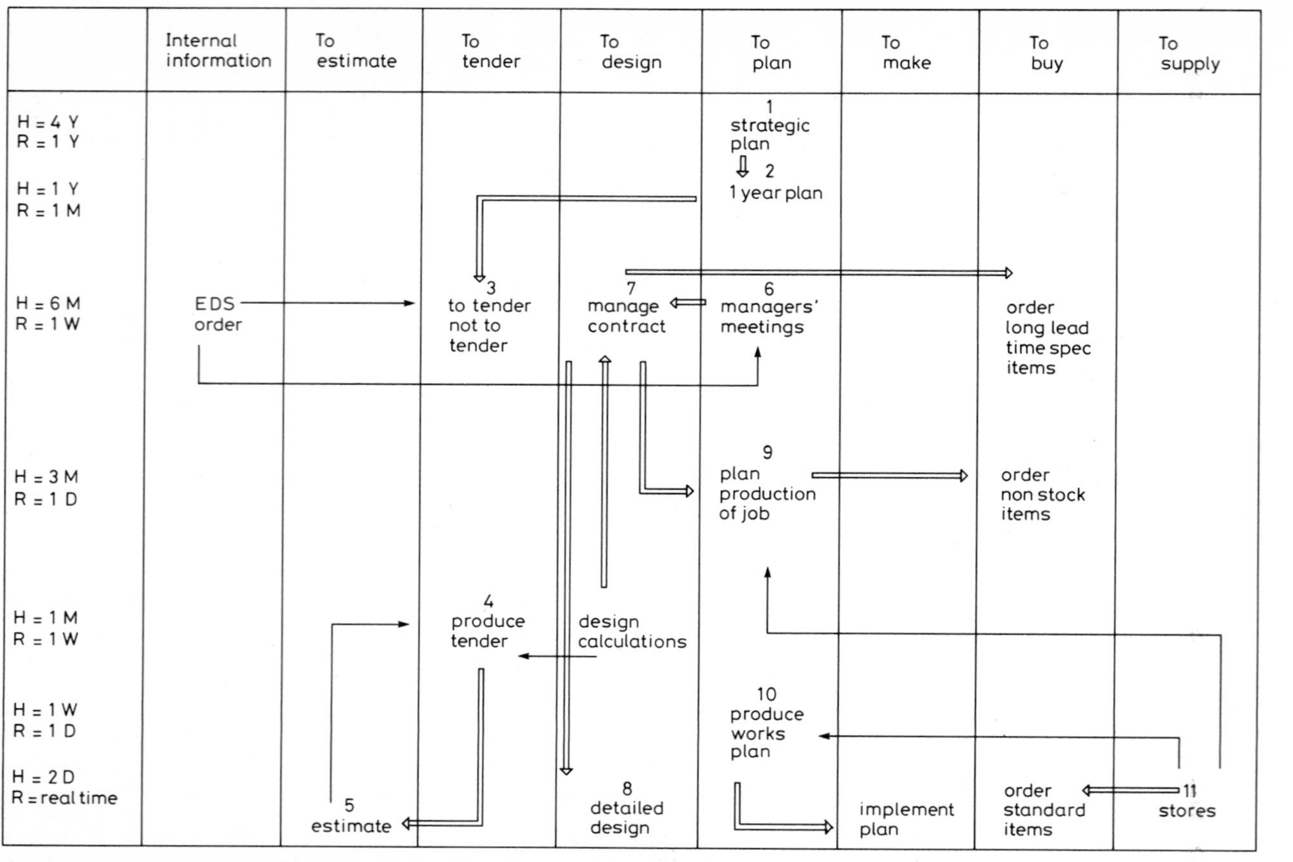

Fig. 8.2 GRAI grid representing the current production management system

phases: The top-down analysis uses information gained from meetings with the synthesis group to construct a GRAI grid. The initial grid represents the decisional hierarchy as foreseen by the synthesis group. The bottom-up analysis looks closely at the decision centres identified during the top-down analysis. Each member of the synthesis group is interviewed by the analyst and a GRAI net is constructed for each decision centre.

The use of the GRAI grid and nets in this manner commonly highlights three major problem areas:

(*a*) Inconsistencies associated with transmission of information:
- no updating of lead times
- no knowledge of large orders
- slow transmission of information
- redundancy of data
- inconsistencies between measuring parameters

(*b*) Inconsistencies specific to individual decisions:
- no account taken of delays
- inventory level insufficiently detailed
- no follow up of low stock levels

(*c*) Inappropriate relationship between decision centres:
- frequency of planning insufficient to cope with frequency of orders.

After using the GRAI grid to identify a decision centre the associated GRAI net can be used to examine the relevance of the information provided. For example, if the decision to tender/not-to-tender (Fig. 8.5) is examined, it can be seen that information concerning the production target, job specification, resources available, delivery data and existing workload are all required. The results of interviews indicate that details of the resources available and existing workload are readily available at the decision centre. After a close examination of the GRAI grid and nets the management structure of information systems can be modified to eliminate the inconsistencies identified. An important element in the use of the technique is the continual evaluation and verification of results by the synthesis group. As the project progresses the group highlights and then corrects the deficiencies within its own structure. This continued involvement is a key to the eventual acceptance of the results of the study and the modifications recommended.

8.3 Application of the GRAI method

8.3.1 Company description

The GRAI method was used to analyse the production management systems of a small company manufacturing sewage purification equipment. The work handled by the company ranges from large contracts to small spares orders. All contracts are managed in-house with about 5 – 10% of the work sub-contracted. The majority of the work done by the company is designed and built for specific contracts.

8.3.2 The study

The GRAI study was initiated with a synthesis group comprising the Technical Director, Marketing/Project Engineer, Production Manager, Contracts Manager and Commercial Manager. The analysis group comprised one analyst and one GRAI specialist. Additional specialist help was provided by the University of Bordeaux 1. After initial discussions with the synthesis group the GRAI grid (Fig. 8.2) was produced. Following this, a series of interviews were performed and GRAI nets were produced for each decision centre. In projects carried out at the University of Bordeaux 1 the synthesis group met regularly to discuss and verify the grid as it was developed. In the small company considered this approach was thought to be impractical and a more informal repeated interview technique was adopted.

8.3.3 Analysis of results

The GRAI grid for the company is shown in Fig. 8.2. To assist in the analysis a second grid identifying personnel at each decision centre was produced (Fig. 8.3). Examination of these grids highlights several inconsistencies:

(*a*) The Contracts Manager plans over a six month period and reviews the situation at monthly intervals. Departmental managers review new orders and hand them over to the Contracts Manager at weekly intervals. This demonstrates that load fluctuations may reach the contracts department between planning intervals.

(*b*) The Production Plan is affected by shortages of stock. Many of these could have been foreseen as they are a consequence of long standing orders. If the grid is examined in conjunction with the GRAI nets further inconsistencies are highlighted.

(*c*) The Master Production Plan does not include feedback from the shop floor and may be unrealistic.

(*d*) Decisions to make/sub-contract (Fig. 8.4) are inaccurate because:
   (i) Inventory levels are not detailed and stock levels can only be accurately determined by physical stock checks.
   (ii) Detailed information on workshop capacity is unavailable.

(*e*) Detailed production planning is not carried out. Jobs are not broken down into their component parts and there is little detailed information on the resources required.

(*f*) The daily work plan does not follow a detailed production plan and

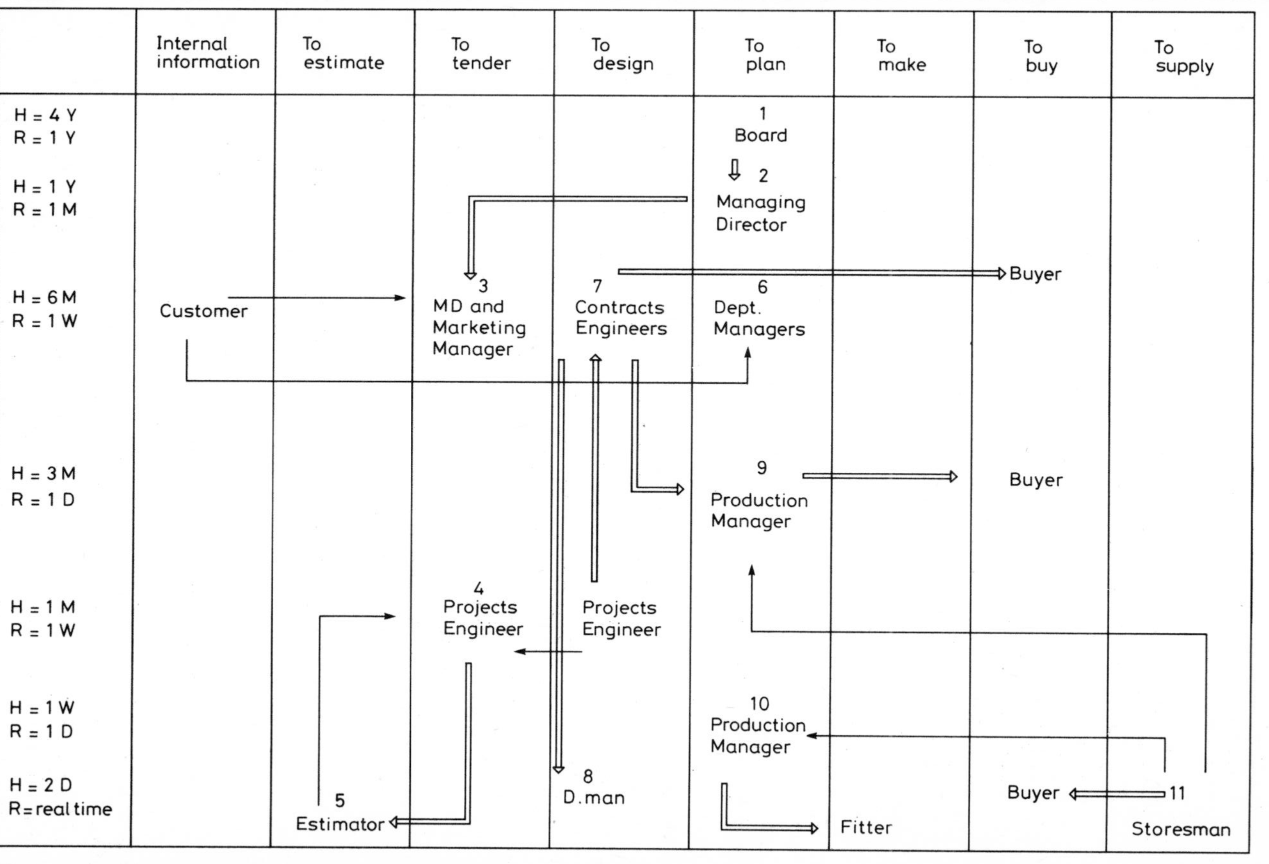

Fig. 8.3 GRAI grid indicating personnel at decision centres

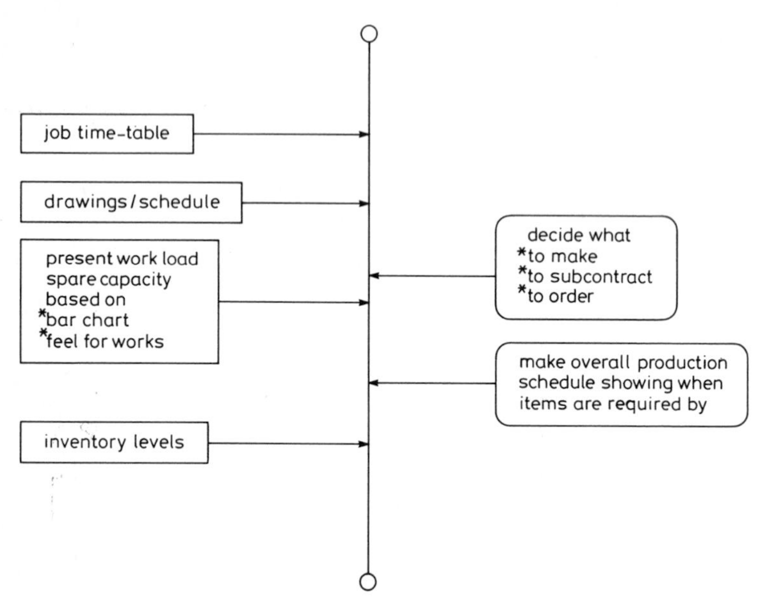

Fig. 8.4 GRAI net for the decision to sub-contract

is heavily influenced by unforeseen events; this leads to reactive activities i.e. fire-fighting.

(*g*) Stock control is inadequate as:
- (i) Stock levels can only be assessed by physical checks.
- (ii) No record of the eventual destination of stock is maintained. This leads to inaccuracies in job costing and affects the accuracy of future estimates.
- (iii) Outstanding stock orders are difficult to obtain and duplications can occur.
- (iv) The purchasing strategy is vague with purchase orders initiated at three points.
- (v) Responsibility for stock control is at a very low level in the hierarchy.

(*h*) The decision to tender is taken without considering available capacity and current work load (Fig. 8.5). This information is not readily available at the decision centre.

8.3.4 Design of a modified system

Many of the inconsistencies highlighted will be eliminated when the CAPM package recently purchased is fully operational. The majority of problems are due to a lack of accurate inventory, costing and resource information. If this data is available, decisions can be made at a more

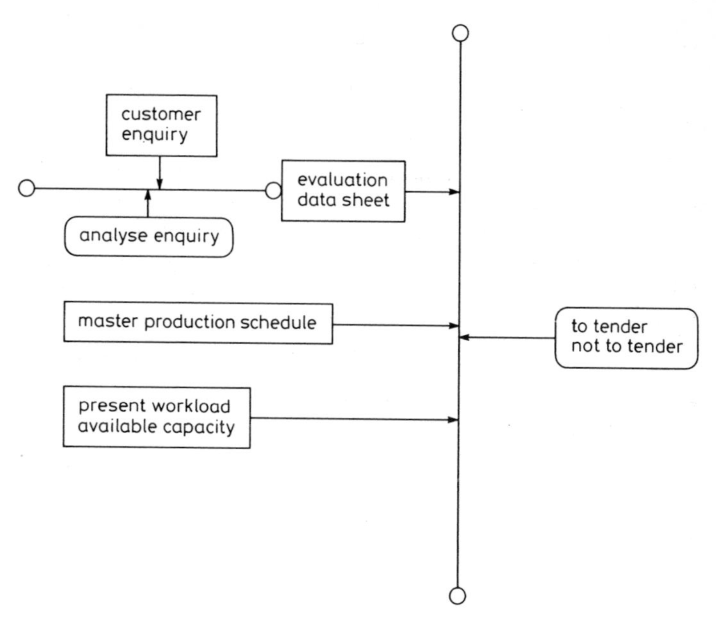

Fig. 8.5 GRAI net for the decision to tender or not-to-tender

appropriate level in the hierarchy and the management structure can be modified. Where the information is not provided by the CAPM system the GRAI method is being used to define the requirements of the decision centre and assist in the specification of modifications. In addition to the improved information system there is also a need to improve the planning function. In particular, a more detailed assessment of resource capacity and a realistic master production schedule is required. With a realistic MPS and accurate work load information available a more informed decision to tender or not-to-tender can be made. Sales forecasts indicate that the proportion of bought out components will increase. It is therefore important that a purchasing strategy is formulated and that a material requirement planning system is introduced.

The GRAI technique is currently being used to design a modified production management system which will eliminate the inconsistencies identified. Fig. 8.6 shows a GRAI grid of the first draft of the modified production management system.

8.4 Evaluation of the GRAI method

The GRAI method has proved to be a valuable tool in the evaluation of the company's production management system. Its use has encouraged a

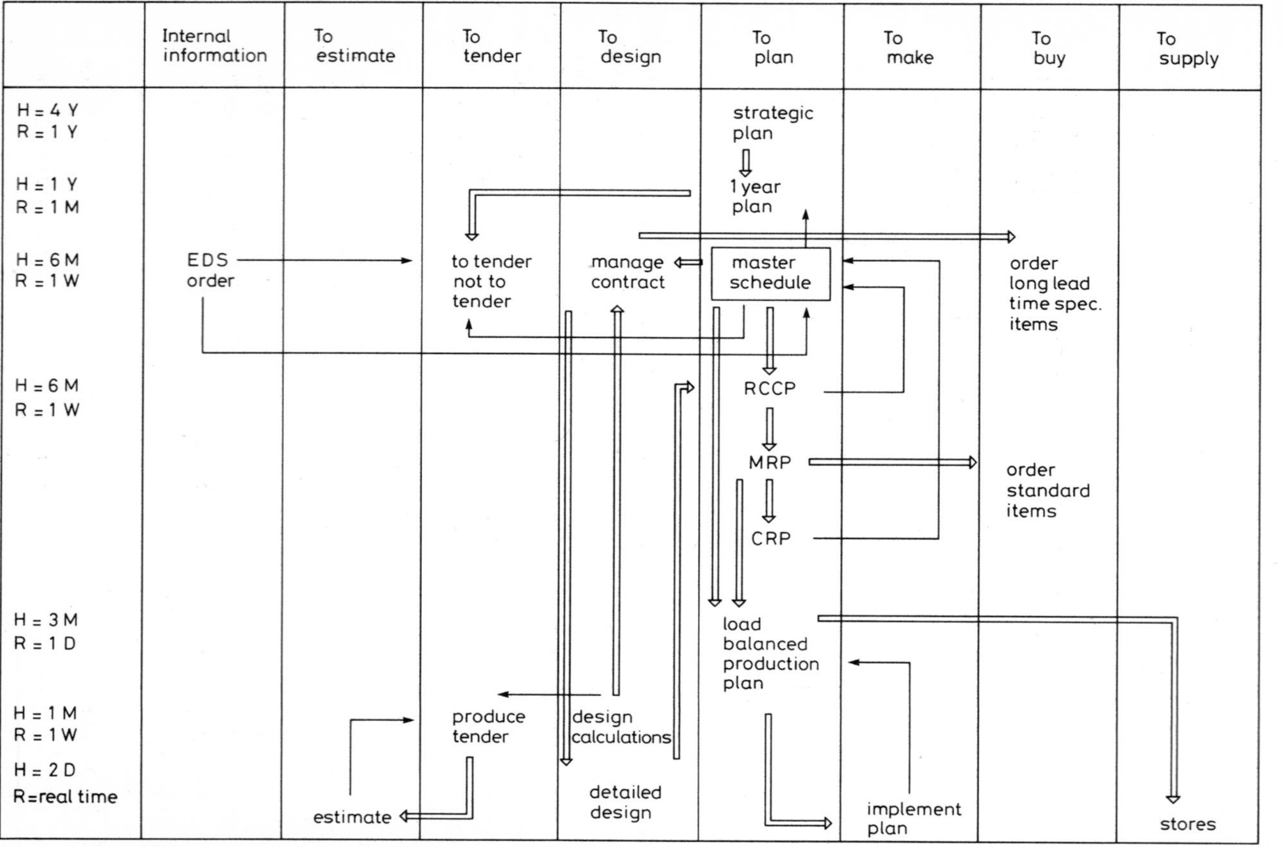

Fig. 8.6 GRAI grid of modified production managment systems

structured approach to the analysis and has assisted in the identification of a number of inconsistencies. In addition, the technique has encouraged the synthesis group to critically examine its effectiveness. The interview and evaluation phase are an integral part of the study and their importance should not be underestimated. In previous studies (Maisonneuve, 1987; Bourely, 1987), the synthesis group met at frequent intervals to evaluate the results. This was thought to be inadvisable in a small company and a repeated interview technique was adopted. This technique gave good results and caused little disruption within the company. In hindsight the original group meeting technique, used in previous applications of the method, has the advantage that the management team more readily adopt the recommendations of the study.

The study has also demonstrated that the GRAI method can be used to assist in the definition of CAPM requirements. When the decision centres have been identified the GRAI nets can be used to document the information required. For example Fig. 8.4 indicates that the Production Manager requires access to the following information:

(i) a breakdown of the machining content of the job submitted,
(ii) the spare capacity at each work centre,
(iii) details of any resources temporarily unavailable, and
(iv) details of any short term orders which may have to be scheduled into the workload.

The CAPM package purchased by the company does not provide information concerning items (i), (ii) and (iii) and the package must be modified to eliminate this discrepancy. The study demonstrates how the GRAI method can be used to analyse decisions and information requirements. It can be seen that the method can be applied in any situation where a structured approach to the analysis of information is required.

8.5 Acknowledgments

The authors are grateful to Tuke and Bell Ltd. for their co-operation throughout the study. Gratitude is also expressed to the staff of the GRAI laboratory, University of Bordeaux 1 for their technical assistance and support. In 1987 and 1988 the British Council funded study visits to the University of Bordeaux 1; this was invaluable in obtaining the necessary background information to utilise the GRAI method. The assistance of the British Council is gratefully acknowledged.

Chapter 9

The technological analysis of possible machine-tool investment options

W. Gerrard
Glasgow College, UK

9.1 Introduction

Recent research has found the manner in which engineers and managers actually approach the investment decision and the criteria they consider important to be in variance with the generally held theories and beliefs postulated by much of the academic literature (Gerrard, 1988*a*). To overcome these shortcomings, a step-by-step methodology has been proposed for the selection and introduction of new technology/machine-tools which more closely satisifies the needs of those making the investment decision (Gerrard, 1988*b*). The overall philosophy of the methodology is shown diagrammatically in Fig. 9.1 and indicates the need for *seven* distinct stages to be considered between the start and finish of any new equipment acquisition. As there are likely to be many equipment options which will satisfy the criteria established in Stage 2, the relative technological merits and demerits of each suitable option must be analysed and the most suitable options noted for further consideration.

9.2 Technical specification analysis

Responses to the survey of industry (Gerrard, 1988*a*) highlighted the relative importance of the many areas which require consideration during overall technological analysis as viewed by those making this analysis. Quality of production, equipment specification and cost were considered the three most important areas for consideration when selecting machine tools, and the full relationship was as shown in Fig. 9.2.

In order to establish a systematic approach to this analysis whereby merit points were awarded by comparing similar features of different machine tools, this data was used as the basis for creating

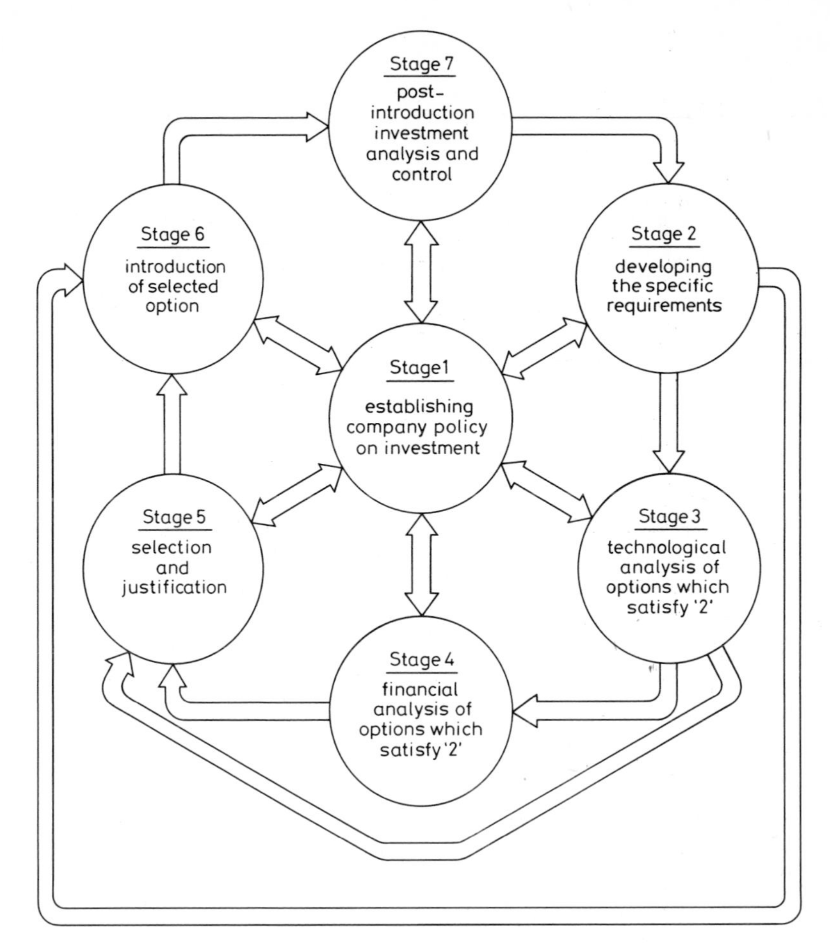

Fig. 9.1 Proposed methodology for the selection and introduction of new-technology/ machine-tools

weighting factors to reflect the relative importance of each broad area under consideration as shown in Table 9.1.

To assess the relative ranking of the different specifications for machine tool options proposed by various manufacturers, a precise checklist is required to allow the merits of each option to be evaluated in a systematic manner. Two different groups of factors for consideration are required depending on whether cylindrical or prismatic components are to be manufactured and examples of each checklist type are shown in Fig. 9.3 and 9.4 respectively.

Some of the factors included in these checklists will be governed by the machine requirements established in Stage 2 and as such certain features

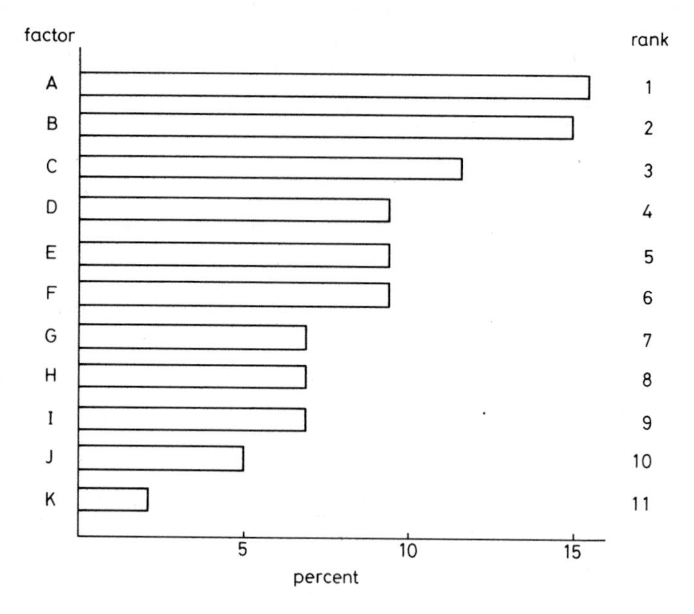

Fig. 9.2 Relativity between factors considered important in the selection of new technology/machine-tools

All Companies Factor		Points	%
A	Quality of production	658	16
B	Equipment specfication	646	15
C	Cost	553	13
D	Service provided	397	9
E	Ease of maintenance	390	9
F	Labour requirements	369	9
G	Delivery promised	314	7
H	Compatibility	308	7
I	Availability of association equipment	303	7
J	Ease of installation	198	5
K	Other factors	97	2

Table 9.1 Feature comparison of machine tools

will be specified in the 'preferred' column by the company; e.g. spindle bore diameter, swing over bed, distance between centres, etc. will all have been defined during Stage 2. Other factors will be specified by the machine tool manufacturers after considering the production requirements indicated them. All factors considered in each checklist relate to the Technical Specification of the machine tool proposal and as such must be assessed by technically competent personnel such as production engineers, manufacturing engineers or industrial engineers. Each feature of the machine tool must be considered and marks awarded out of ten to reflect suitability. All these factors, after totalling up all marks awarded, will be given an overall score which will then be converted to a mark out of ten to produce a score for inclusion in the overall technical analysis.

9.3 Overall technological analysis

Factors relating to the Technical Specification of any proposed machine tool are not the only ones considered when assessing its suitability, as has been shown from responses to the survey of industry. All categories outlined in Fig. 9.2 must be considered and weighted to emphasise their relative importance; to this end an Overall Technological Analysis Checklist as shown in Fig. 9.5 should be used to arrive at a basis for comparison reflecting the overall suitability of proposed options. This task can be undertaken by a panel of people at different levels within the company to avoid the influence of biased views and, ideally, should comprise persons who will ultimately be involved with the operation of the equipment once it has been installed, as this will have a positive effect by further encouraging a team approach. The panel should therefore be representative of:

- Production management
- Production engineering
- Production planning & control maintenance
- Shop floor supervision
- Shop floor operators/Trade Union representatives.

Machine			Sheet 1 of 3
Technical specification checklist – cylindrical components			
Specification feature	Preferred	Proposed	Merit points/10
Basic machine configuration			
1. Chuck capacity/size			
2. Spindle bore			
3. Swing over bed			
4. No. of axes			
5. No of speeds			
6. No. of speed ranges			
7. Actual speed ranges			
8. Feed rates			
9. Motor type			
10. Motor power			
11. Distance between centres			
Tooling specified			
12. Tool changer used			
13. No. of tool stations/turret			
14. Tool bar diameter			
15. Setting facility for O/D tooling			
16. Setting facility for I/D tooling			
17. Gauging facilities			
18. Tool holding facilities			
19. Turret anti-collision safety features			
20. Coolant facilities			
21. Tooling system/supplier			

Fig. 9.3 Technical specification checklist for cylindrical components

One limitation to this analysis would be that marks awarded for Technical Specification would not already be included on the Overall Technological Appraisal Forms given to panelists so as not to influence their opinion, but added later after completion of the forms. As before, points should be awarded out of ten for each category on the appraisal form and multiplied by their weighting factor to arrive at a total weighted assessment which will

Machine			Sheet 1 of 3
Technical specification checklist – prismatic components			
Specification feature	Preferred	Proposed	Merit points/10
Basic machine configuration 1. Worktable size 2. Worktable type 3. Worktable load capabilities 4. Table travel (x-Axis) 5. Table travel (z-Axis) 6. Spindle travel (y-Axis) 7. Gauge point position 8. Line of centre of worktable 9. Spindle design 10. Spindle speed range 11. Spindle drive motor type 12. Spindle drive motor power 13. Feed motors types 14. Feed motors power *Tooling specified* 15. Tool changer capacity 16. Tool shank type 17. Maximum tool diameter 18. Maximum tool length 19. Maximum tool weight 20. Tooling system/supplier 21. Tool selection i.e. random/fixed sequence 22. Programming required 23. Coolant facilities 24. Tool presence safety checks			

Fig. 9.4 Technical specification checklist for prismatic components

Machine			
Overall technological analysis checklist			
Machine feature	Category	Weighting	Merit points/10
1. Quality of production	A	× 16	
2. Quality of machine tool i.e. state of art etc.	A	× 16	
3. Machine tool specification	B	× 15	
4. Materials handling i.e. to/on/from machine	B	× 15	
5. Cost i.e. Machine, Installation/services, Foundations, Software packages, Tooling/workholding	C	× 13	
6. Service i.e. Back-up offered, Knowledge of supplier	D	× 9	
7. Maintenance i.e. Ease of access, Routine/planned, Availability of spares	E	× 9	
8. Labour i.e. Skill levels required, Training provided	F	× 9	
9. Delivery promised	G	× 7	
10. Compatibility i.e. with existing equipment	H	× 7	
11. Availability of associated equipment	I	× 7	
12. Ease of installation	J	× 5	
13. *Other features*	K	× 2	

Total marks for ranking purposes .... (*t*)

Fig. 9.5 Overall technological analysis checklist

be out of a total of 1300 points. By conducting this exercise and producing an average of all opinions for each proposed machine tool option, a ranked list of the options will be created to reflect the relative technological suitability of each machine tool proposed by different machine tool manufacturers.

9.4 Other technologial considerations

When considering the design of machine tools, it must be borne in mind that human operators will be involved in operating them unless the machines are fully automatic. Some other features of each machine tool option under consideration must therefore also be assessed under the broad headings of:

(i) ergonomic factors
(ii) control factors.

9.4.1 Ergonomic factors

Ergonomics tends to consider the appropriateness of the design of the man/machine system with respect to its operation, and when considering the suitability of each machine tool proposed, the following factors may require to be assessed; namely:

- overall layout of machine and total floor space required
- access to working area
- compactness of the machine tool
- position of controls with respect to the operator
- ease of loading/unloading
- additional cranage, e.g. capacities, access, operations etc.
- lighting and heating requirements
- safety requirements, e.g. provision of emergency stops, safety barriers, screens, etc.

9.4.2 Control factors

The design and positioning of the controls for a machine tool will depend largly on the type and physical size of the machine tool under consideration. If, for example, a large machining centre is being considered, an overhead swinging pendant carrying the necessary control console will be incorporated to allow the operator a degree of mobility whilst still being in control of the machine. In a smaller machine tool, e.g. a CNC lathe, there is no need for the same degree of mobility of controls, and hence these tend to be mounted on the machine in a position suitable for operator vision and access. The control system used for CNC will possess many features and these will be well documented in the machine tool manufacturers' literature. It is therefore important at this stage to recognise the likely effects these features will have on future operations and note these for consideration during the Justification/Selection Stage 5.

Factors which may require further consideration are likely to relate to:

- manual data input (MDI)
- tape preparation
- test piece manufacture
- simulation/diagnostics.

9.5 Summary of Stage 3 technological analysis

To sum up the procedure for analysing the technical suitability of the many options which may be proposed by different machine tool manufacturers, a diagrammatic representation of the overall process is shown in Fig. 9.6 and the basic concept may be defined as follows:

- The technical specification of each machine tool option under consideration must be assessed; this is achieved by completing the Technical Specification Checklist shown in Fig. 9.3 for cylindrical components or Fig. 9.4 for prismatic components with marks awarded out of 10 for each specification feature.
- The total marks awarded for technical specification are next converted to an overall mark out of 10 and this score is held on file for later inclusion in the Overall Technological Analysis Checklist.
- The overall suitability of each machine tool option under consideration must next be assessed; using the Overall Technological Analysis Checklist shown in Fig. 9.5, marks are awarded out of 10 for each technological feature by members of an appropriately balanced panel.
- The marks awarded in Step 3 are then multiplied by their relative weighting factor and a total mark arrived at which is out of 2100.
- The marks calculated in Step 4 for each machine tool option are averaged by adding all marks and dividing by the number of panelists to arrive at a basis for ranking the perceived suitability of each option.
- Finally, other factors relating to the introduction of each specific option are noted for further consideration.

9.6 Interactive computer technological analysis

The steps required to perform the Stage 3 Technological Analysis can, however, be laborious and time consuming, and consequently there is a need for a more succinct method of calculation. This has been achieved by

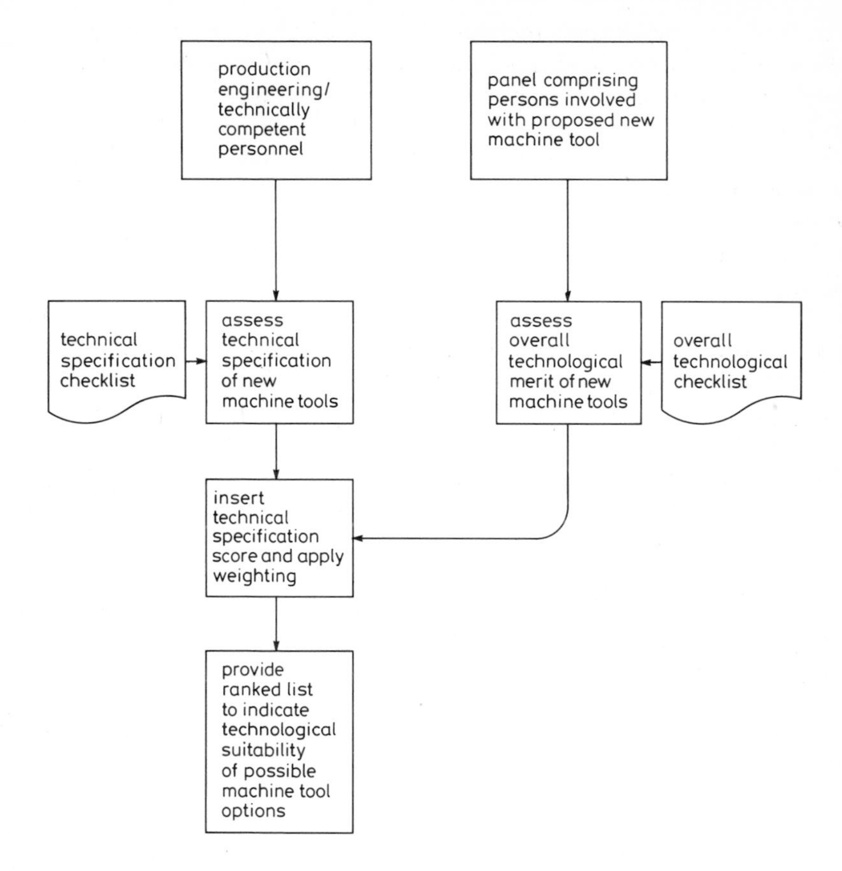

Fig. 9.6 Diagrammatic representation of technological analysis of possible machine tools investment options

the application of a computer, and to this end an interactive computer program has been developed using Fast Basic and any Atari ST computer. When considering a possible machine tool option, the program asks the user to award marks out of ten to reflect the suitability of the various features considered in line with the general methodology previously outlined. Two programs have been developed for Technological Analysis of machine tools for the manufacture of

(*a*) cylindrical components
(*b*) prismatic components. For each machine analysed, features are broken into five different categories, namely:

(i) machine configuration
(ii) tooling specified

(iii) general machine features
(iv) controls
(v) other features

And for each of the five categories, the average is calculated and displayed along with the overall average of the particular machine tool which is given at the end. The results of this analysis are displayed on the screen with the table formatted to provide an overall ranking of the marks awarded for different options. A third program to reflect the Overall Technological Analysis for each machine tool option is also organised in a similar fashion and results in a ranked listing which reflects the overall suitability of possible machine tool investment option. Space limitations prohibit the listing of this program, but typical hard copy results of a trial analysis of four machine tools are shown in Fig. 9.7. This was achieved by the random selection of values for each machine tool feature and entering these into the computer upon the appropriate prompt.

Press ANY KEY to continue

Machine	Basic machine configuration	Tooling spec	General features	Controls	Other features	Overall
4	7.00	7.10	6.83	7.22	7.00	7.05
1	7.55	7.30	6.67	6.78	6.20	7.02
3	7.27	6.70	6.50	7.22	6.60	6.93
2	7.00	6.00	7.00	6.44	7.40	6.68

Machine 4 is YASDA
Machine 1 is CINCINATTI
Machine 3 is YAMASAKI
Machine 2 is SMT

Fig. 9.7 Typical hard copy results of a trial run analysis

9.7 Summary and conclusions

Thus, in Stage 3, the analysis of the many technological features associated with the selection of any new machine tool is considered in a systematic manner with due weighting applied to reflect their relative importance. In this way, a ranking of the relative technological suitability of the various proposals which satisfy Stage 2 of the methodology is achieved. This ranking is considered in the Selection and Justification Stage 5 in conjunction with the results of the Financial Analysis performed in Stage 4 which forms the subject of Chapter 10. By employing an interactive computer program, the Engineer/Manager is reminded of the features he should be considering. The program guides the user through each step, allowing for input errors, and provides a concise table of results which clearly indicates the machine tool most suitable along with specific areas of strength/weakness. These features make the methodology/program a most useful tool for any company involved in investments in new equipment. By removing the vagueness which so often surrounds the selection decision, a clearer picture of the technological suitability of possible investment options is achieved.

Chapter 10

The financial analysis of possible machine-tool investment options

W. Gerrard
Glasgow College, UK

10.1 Introduction

A methodology has been proposed which guides the engineer or manager through the various steps required to select and introduce new technology/ machine tools (Gerrard, 1988*b*). Following Stage 3 of that methodology it is necessary to analyse the financial viability of each of these options.

From responses to a survey of industry (Gerrard, 1988*a*), it was clear that in the companies surveyed, the Payback model played a dominant role in the financial appraisal of new technology/machine tools. Of all companies surveyed, 55% considered that Payback was the main model used for assessing the relative financial ranking of different machine tool investment options; when considering the use of Payback at any time, this figure rose to 75%. However, the Payback model has certain inherent theoretical deficiencies. Not the least of these deficiencies is its total disregard for the 'time value of money', i.e. the fact that £1 today is worth more than £1 in a year from now. Thus the shortcomings of the Payback model are considered in detail below, as are the advantages of those appraisal methods which have their theoretical basis in discounting techniques; following this, a methodology is proposed for systematically assessing the relative financial merits of different investment options which will more closely meet the needs of industry.

10.2 The Payback Model

The Payback Model calculates the number of months or years required to recover the capital outlay for a new investment.

In its favour, basic advantages quoted by those who advocate its use include:

- It is readily understood by all involved in the financial appraisal/selection process.
- There is a certain logic in attempting to recover the investment outlay as soon as possible.
- It is a useful tool when accountants talk to non-accounting functional managers.
- It is readily adaptable to changing economic conditions: i.e. in times of booms — extend the acceptable payback period; in times of slumps — reduce the acceptable payback period.
- There are minimal assumptions required as to future events, e.g. interest rates, etc.

When considering those employing the Payback model at any time, the usage increased to 75% of all companies surveyed; with companies tending to use more than one technique when financially assessing the worth of different investment options.

From those advocating its use at some time, the following responses were noted:

- Payback tends to be used as a back-up to a DCF technique.
- It is considered an aid for reducing risk; Payback tends to highlight those machine tool options which recover the investment outlay in the shortest time — arguably reducing risk.
- Payback tends to be used to differentiate between two machine tool options considered very closely matched after employing a DCF technique, with Payback analysis tipping the scale in favour of one.
- Payback can be used to rank the various machine tool options in times of investment capital shortage.

However popular the Payback model may be, it has been well documented in the relevant literature that, as a basis for financial appraisal of investment options, the model is inadequate. It does not give any indication of profitability nor does it take cognisance of differing economic lives of alternative investment options; as a result, it is argued that the Payback model has an inherent inability to assess the economic worth of an investment. However, by far the most serious omission of Payback analysis is that it does not take into account the time value of money, in that it treats all revenues similarly irrespective of when they are accrued. To conclude, therefore, it is apparent that a vast section of management opinion tends to favour a Payback model approach for the financial appraisal of possible investment options; not through any theoretical or academic correctitude, but more due to the fact that it is very alluring to managers in search of a simple answer to very complex questions. The use of a discounted cash flow technique resulting in the net present value

(NPV) of a flow of future revenue is, however, theoretically correct and will be discussed in the following section.

10.3 Discounted cash-flow models

As has been stated, the shortcomings of the Payback model have led to the application of Discounted Cash Flow (DCF) based techniques. It is generally considered that these provide a more objective basis for evaluating possible machine tool investment options both in terms of magnitude and in terms of timing of expected cash flow over the anticipated economic life of the investment option. The two DCF models in use in industry are the internal rate of return (IRR) and the net present value (NPV) techniques, the mechanics of which are as follows:

10.3.1 Internal rate of return (IRR)

The IRR for an investment option is the discount rate that equates the present value of expected future cash outflows with inflows, and is represented by the rate r such that:

$$\sum_{t=0}^{n} \frac{A_t}{(1+r)^t} = 0$$

where A_t

n = last period in which a cash flow is expected

Thus r is the rate which discounts a stream of future cash flows from A_1 to A_n to equal the initial outlay at $t = 0$. Investment options are selected using the IRR criterion by comparing the IRR criterion by comparing the IRR with the required rate of return for the company which is set as a matter of company policy. If the IRR exceeds the required rate of return, the investment option may be deemed acceptable and the different rates of IRR established for each machine tool option can then be used as a basis for ranking the different investment options under consideration.

10.3.2 Net present values (NPV)

Like the IRR method, the NPV model is a DCF technique used for capital budgeting and financial appraisal. With NPV, all cash flows are discounted to their present value using the required rate of return such that

$$\text{NPV} = \sum_{t=0}^{n} \frac{A_t}{(1+k)^t}$$

where k = required rate of return and if the sum of these cash flows is equal to or greater than 0, then the investment option is deemed acceptable, for, if the required rate of return is the return investors expect the company to earn, it is accepted that investment proposals with an NPV > 0 will maintain or increase the shareholders' wealth. Although the levels of NPV of different investment options can then be used as a basis for ranking the suitability of investment option, it must be noted that different values of NPVs will accrue by applying different required rates of return, and, as such, the company's investment policy regarding the acceptable rate of return plays a crucial role in the final investment decision made using this DCF technique. In general, IRR and NPV methods lead to the same acceptance or rejection decisions. However, important differences exist between the methods and, in some cases, this can result in conflicting answers with respect to acceptance or rejection of a particular investment option. This discrepancy comes about owing to the fact that the IRR method implies a re-investment rate equal to the IRR whereas the NPV method implies a re-investment rate equal to the required rate of return, which is also used as the discounting factor. Consequently, the NPV model is generally accepted as being theoretically superior to IRR and for the reason cited, the financial justification criterion used as the basis of the proposed financial appraisal methods for possible machine tool options is based on the NPV model.

10.4 Proposed financial appraisal methodology

Any manager given the task of financially assessing the appropriateness of a particular investment option must be aware of the different purchasing policies which can be adopted in order to acquire the machine; namely outright purchase, hire purchase or by leasing the machine. To buy outright is the most common policy, although with hire purchase there is the aspect of not totally committing the company's liquid resources and hence the advantage to the company's case flow position. Furthermore, with hire purchase the machine is legally the property of the company whereas, with leasing, the machine is always owned by the lessor, and as such any investment grants or depreciation tax benefits remain with the lessor. Leasing may not always be an option which is available for new technology/machine-tool purchases, and generally only when a company is paying very little corporation tax with few ownership rights does leasing become attractive. In the manufacturing industry, this situation does not normally exist and as such no further consideration will be given, and discussions will concentrate on outright purchase and hire purchase approaches.

10.5 NPV techniques for financial appraisal of possible investment options

10.5.1 Assumptions for proposed NPV appraisal methods

When considering the financial appraisal of new-technology/machine-tools, the dynamic nature of technological developments along with possible changes in product markets must be borne in mind. Owing to the uncertainties relating to these factors, it is only just possible to forecast events over the next five years or so with any degree of confidence, and beyond this horizon predictions become fraught with uncertainty. Consequently, the costs associated with the benefits likely to accrue from any new capital equipment purchases can be reasonably clearly defined over a five year span, but beyond this tend to be difficult to forecast. For these reasons, it is argued that any proposed appraisal technique should consider only those costs and benefits which can be defined with a fair degree of confidence. Consequently, the methods proposed for the financial appraisal of both outright purchase and hire purchase of capital equipment consider only those costs and benefits accrued in the year of introduction along with a further five full working years. The methods also assume the notional sale of the new equipment at the end of the sixth year and use the NPV calculation as a basis for comparison between different possible investment options.

10.5.2 NPV technique for financial appraisal of outright purchase

The worked example shown in Fig. 10.1 considers the financial appraisal of the outright purchase of a CNC machining centre at a basic cost of £350 000. The overall cost, anticipated savings, capital allowances and effects on taxation are included in the calculation of net cash flows over the subsequent six years after introduction, and these are discounted to provide a DCF as shown.

10.5.3 NPV technique for financial appraisal of hire purchase

The worked example shown in Fig. 10.2 considers the financial appraisal of the hire purchase of the same CNC machining centre at a basic cost of £350 000. The overall cost, down payment, anticipated savings, subsequent payments, capital allowances and effects of taxation are

Ref. Note	Item / Year	1	2	3	4	5	6	7
a	Purchase Cost	(350,000)						
b	Sale of Old Equipment	10,000						
c	Other Costs	(5,000)						
d	Investment Grants	53,250						
e	Total Costs	(291,750)						
f	Anticipated Savings	50,000	100,000	100,000	100,000	100,000	100,000	
g	Capital Allowances	(85,000)	(63,750)	(47,813)	(35,859)	(26,895)	(20,171)	
h	Net Increase in Taxable Profits	(35,000)	36,250	52,187	64,141	73,105	79,829	
i	Tax Due on (h)	12,250	(12,688)	(18,265)	(22,449)	(25,587)	(27,940)	
j	Tax Payable		12,250	(12,688)	(18,265)	(22,449)	(25,587)	(27,940)
k	Sale of New Investment							50,000
l	Net Cash Flow	(241,750)	112,250	87,312	81,735	77,551	74,413	22,060
m	Discounting Factor	1.000	.833	.694	.578	.482	.402	.335
n	Total NPV Stream	(24,750)	93,504	60,595	47,243	37,380	29,914	7,390
o	NPV of Investment							34,267

Fig. 10.1 Worked example showing NPV method for financial appraisal of outright purchase of £350 000 CNC machining centre using a discounting factor of 20%

included in the calculation of the net cash flows over the subsequent six years after introduction and these are discounted to provide a DCF as shown.

10.5.4 Summary

By applying the methods outlined, financial appraisal for both outright purchase and hire purchase of the three or four best options highlighted in the Stage 3 Technological Analysis phase can be carried out. It will thus be possible to create a ranked listing of investment option representing their financial contribution to company profits if purchased outright or by means of a hire purchase agreement. This listing can be used in the Stage 5 Selection & Justification phase as a basis for comparison between different options; however, up to this point no consideration has been

Ref. Note	Item / Year	1	2	3	4	5	6	7
a	Purchase Cost	(350,000)						
b	Sale of Old Equipment	10,000						
c	Other Costs	(5,000)						
d	Investment Grants	53,250						
e	Total Costs	(291,750)						
p	Down Payment	(58,350)						
q	Annual Repayments	(64,768)	(64,768)	(64,768)	(64,768)	(64,768)		
f	Anticipated Savings	50,000	100,000	100,000	100,000	100,000	100,000	
g	Capital Allowances	(85,000)	(63,750)	(47,813)	(35,859)	(26,895)	(20,171)	
r	Interest on HP	(18,088)	(18,088)	(18,088)	(18,088)	(18,088)		
s	Net Increase in Taxable Profits	(53,088)	18,162	34,099	46,053	55,018	79,830	
t	Tax Due on (s)	18,581	(6,357)	(11,935)	(16,119)	(19,256)	(27,941)	
u	Tax Payable		18,581	(6,357)	(11,935)	(16,119)	(19,256)	(27,941)
k	Sale of New Investment							50,000
v	Net Cash Flow	(73,118)	53,813	28,875	23,297	19,113	80,744	22,059
m	Discounting Factor	1.000	.833	.694	.578	.482	.402	.335
n	Total NPV Stream	(73,118)	44,832	20,039	13,466	9,213	32,459	7,390
o	NPV of Investment							54,281

Fig. 10.2 Worked example showing NPV method for financial appraisal of hire purchase of £350 000 CNC machining centre using a discounting factor of 20%

given to the risks associated with each capital investment option and this will be examined next.

10.6 Consideration of risks associated with capital investment

Every investment in capital equipment has an element of risk associated with it. When the equipment finally selected arrives on the factory floor, it may take longer to install than was anticipated and, as a result, the expected savings may take longer to accrue. When the equipment eventually comes into full production, sales prices and demand may differ significantly from those anticipated, resulting in a reduction in expected profits. For these reasons, it must be recognised at the outset that there is always a possibility that the actual financial returns from any investment may be lower than those anticipated during the formal analysis of investment options. Consequently, some index reflecting the susceptibility of each investment to risk is required and a Discounted Payback Method is proposed.

10.6.1 Discounted payback risk analysis

As was indicated earlier, the Payback model has many shortcomings as a means of financial appraisal. However, as a means of assessing the risks associated with an investment, these shortcomings become advantages in that Payback analysis highlights those options which recover their investment outlay in the shortest time — thus reducing risk.

The formula proposed determines the average annual returns over the six production years assumed for the machine tool under consideration and is defined as follows:

Discounted payback period (DPP)

$$= \frac{\text{Initial outlay}}{\text{Average annual returns}}$$

For outright purchase in Fig. 10.1, this becomes

$$\mathrm{DPP} = \frac{(n)^{1}}{{}^{7}_{2}\;(n)\,/\,6}$$

$$= \quad 5.255 \text{ years}$$

For hire purchase this becomes 3.432 years.

Clearly, the measure of risk associated with the outright purchase of the machining centre appears to be greater than that associated with hire purchase. However, it must be borne in mind that hire purchase repayments require to be honoured even if no further use for the machine existed after the year of its introduction. Thus the DPP method of assessing the risks associated with different investment options should be used to compare these options under similar purchasing conditions; i.e. to provide an indication of risks associated with the outright purchase of all options under consideration *or* to provide the same indication for all options if purchased under a hire purchase agreement.

10.7 Summary

The shortcomings of financial appraisal of capital investment using payback criterion have been discussed and in the light of this, two financial appraisal models based on DCF have been proposed. The models assess the relative financial contribution to profits of each investment option under consideration if purchased outright or under a hire purchase agreement. The risks associated with each investment option have also been considered and a Discounted Payback Method proposed as a means of assessing the sensitivity of each possible investment option to changes in expected operating conditions.

Thus, after the Stage 4 financial analysis phase, enough information on possible investment options with regard to technological suitability, contribution to profits and susceptibility to production related risks should exist to allow progress to the next Stage 5 Selection and Justification phase.

Part 2:

Management strategies

Chapter 11

Introduction

H. M. Ryan
Sunderland Polytechnic, UK

The chapters in Part 2, while grouped within the framework of management strategies, cover a wide range of subjects including: the challenge of new technology to conventional manufacturing (thinking) strategy and the obligation to embrace it if the UK is to stay competitive in world markets; effective organisation of high technology environment and its implementation in manufacturing companies; effective office support for redesigned manufacturing systems; implementation and development of CIM from a managerial perspective and as a strategic management issue or form of business control; flexibility through the design of manufacturing infrastructures; modular-assembly/systems-integration for flexible manufacturing; while one chapter poses the question — Management information — is it worth it?

Chapter 12 by C. C. New provides a valuable overview of the challenge of incorporating new technology to conventional manufacturing, ranging from a discussion on alternative approaches to improving time productivity in manufacturing; consideration of CIM; new technology investment in the UK; the reasons for the UK investment failure; from islands of technology to CIM; and finally presenting an action plan for management.

Chapter 13 by Wilson and Whitehurst considers effective organisation of a high technology environment, namely Westland Helicopters Ltd. The chapter outlines the company's strategy which identified an essential need for dramatic reductions in lead time and working capital. The chapter describes one of the specific initiatives under this programme.

Johnson, Lucas Engineering and Systems Ltd., UK, (Chapter 14), discusses effective office support for redesigned manufacturing systems. The chapter opens by referring to an earlier companion paper by Parnaby (1988*a*) which highlighted, by means of a Performance Comparator, the gap in performance between a typical European and a typical Japanese engineering manufacturing company. The object of Johnson's paper is to explain two key concepts:

(i) the elimination of waste
(ii) natural groups which apparently are fundamental to the Lucas approach to redesigning a total business.

These two aspects, together with several other important concepts also detailed in the reference literature quoted, make up the total systems engineering approach to redesigning a total business favoured by Lucas.

A managerial perspective relating to CIM implementation is presented by Prabhu and Kimble (Chapter 15) who make reference to a 1988 regional survey highlighting managerial experiences of CIM reported by some 40 manufacturing companies in the north east of England. Some of the issues considered included: what choices of CIM are they making; why are they implementing CIM; what problems do they face in implementing CIM; what changes do they foresee; implications of users, practitioners, CIM vendors and advisors. This chapter attempts to draw together issues critical to the successful implementation of CIM both from users' perspective and external advisers who recommend CIM systems.

In Chapter 16 Chuah considers CIM development as a strategic management issue. He considers that: 'CIM should not simply be a technology driven system. Its development and implementation must be based on business needs. It is a strategic management issue which requires a correct approach and the right attitude from the outset. Physical integration itself is often not enough. CIM will only be effective and efficient if the information needed to direct and control all the business functions and production processes has been designed to flow through the system smoothly and efficiently'. In order to do this Chuah considers that there must first and foremost be a thorough review of the company's business, product, manufacturing and marketing strategy and practice in all aspects. 'There is no generally agreed CIM module or sub-system structure. Management has to decide whether total or partial integration of these modular functions and activities is more appropriate to the company's needs. It is important that the information and communication requirements of these inter-related functions are clearly defined at the earliest stage of any proposed CIM development. Effective communication, with or without the aids of computer systems, is a pre-requisite for successful company operation'.

The design of such an information system is often a difficult task. Long established company culture and organisational structure may not be conducive to changes in such direction. Moreover, existing systems may be difficult or expensive to modify to complement new set-up. The general problems of CIM development are discussed, but the main aim of Chapter

16 is to emphasise the need for a strategic approach by top management to establish the overall objectives and framework of the CIM system.

Chapter 17 discusses the implementation of an IT strategy in a manufacturing company (ICL). The chapter describes clearly how ICL faced this issue, evolved a strategy and worked it through (identifying key issues/elements), to become a highly successful example of IT use in manufacturing.

Chapter 18 describes flexibility through the design of manufacturing infrastructures. The chapter, which presents the six basic functions of the production management infra-structure (e.g. designs, materials, methods information, orders, facilities and trained/motivated personnel) stems from an ACME sponsored research grant held jointly by the CIM Institute, Plymouth Polytechnic and the Change Management Research Unit, Sheffield Business School, entitled 'The development of a user-led implementation methodology for integrated manufacturing systems within the electronics industry'. The chapter is based upon the early results of this work.

The need for flexibility in assembly operations in the sector of electronics manufacturing is well known, as is the development of 'market driven' strategies for electronics component production which will enable companies to compete on the basis of responsiveness both in terms of delivery and the ability to provide precisely what the customer wants. Chapter 19 presents a valuable contribution describing modular systems integration for flexible manufacturing — the case of low volume, high variety assembly. The authors discuss the DRAMA methodology (design routine for adopting modular assembly) and indicate the present state of its development/application.

Chapter 20 discusses computer integrated manufacturing (CIM), its application within a manufacturing environment and its relationship to business control. It argues that the current view of CIM can be considered parochial, simplistic and limiting, and has contributed to the widespread confusion, and misunderstanding surrounding this subject. This, it argues, is one of the main contributing factors in its slow take up. It concludes that CIM is a philosophy which sets a direction towards total business integration, where all elements of the business are integrated to optimise performance. It also argues that there is a close relationship between CIM and business control; a control hierarchy is proposed which supports this argument and concludes that there is a need for an integrated approach to business control.

It suggests that level 2 of the control hierarchy should be responsible for the integration of systems into the overall business process and would provide great potential for redesigning the business to respond to high level inputs. The paper concludes that there are eight business elements which require integration during a CIM undertaking.

The modern manufacturing managers have a vast amount of data available to them. Chapter 21 presents a brief contribution which attempts to answer the question: is the cost of producing management information value for money? Retrospective and forward projections are made to develop this theme. The final chapter in Part 2 (Chapter 22) by Davis, John Brown Automation Ltd., UK, is in the nature of a case study relating to integrated assembly for flexible manufacture. It outlines:

(i) the wide mix of technologies involved to achieve a high level of productivity
(ii) the integrated quality assurance procedures and relevant management control and progress monitoring strategies.

Chapter 12

UK manfacturing: the challenge of transformation

C.C. New

Cranfield School of Management, UK

12.1 The challenge of new technology

The challenge of incorporating new developments in manufacturing technology into manufacturing strategy can be seen as the laying down of a series of major challenges to conventional manufacturing thinking:

(i) Reduce work in process by 50% or more.
(ii) Reduce lead time by 50% or more.
(iii) Facilitate the introduction of new products at two to three times the existing rate on half the current design/development lead times.
(iv) Reduce 'support' labour by 50% or more.

It is not enough to think of mere 'tinkering' with existing systems and technologies; such radical changes require a totally novel approach. Yet such an approach is perfectly feasible if one examines the current 'norms' for manufacturing in most engineering companies, even those with relatively high volume final assembly processes. The normal conditions are:

- bought out content 50%
- throughput efficiency in component manufacturing, that is the ratio of work content to total lead time, of 20% (or 80% of the time queuing or idle)
- throughput efficiency relatively high in final assembly
- additionally we need to recognise: extensive periods in materials stores, finished component stores and often finished goods stores.

Any approach which can reduce the 'idle' time spent in the system will in fact achieve the first three objectives automatically and is likely to make labour reductions far easier to accomplish. However, instead of concerning itself with the 'idle' portions of time productivity, most Western management seems to have been concerned with the (much smaller) 'busy' periods, that is with concentration on the actual work task. While this is clearly not to be ignored it seems to have been a case of not seeing the wood for the trees.

12.2 Some alternative approaches to improving time productivity in manufacturing

12.2.1 The Japanese (Just-in-Time) approach

This involves tight control over the flow of orders through the system to maintain very low queues. It requires considerable delegation to operators, a very directed application of manufacturing engineering to reduce set-up-time and high levels of commitment to continuous and relatively constant production. Even the Japanese have only been able to apply this approach in a limited range of industrial environments but these have shown major benefits.

A European version of this courage + control + people-power approach can be seen in the application of Group Technology principles, which incidentally are very much the current vogue in the USA despite their relatively unfashionable image in the UK.

12.2.2 The systems-intensive approach

The more usual Western approach has been through sophisticated data processing systems in order to 'solve' the manufacturing problem. This involves extensive data collection and daily scheduling of every operation. While this approach seems to have been more broadly applicable than the Just-in-Time (JIT) approach, its effects have been far less dramatic: a move from 'one operation per week to two or three operations per week' is certainly an advantage but it may still mean only say 10 hours work in a 40-hour week, a throughput efficiency of only 25%.

12.2.3 The capital-intensive approach

The use of capital investment in high technology manufacturing equipment in order to effect time productivity is not new. It started with multi-axis machining centres and other multi-operation machines, the objective being to reduce the number of operations; the ultimate aim being one operation per part. Too often, however, the promise was unfulfilled because, although 10 components of a 100 component assembly were produced in 4 weeks instead of 14, the other 90 still took 14 with no effect on the customer lead time. Alternatively, the planners went overboard in putting as many components as possible on such machines only to generate bottlenecks ahead of them. The successful application of the capital intensive approach requires a plant-wide view to be taken and careful planning of capacity.

12.2.4 The time-intensive approach

This simple approach has often been neglected; multi-shift operation is a very effective method for shortening the total elapsed lead time, since throughput efficiency relates hours of work to available hours. Thus at an efficiency of 25%, 100 hours work would take 10 weeks at 40 hours per week but only 5 weeks at 80 hours per week on a two-shift basis.

The likely impact of the newest manufacturing technologies can best be understood by seeing them as providing the vehicle through which all of the four fundamental approaches described above can be implemented simultaneously:

(a) Just-in-time production is made possible through the few operations, tight deterministic scheduling and minimum tool change times.
(b) Again, because of the high level of predictability obtained and the reduction in lot sizes, close co-ordination of component requirements is not only necessary but is also significantly facilitated.
(c) The solution is clearly part of the capital intensive approach but it is important to re-emphasise the point that it must be a plant-wide view that is taken in the overall system design.
(d) Multiple (unmanned) shift operation becomes a reality.

12.3 Computer-integrated manufacturing (CIM)

The complete CIM model includes at least the following major modules:

12.3.1 Computer-aided design (CAD)

This includes the production of schematic drawings and detail dimensioning and may include finite element analysis and simulation capabilities.

12.3.2 Computer-aided production engineering (CAPE)

The CAPE module in general accepts data from the CAD module, and from this process routings, operation lists and times can be produced. In addition it will normally include provision for tooling and the design of jigs and fixtures. CAPE would also usually produce the necessary computer control instructions/tapes for both manufacturing machines and test equipment.

12.3.3 Computer-aided production planning (CAPP)

The CAPP module is probably the most familiar and has been around for some time in the form of computer scheduling systems. CAPP takes inputs from CAPE and the order acceptance systems, and using its own basic data carries out load and capacity planning and detailed scheduling.

12.3.4 Computer-aided manufacturing (CAM)

Although used in a general sense this may include everything except CAD; it is convenient to regard the CAM module as only carrying out those activities directly related to machining processes. CAM therefore operates the production machines and carries out real time job sequencing.

12.3.5 Computer-aided storage and transport (CAST)

The final basic module is concerned with materials handling. This may include warehousing, automatic stock picking systems and materials conveyancing. It would also control any automatic guided vehicles (AGVs) used within the manufacturing system.

The basic links between these modules are illustrated in Fig. 12.1.

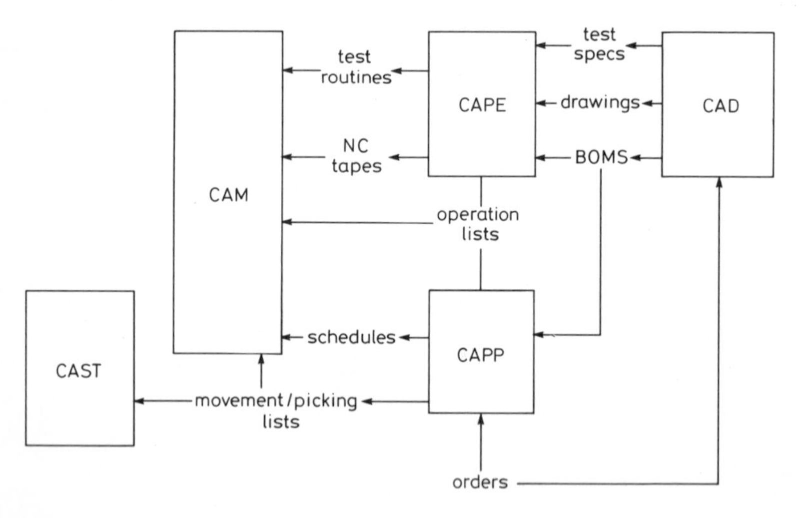

Fig. 12.1 Computer-integrated manufacturing

12.4 New technology investment in the UK

In a recent survey of UK manufacturing operations (New, 1986), plants were questioned concerning their current and future plans relating to investments in the new manufacturing technologies.

For many of the plants the perceived payoffs to date from new technology seem to have been low or even non-existent. While this in itself should not be surprising what is more worrying is that the failure to obtain short term results may be discouraging companies from pursuing these technologies further. If this view is carried through it spells disaster for the UK economy in the long run.

The 240 sample plants included 155 in engineering and related activities — the obvious major users of much of the available new technology. However, of the 64 plants reporting experimentation with Flexible Manufacturing Systems (FMS) technology, two thirds reported low or negative payoff to date. Of the 69 experimenting with robotics over three quarters reported low or negative payoff. Taking CAD and CAM (Computer aided design and manufacture) together, while two thirds of the companies who could possibly make good use of the technologies did in fact report using them, almost half did not think they had made any significant gains from their introduction. Tables 12.1 and 12.2 show for comparison the plants' ranking of the specified technologies on the basis of payoff to date and future emphasis.

Table 12.1 New technologies: Payoffs to date, 1985 survey

	Payoff	
	Negative, none or low	Moderate to high
CAD	46%	54%
CAM	46%	54%
MRP	19%	81%
FMS	67%	33%
Robots	76%	24%

Table 12.2 New technologies: Future emphasis: Next 2 years

	None to moderate	Fairly high or high
CAD	54%	46%
CAM	54%	46%
MRP	21%	79%
FMS	75%	25%
Robots	85%	15%

In relation to planned future emphasis the results do appear to be slightly disturbing. Less than half the respondents intended to put high or even fairly high emphasis on CAD/CAM and this proportion dropped to 25% for FMS and a meagre 15% for robotics — in fact around one in three reported *no* emphasis on the latter two. Only the computerised production planning and control technology represented by MRP showed any real payoff to date or indication of extensive future emphasis: 56% of plants reported fairly high or high payoffs already and 79% intended putting that level of effort behind such systems over the next two years — good news for the computer companies and software houses.

Table 12.3 New technologies: growth in user base

	No. of user plants	No. of plants indicating future use	% increase
CAD	106	167	57%
CAM	102	165	62%
MRP	167	217	30%
FMS	64	115	79%
Robots	69	111	61%

Total sample size: 240

The slightly encouraging side of the results is seen best from Table 12.3 which shows that there is at least a fairly dramatic increase in the number of plants intending to pick up the new technologies over the next two years. For example, while only 64 plants reported themselves as using FMS technology to date, 115 reported some future emphasis (that is, ranking its importance above 'none') indicating a 79% increase in take up. Similar increases are apparent for the other technologies. The technique showing the least percentage increase, MRP, was already being widely used.

Table 12.4 Competitive priorities of UK manufacturing plants

Rank	
1.	Consistent quality
2.	Dependable delivery
3.	Low costs
4.	High performance products
5.	After sales service
6.	Fast deliveries
7.	Rapid product design changes

12.4.1 Competitive priorities

In terms of competitive strategy in the marketplace the current ranking of the 'competitive edge' criteria planned by the UK respondents is shown in Table 12.4. The degree of importance apparently attached to the top ranking items — consistent quality and dependable delivery — is impressive. Almost three quarters of respondents ranked consistent quality as 'high', rising to 97% ranking it fairly high or high. Similarly 58% ranked dependable delivery as of high importance rising to 92% fairly high or high. There is of course some element of 'motherhood' in such results — no one is likely to say such things are unimportant. However, if we compare the UK results with those from the European Futures Study (Ferdows, 1985) and from the North American Manufacturing Futures Survey (Miller and Vollman, 1985), as shown in Fig. 12.2, some interesting comparisons emerge.

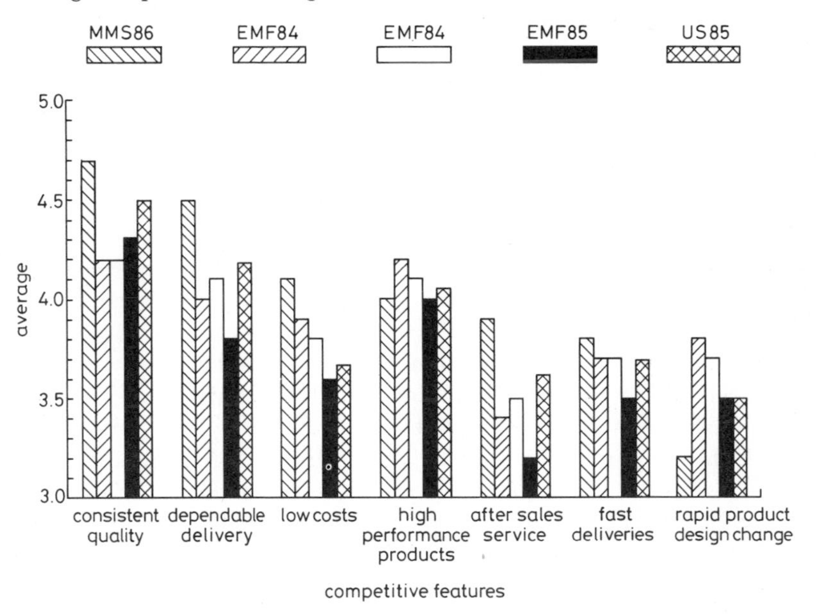

Fig. 12.2 Competitive priorities: Comparison between UK, USA and Europe

The similarities are not surprising; consistent quality is clearly a number 1 priority in the UK as it is in the USA and as it is in Europe. In second place for the UK and the USA comes dependable delivery, but for the European sample as a whole the ability to produce high performance products was ranked higher than dependable delivery in 1983 and 1985 and equal with it in 1984. In fact, as can be seen clearly in Figure 12.2 the UK rankings in absolute terms tended to be higher than for the US or European sample except in two directly related areas: the ability to produce high performance products and even more markedly the ability to implement rapid product design changes. For the UK sample the ability to produce at low cost ranked higher than the ability to produce high performance products whereas in every case for the European surveys and for the US this ranking was reversed.

This is a strange and worrying finding for a sample heavily biased towards the UK exporting industries. The emphasis in UK manufacturing is still apparently seen to be on the production of fairly standard products at low costs. It is clear that the developed economies of Europe are much more concerned with the rapid introduction of new high performance products than with low cost production — and this is almost certainly the only viable long term strategy for a highly developed economy. To be fair this trend was more marked in the consumer and intermediate products plants than in those producing capital equipment: while 74% of all plants rated low costs 'fairly high' or 'high' only 63% of the plants producing capital goods did.

The other striking feature of Fig. 12.2 is the difference in the rating assigned to 'after sales service' between the UK and US plants, who rated it fairly important, and the European plants, who generally ranked it below any of the other criteria. Does this indicate a competitive weakness of European companies? This could possibly be exploited by UK manufacturers who are clearly aware of this factor's crucial importance, particularly in the capital goods field. In fact in the UK survey, four out of five plants in the capital goods sector rated after sales service at the fairly high/high level.

One could perhaps argue that with less emphasis on new product development and design flexibility the UK plants perceive less of a requirement for some of the new technologies. However, some of the most important potential benefits to be gained from CAM, robotics and FMA are likely to be in the areas of quality and delivery reliability, so this argument seems to be difficult to support in practice.

Returning to delivery reliability, it is clear that the plants all rated the ability to deliver on time to quoted delivery dates as a major competitive factor. Yet the reported delivery performance of the participating plants was actually fairly poor: almost one in four plants reported delivering less than half of their customers' orders on time. Setting a very modest target of '75% or better on time delivery' only 46% of the plants achieved this. Moreover, the proportions as we have seen are remarkably consistent between 1975 and 1985: in 1975 only 44% of plants achieved an on-time delivery target of 80% or better — plus ca change c'est la meme chose!' Again, in 1975 40% of plants reported that they had no formal system for measuring performance against promised delivery dates and this figure appears if anything to have gone up (to 46%) rather than down.

The situation which seems to emerge is one of 'lip service' to delivery reliability: the plants almost unanimously rate it as important but: one in four plants deliver more orders late than on time and the median plant in the sample delivers one order in every four after its originally quoted delivery date. Finally, only about half the plants even bother to monitor their actual delivery performance. The strategy is good, the tactics simply don't deliver.

12.5 Reasons for the UK investment failure

It is clear from the survey results reported above that there is a reluctance in the UK to invest even in the 'Islands of Technology' and consequently even more reluctance to move towards CIM.

If we consider a typical FMS capital expenditure proposal it is not hard to understand the reluctance of management to take up such investment:

	Conventional	*FMS*	*difference*
Number of machines	68	18	-78%
Number of operators	215	12	-94%
Floor space (m^2)	9750	3000	-69%
Lead time (days)	80	3	-96%

This all looks very impressive and certainly from a labour productivity point of view it *is*. However, when we examine the expenditure requirements we find that the FMA requires something like an additional £13 million up front capital investment and shows savings of the order of:

One off (inventory reduction)	£2.8 million
Annual (labour savings)	£1.0 million

The result? A DCF rate of return of less than 10% and a Payback period of 10 years. For a project which requires main Board approval this is clearly an unlikely proposition. The bald outline figures of this proposal provide all the clues as to *why* such investments appear unattractive and why UK management in particular have been slow in putting money into radically new technologies. Let us examine each of these in turn.

12.5.1 Inadequate tools of financial analysis

(*a*) *Use of excessively high hurdle rates or short payback periods*

Typically, companies require applicants for investment funds to produce a detailed DCF analysis for the proposed project and to submit this for consideration alongside other proposals. A 'hurdle' rate is established by the controlling function (usually the financial controller or accountant) and projects not getting over the hurdle are rejected. A less sophisticated management might require payback within a given time.

On the assumption (not necessarily true but not for discussion here) that capital is in short supply, management tend to set high hurdle rates in order to 'allocate' the available funds — to high return projects. Such hurdle rates are often of the order of 15 — 30% and may reflect the current return on assets being achieved.

Technically there are a number of problems with this approach. The major weakness is that a company's true hurdle rate is in fact its marginal opportunity cost of capital — in other words, the real rate of return which could be obtained by investing the money in something else. This long term rate is around 8 – 10% *not* 15 – 30%. Moreover, the problem is compounded by setting a high discount rate (which includes inflation) but using constants in the cash flow projections — a practice which is of course totally inconsistent. Such errors, even assuming the cash flow projections are complete, introduce a heavy bias against all forms of advanced technology investment because of:

- long start-up periods,
- long project lives, and
- delayed returns.

Similar conditions apply in the case of short payback periods. How can an FMA which will take up to two years to fully implement payback within three years from start-up – which actually implies a payback closer to one year.

The return on current assets is often used as the basis for the hurdle rate on the reasoning that any investment which does not show this return will dilute current return on assets. This is certainly true but is also irrelevant. If a company is making say 30% return on £1 million of assets and invests a further £100 000 at, say 10%, its return on assets will drop to 28%, but if in order to do this it *borrowed* the £100 000 at a *real* cost of say 7% it is now making 10 000 – 7000 = £3000 extra profit per annum. In other words, companies give up extra profit in order to maintain an artificially high return on assets. In any case, the main reason for such returns has nothing to do with the profit being earned but rather with the low valuation of the current asset base.

(*b*) *Use of inappropriate alternative investment assumptions*

In preparing an investment proposal for new technologies it if often the case that the 'do nothing' option assumes future conditions of stable market share, selling prices and costs. Such assumptions clearly fail to take account of possible competitive action. If any one competitor in the market *does* make such new investment, over time such assumptions are unlikely to remain true. In most cases, the new technologies provide a substantially lower marginal unit cost. In order to achieve a better return on the initial investment a competitor using the new technology is likely to use such cost advantage to increase market share. It is obvious that such status quo assumptions make new technology investment look less attractive. The true base-case comparison should probably be falling market share, falling sales and rising unit costs.

In the same way, comparison of new technology investment with 'like-for-like' replacement often assumes an inappropriate life for the 'old' technology. Flexible manufacturing systems by their nature may be flexed to meet changing product requirements. Traditional forms of product specific investment may often have their true life limited by product changes — this is of course particularly important for high volume manufacturing businesses.

We have already discussed the use of high cut-off or hurdle rates but we should repeat again the fact that using a hurdle rate of 30% implies that there are alternative investment opportunities available which do give a *real* rate of return of 30%; this is simply not true.

c) Bias towards incremental investment

In most companies the capital expenditure proposal system is set up around authorisation levels: up to £10 000 might require department head approval, £30 000 plant manager approval and so on. It is therefore not uncommon to see clusters of proposals which manage to come in just below the established ceilings. This practice sets arbitrary limits on incremental investments and forces many junior managers (who raise the proposals) to think only in terms of such incremental investments.

It may well be that each incremental investment may *apparently* be justified in its own right; the problem is that added together the increments actually cause dysynergy and are certainly not as effective as a total innovation could have been. It is also not uncommon to discover that if, for example, all the inventory reductions claimed for each incremental investment were added up the company would have a negative inventory (!), as much post-investment auditing often shows the savings claimed in order to meet high hurdle rates simply do not materialise in practice.

Perhaps an even worse consequence for the company is that the existence of a whole set of ongoing incremental investments actually precludes consideration of revolutionary investment proposals. At any point in time, there are a number of investments whose benefits are still outstanding, and to abandon them in favour of a radical change clearly indicates that such investments should not have been started — that is an admission of error by the managers concerned. The correct strategy would appear to be to assess the technological life of the existing plant and only accept investments which show a return within that period, say three years — the intention of course being to replace the existing plant with a radically new technology at that time. Is that what UK plants are currently doing with their short payback requirements? Do they really intend to scrap their existing plants in three years anyway? I wish I believed that that were true.

(*d*) *Failure to include all tangible benefits*

Typically investment proposals have included only the most obvious of the tangible benefits expected and only those which it has been possible to quantify easily. As it happens this did not matter very much before the advent of CAM/FMS systems simply because the changes were not dramatic enough. If a new machine requires say 25 m^2 instead of 30 m^2 the cost of the factory space was hardly very relevant. With FMS type systems, however, we are talking of space reductions of the order of 70% and inventory reductions of a similar order — these offer very real opportunities for cost reduction.

The tangible benefits which need to be accounted for certainly include:

- *Inventory reductions* in terms of investment savings and space and warehousing costs.
- *Floor space reductions* taking into account the true opportunity cost of new space and such items as heating/lighting etc.
- *Quality improvements*, including the benefits of process repeatability and monitoring, waste reduction, inspection reduction and reduced warranty payments to customers.
- *Wage inflation protection*: The far lower number of people involved in running the new technologies clearly provides considerable protection against general wage inflation. In a sense, the plant is able to 'freeze' much of its unit cost through the capital investment. This need not, however, be seen as exchanging variable (labour) cost for fixed (machine) cost because these concepts become irrelevant. In practice, much labour cost has become fixed and different depreciation policies which relate to output volume rather than simple passage of time make much more sense for the new technologies.
- *Higher equipment utilisation*: Even when fairly substantial increases in downtime are allowed for technical problems, the new manufacturing

technologies can still provide dramatic increases in *real* utilisation through unmanned operation and flexible scheduling.

(*e*) *Failure to include soft or intangible benefits*
Even when all the possible tangible benefits are included in the analysis, there are still a lot of extremely important factors which remain unaccounted for:

(i) higher market penetration due to short reliable lead times
(ii) schedule dependability due to deterministic scheduling
(iii) the learning curve benefits which can only come from adoption of the technology
(iv) the volume flexibility in terms of output for the market which is possible through the use of unmanned operation
(v) labour stability *because* of volume flexibility with a given labour force
(vi) product flexibility to move quickly as the market changes.

Most of these characteristics are *revenue enhancing* rather than *cost reducing* and for this reason are regarded, particularly by accountants, as highly subjective and therefore of dubious value in the analysis. It seems to be much easier to accept the idea of a 10% cost reduction based on a well established and detailed cost statement than the nebulous idea of a market share increase due to better delivery performance. It is obviously difficult to value such benefits but that does not make it correct to assume that they are all zero! The trouble with much 'traditional' accounting is that it prefers *precision* to *accuracy*; it would rather be precisely wrong than vaguely correct. One useful way of including such factors in the analysis is to establish what the annual return from such soft benefits *would need to be* in order to reach the required hurdle rate and then ask: 'Is this a feasible value to put on such soft benefits?'

12.5.2. The management communication gap

The gap in communication between manufacturing engineering and senior management has always been a problem but the new technologies have brought this to a new level of significance. To be fair, senior management have been right to be scpetical in the past and their faith has hardly been restored by the failed promises of the benefits form individual machining centres put into traditional environments. What is needed today are manufacturing engineers with vision and a wide market perspective who are not afraid to raise capital expenditure proposals for figures 10 or even 20 times the corporate norm. On the management side too we need a long term perspective, a commitment to major changes in organisational structure and the will to make it work.

There are usually three clearly identified management groups in many companies faced with the new technologies:

- *Senior management*: They know it is necessary to do *something* for strategic reasons and they are worried about what the foreign competitors are doing.
- *Junior manufacturing engineers*: They (if trained properly) know the technologies and have the expertise but they have little influence and find it difficult to relate to the competitive strategy of the business.
- *Middle manufacturing management:* They, too often, do not know the technologies and feel uncomfortable about them and, in any case, are being measured on short term results.

The existence of these three different positions often leads to considerable frustration for both senior management and the manufacturing engineers, and considerable stonewalling by manufacturing management.

12.5.3 Risk aversion and the fear of failure

Most European and particularly most UK managers suffer from a severe form of risk aversion. The reasons for this are complex and cultural but the underlying problem is fear of failure on the part of the individual. In the United States, it is readily accepted that if you try lots of innovations some of them will fail, but there need be no stigma attached to the individual managers involved in the failure (unless it is clearly caused by incompetence or mismanagement) — it is better to have tried and failed than not to have tried at all. In Japan, individuals are perceived to be carrying through a consensus decision so that failure (or success) is shared by all and again no stigma attaches to the individual. In the UK, however, two factors seem to govern many aspects of managerial behaviour particularly in manufacturing:

(*a*) Most managers see themselves as being in their current job for a very limited time span (2 — 3 years).
(*b*) Visible failure in their current position will prevent further progression whereas visible success will guarantee their onward progression in the organisation.

The snag with having these two factors together is obvious — it leads to short term decision making and strong risk aversion.

Of course there will be difficulties with implementing new technologies, of course there will be teething troubles and loss of output and of course there will be extra costs not known at the start of the projects. However, how many new investments can you think of that have not eventually met most of the initial objectives? Even though individual managers may only be in place for a relatively short period the organisation is intended to continue indefinitely — new technology is a long term commitment.

A manufacturing manager investing substantially in new technology raises his asset base, lowers his return on assets and under current accounting conventions raises his fixed cost — small wonder so few are keen to do it when they are measured on *annual* performance.

Perhaps the most relevant insight here is a general failure to differentiate between 'sins of commission' and 'sins of omission'. A manager is responsible for a sin of commission if he/she does something which turns out in retrospect to be an error — this is to be expected occasionally if the manager is operating in an innovative way. Sins of commission are therefore unfortunate but should hardly be penalised unless regularly repeated. By comparison, a manager is responsible for a sin of omission if something goes wrong as a result of something the manager did not do (but by implication should have done). This is serious and should be strongly penalised in terms of future career progression — it implies that the manager is failing to exert proper custodial control over the business. Applied to the area of new technology investment this idea has much to commend it. It is clearly possible (even inevitable) to commit sins of commission. It is, however, absolutely certain that the 'do nothing' option will in most cases be a serious sin of omission.

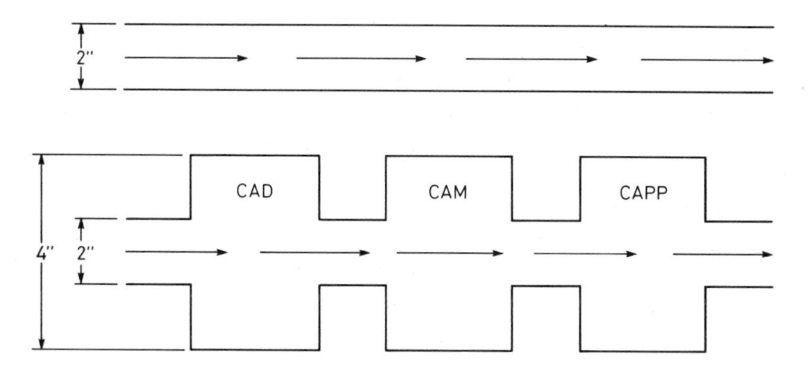

Fig. 12.3 Independent computer modules do not necessarily improve productivity

12.6 From Islands of Technology to CIM

If, as we have discussed, there are considerable barriers to investment in the individual Islands of Technology there are even more barriers to integrating the islands together to achieve a CIM environment. In justifying a CAD investment, for example, the design office may do this based on such benefits as design productivity, component standardisation etc. In justifying a CNC machine the manufacturing manager may do this on cost reduction, inventory reduction and better quality. Each of these investments may apparently increase productivity in their own area but the real question is whether the business *as a whole* is actually better off. One useful analogy to illustrate this is to imagine the business as a fluid flowing through a pipe (Fig. 12.3). We started with a 2 in diameter and an appropriate flow level. The CAD investment expanded the diameter pipe of the pipe in design to 4 in, the CAM investment expanded the diameter in component machining to 4 in and similarly a new CAPP (MRPII) system was installed which made it possible to double productivity in planning and control — to a 4 in pipe. The snag? The CAD interface with the CAM systems still created a bottleneck of a 2 in pipe and similarly information availability between the CAM system and production control still restricted flow to a 2 in pipe. The result — no real benefit to the business as a whole.

My recurring nightmare of lack of system integration is a designer using a CAD system to produce new drawings (twice as fast as he used to) so that he can fling them 'over the wall' into manufacturing that much faster. When they arrive in manufacturing, the manufacturing planning engineer's first task is to feed back into his computer all the necessary data from the drawing in order to use his own CAPE system. This of course prints out operation lists and possibly Bills of Material (BOM), which he posts to the production control department, and tooling and fixture requirements, which he posts to the tool drawing office. Production control enter the operations lists into their routing file (for CAPP) and the BOMs into their product structure file. The tool drawing office of course uses a different CAD system chosen by the chief tool designer and so re-enter the detail part co-ordinates before commencing tool design. Meanwhile the CNC programmers are of course preparing from scratch the machine tapes needed for CAM and the plant engineers are similarly transferring the data into the automatic warehousing system which uses customer identification numbers instead of the internal part number.

The dream is of an integrated set of systems each able to operate independently but each able to draw off or supply appropriate interface data at the touch of a button.

There are, however, two very real and very distinct barriers to such integration: one technical and the other behavioural.

12.6.1 The technical problem

I am (I think reliably) informed that — given time, expertise and money — you can in software terms interface anything with anything. In fact, despite all the (to me largely incomprehensible) technical problems, the

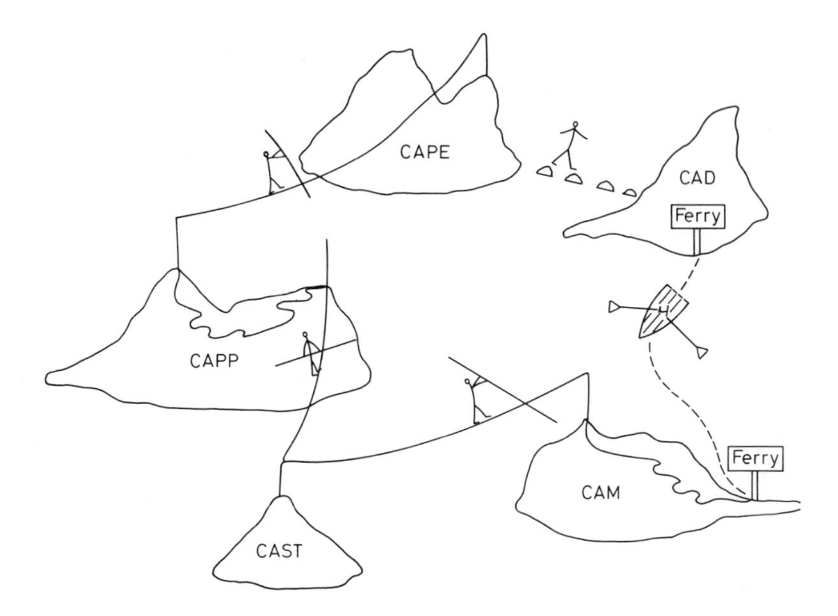

Fig. 12.4 Islands of Technology: The 'fix it' software solution

basic requirement is quite straightforward. Each Island of Technology needs a 'mailboat' package of data to be transferred between islands — the difficulty comes of course in specifying the data package in the first place, having transport which is suitable and arranging for a frequency of service which matches the needs of the organisation. The resulting 'integration' more often that not takes the form of Fig. 12.4 with various methods of transfer ranging from the 'daily ferry service' through the 'DP highwire Act' to the 'stepping stones only usable at low tide'.

The technical problem is non-trivial and there are any number of cases in which a design office has invested in a particular CAD system in order to improve design productivity only to discover that appropriate post-processors were not always available to enable direct links with CAM.

If the technical problem is daunting, the behavioural problem is even worse.

12.6.2. The behavioural problem or 'keep off my patch'

The behavioural problem is that the integration of the systems described requires a correspondingly integrated approach from management, Design engineers can no longer sit in splendid isolation in the drawing office and fling designs 'over the wall' into manufacturing without regard to available tooling, machine capability or capacity. Design, operation planning, machine choice and capacity planning are strongly interrelated and the 'design to production' transfer is a continuous (and interactive) process. The key point, however, is that optimisation of the whole process is unlikely to come from optimisation of the individual Islands of Technology — the individual functional managers cannot be allowed to run their own show as if the other areas did not matter. In practice, this means that there will be limits on system choice placed on the individual areas in order to enable overall integration to be achieved. There are a number of possible ways in which this problem can be tackled and we return to these below.

12.7 Action plan for management

It is always dangerous to recommend 'global' action plans, but from our discussions a number of clear points do emerge and it is appropriate to summarise these here under a number of headings.

12.7.1 Major changes in financial evaluation procedures

- realistic hurdle rates (8%)
- high for fix-its
- low for long-term high-tech
- premium for 'as-is' technology
- include all tangible benefits
- allow for intangible benefits: use 'hurdle' annual benefit requirements needed to make investment worthwhile
- compare with realistic alternatives assuming competitive action
- top-down authorisation
- abandon ceilings?

12.7.2 Set up strategic technology teams

- senior management
- middle manufacturing management
- technically qualified manufacturing engineers
- information technologists
- engineering designers.

To address the problem: 'Given that with no action we will cease to be competitive, what bundle of investments will maximise competitiveness on the following dimensions . . .'

To answer such a question the team will clearly also require inputs from marketing and product development.

12.7.3 Initiate a major education and training programme

For
- senior
- middle
- supervisory management
- accountants
- financial controllers.

In
- manufacturing systems engineering
- technologies
- information processing
- financial justification of high-tech investment.

12.7.4 Incorporate long-term strategies into short-term performance measures

- teething problems
- productivity variances
- indirect support costs
- training costs.

12.7.5 Establish a specific technology input to the manufacturing strategy

- 5 year manufacturing technology plan
- 1, 3, 5 year implementation plans.

12.7.6 Organisational structuring

- creation of design to production (DTP) teams.

This is the major feature of most attempts to overcome the problems of organisational boundaries. Fig. 12.5 illustrates three approaches which have been used:

(i) specialist designers and DTP teams to take the basic design from concept to finished item
(ii) DTP teams which are product (or product group) specific but which can call on expert 'staff' help when necessary from consultant designers, manufacturing engineers or part programmers
(iii) totally dedicated DTP teams by product group. This is a form probably most suitable to the larger company with few high volume product lines.

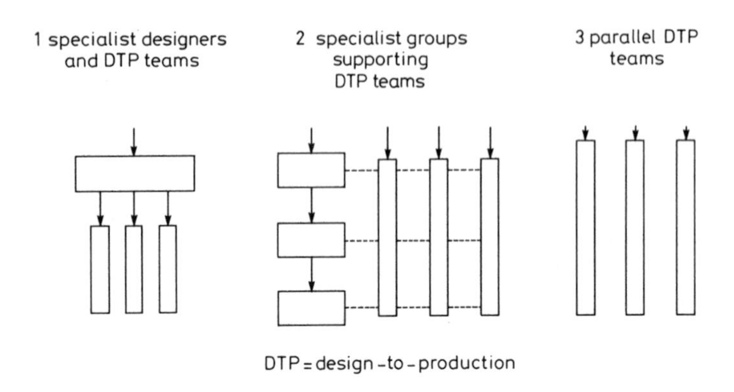

Fig 12.5 Role convergence under CIM

DTP: Design to production

12.8 Conclusion

The challenge of transformation is there but it is necessary to see the new technologies as an opportunity not a threat. Consideration of whether to use the new technologies is no longer the issue. It is not an option but an obligation if the UK is to stay competitive in world markets. We are in danger not merely of falling behind in the competitive race but of dropping out of it altogether.

Chapter 13

Effective organisation of a high-technology environment

R. J. Wilson and J. Whitehurst
Westland Helicopters Ltd., UK

13.1 Introduction

The name Westland became commonplace owing to media coverage of a highly publicised debate about the pros and cons of European or American money being used to support the financial reconstruction of the company in 1985. However, the subject of this chapter concerns only one of the three subsidiary companies owned by the Westland Group, namely Westland Helicopters Limited, who employ 4200 personnel mainly in Yeovil, Somerset.

In the context of the book, the background is simply a low volume high variety manufacturing operation with high levels of product engineering change and consequently a complex set of interactions which impinge on manufacturing, and therefore business, performance.

The company had made progressive improvements in manufacturing performance, but determined that if both margins and competitiveness were to be improved significantly, there was a need for a step change. Dramatic reductions in lead time and working capital were essential elements of overall strategy. The cornerstone of these improvements was to be organisational technology rather than process technology and the chapter describes one of the specific initiatives under this programme.

13.2 Production systems

Lynx and Sea King helicopters have around 20 and 30 thousand parts respectively and control of assets is consequently a significant task for which a suite of computer systems is used. At the heart of this suite is Westland's MRP system, the PRR (Production Requirements Reporting) (see Fig. 13.1).

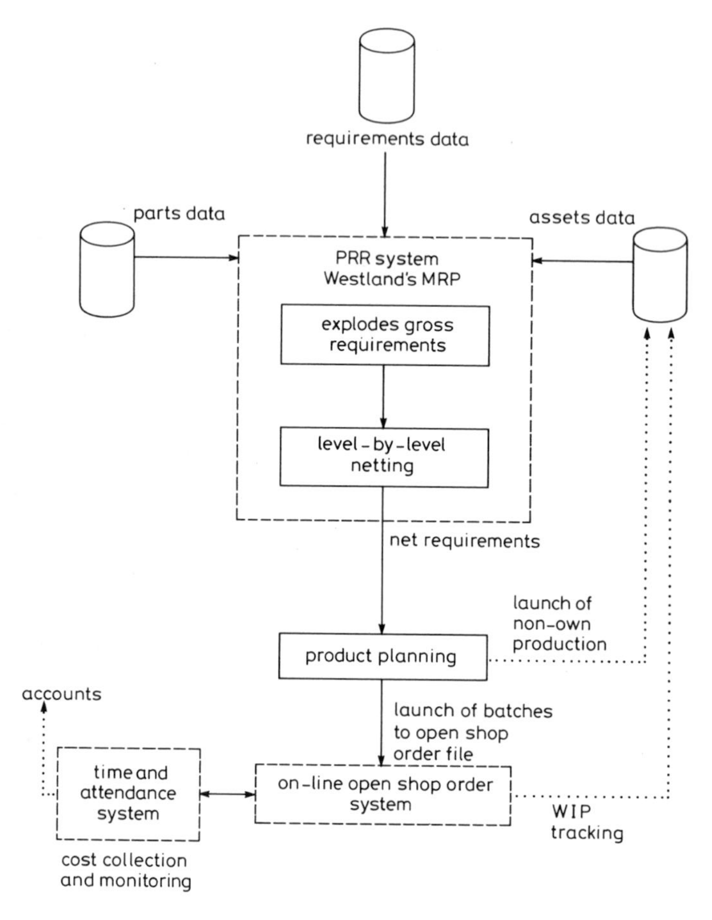

Fig. 13.1 Current production control system

The PRR determines from the 'order book' gross requirements in terms of quantity and delivery for aircraft, or spares assemblies. These are then 'exploded' via the Bill of Materials which breaks them down into detail parts. A level by level netting process is then carried out whereby assets (in the form of WIP and stock) are subtracted from the gross requirements at each assembly level to give the net requirements. The result of these calculations is that a series of reports are printed which enable the appropriate planning areas to launch work to the shop floor and to initiate purchase orders for equipment, hardware and sub-contract work.

There are some very real problems with such a complex planning system; these include:

- poor data integrity (as with many centralised systems with no ownership of data)
- no finite scheduling capability, often leading to unachievable schedules
- constant rescheduling of due dates with little or no consultation with those affected by them.

The above have led to results with which many readers will be familiar: long lead times, high levels of work in progress, poor adherence to schedule and a general lack of visibility of the manufacturing process. In parallel with the development of these complex systems, there has been a tendency to invest in very high value capital equipment leading to the creation of a number of Islands of Technology. Our experience of this approach has shown that, whilst individually many islands have been extremely successful, collectively, they have had minimal impact on total factory performance, owing to a lack of focus on overall organisational improvements.

It was therefore imperative that we began to take a 'total systems' approach, whereby we redress the imbalance of previous years. Consequently, we have taken a broader view over the past year and have embarked upon a Systems Engineering Programme whereby the total factory system is being redesigned using concepts such as simplified material flows, natural groupings, team working, ownership and accountability for the process, and an embracing of the Just-In-Time philosophy.

13.3 The process of change

In November 1987, we took the Board of Directors and a few senior managers off-site for a two day seminar on the systems approach. The objective was to convince the Board that a process akin to that being undertaken within Lucas Industries PLC, adopting world best practices under the umbrella of Manufacturing Systems Engineering, could provide a route by which the company might make a step change in performance and competitive position.

Having gained general agreement and varying levels of commitment, the task of deciding where to begin lay before us. The change in working methods and practices that we envisaged for manufacturing was to be fairly dramatic. The first project had to be, and had to be seen to be, an unqualified success if further projects and redesigns were to be undertaken. What we required was an area in the company with a management receptive to change, and which was also largely self-contained, thus minimising the impact on the rest of the company. Gear manufacturing stood out as an ideal candidate, and on further examination it transpired that within our strategically important transmission area, the gears not only added the highest value, but the longest lead time and high cost of quality. This combination of factors convinced us that the gear shop could provide the success story and the necessary 'step change' in performance that we were looking for.

13.4 The gear facility

Our gear production facility is a £10 million business contributing significantly to the profits of the company. Gear production is split evenly between spiral bevel and parallel axis gears and we manufacture around sixty different types of each. The environment itself is a highly complex and demanding one, both technically and in terms of production control. For example, many of our gears conform to AGMA 13 standards, whereby tolerances on key dimensions such as tooth spacing and involute profile can be as low as 5 microns (see Fig. 13.2).

Involute spur	
pitch circle diameter	300
module	2.5
number of teeth	120
pressure angle	20°
face width	100

	AGMA quality number			
units μm	10	11	12	13
single pitch	12	9	6	4
cumulative pitch	36	27	18	12
profile	15	11	8	5
lead	23	18	15	10

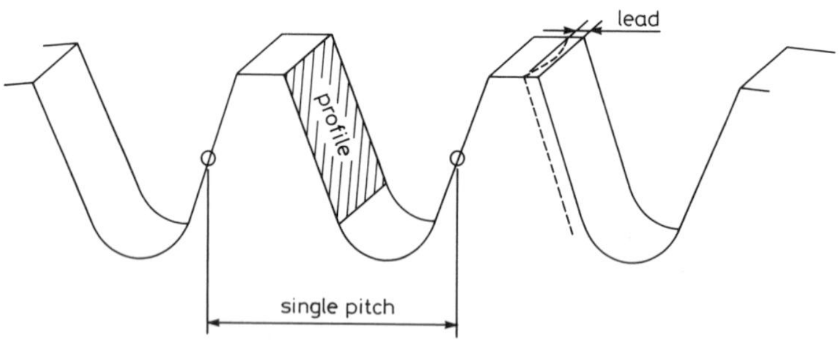

Fig. 13.2 Example of AGMA manufacturing tolerances

Additionally, smooth material flows are hampered by the need to schedule around 11 500 part number operations on over 100 machines. Batch production requires that each of these machines undergoes a set-up changeover after every batch, and the characteristics of the machines dictate batch sizes of typically 20 — 25 gears. These changeovers are often lengthy and can take up to 24 hours on some of our high precision, spiral bevel grinders. Combining this with the scheduling of complex material flows around the functional layout results in average lead times currently in excess of nine months.

13.5 The gears taskforce

Following the decision to redesign the gears facility, a full time taskforce was set up to carry out the project. The taskforce was multi-disciplinary in nature and included a shop floor representative, a supervisor, a technical specialist, four newly appointed Manufacturing Systems Engineers and a Manufacturing Systems Engineer from Lucas Engineering and Systems Ltd. Although the taskforce was allowed great freedom to enable a 'bottom-up solution, they were given some clear terms of reference. We knew that we wanted shorter and simpler material flow paths to give us enhanced visibility of the process. Above all we wanted to put the power to effect change and initiate improvements back to where it belonged — on the shop floor. Thus we could begin to break down traditional 'not my problem' or 'not my job' attitudes by encouraging shop floor responsibility and accountability for process and product integrity.

Conceptually, we believed that some form of cellular structure would provide the solution. The dedication of a family of parts and machines to a specific cell would give us shorter and simpler material flow paths. Furthermore, the combination of team working and better organisation within the cells aimed to foster the right attitudes amongst the workforce. Therefore, the brief to the taskforce was to examine the feasibility of forming cells within the gear shop and to create a new, appropriate job structure to support them. The taskforce was also given a set of demanding objectives and targets to meet, based on an anaylsis of current performance levels. These are shown in Fig. 13.3

	Now	1990	Target
Stockturns	2	6	12
Schedule adherence	4.5%	100%	100%
Leadtimes	9 months	3 months	2 months
Quality defects	156	5%	0%
Effectiveness	25.7/37.5	32/37.5	32/37.5

Fig. 13.3 Gear shop objectives

13.6 Methodology employed

The first activity in the redesign process was to plan the project stages in detail (see Fig. 13.4 for outline). The following is a commentary on the methodology employed, and the results obtained from each of these stages.

13.6.1 Data collection

This phase was a data collection exercise during which the aim was to collect a variety of product, process and market data which would be required throughout the project. Typically items of data collected included routing information for each part, an understanding of the market issues until 1992, a scrap and rework history, and part/machine characterisations.

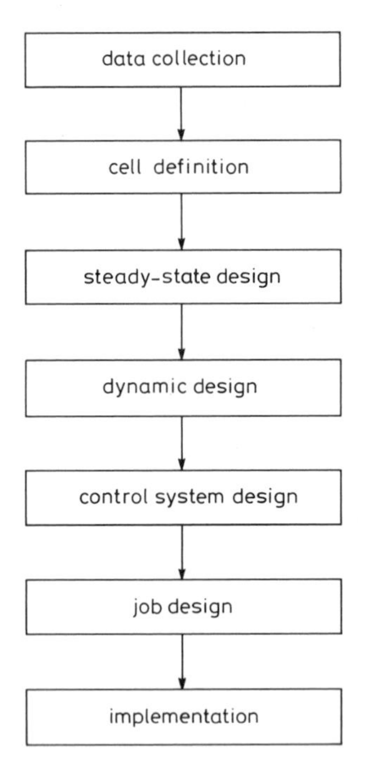

Fig. 13.4 Outline project plan

13.6.2 Cell definition

Cell definition is the process of finding part and machine groupings to form cells and was the key to the success in increasing competitiveness and meeting market needs. The cells formed would lay the foundations for the rest of the manufacturing system redesign and it was important to bear in mind several key/strategic issues (see Fig. 13.5) when deciding upon the approach and techniques to use.

We eventually chose to form cells around the main gear cutting and grinding machines because of their very high value. Primary cells were formed around these key machine groups using both the knowledge of the families of parts which existed and Modroc software which uses an algorithm based on rank order clustering. Individual machine identities for the key machines were then assigned and the primary cells created by allocating support machinery such as lathes, drills and general grinders. Resolution of any overlap between the cells was obtained by making use of:

- approved machine alternatives
- accepting some movement of parts between cells
- purchasing duplicate equipment.

The output from the cell definition stage was six machining cells (three spiral bevel and three parallel axis) and two support cells; a heat-treatment cell and a finishing cell for the final non-destructive test and surface treatments which are common to all parts.

- Need to satisfy WHL own production
- Market segmentation strategy
- Close relationship with Sikorsky
- Volume flexibility
- Generic parts families exist
- Very high value gear hobbing, shaping and grinding machines

Fig. 13.5 Key issues

13.6.3 Cell design

During the cell design process (see Fig. 13.6) the part/machine groupings from the cell definition phase were validated and the cells were tested for size using the volume and demand data gathered during the data collection phase.

The robustness of individual cells was assessed via the process of dynamic design, whereby the cells were modelled using spreadsheets to analyse the effects of changes in design variables such as levels of demand, scrap, rework etc. Other outputs from this stage included manning requirements for each of the cells and a nominal layout of the cellularised facility.

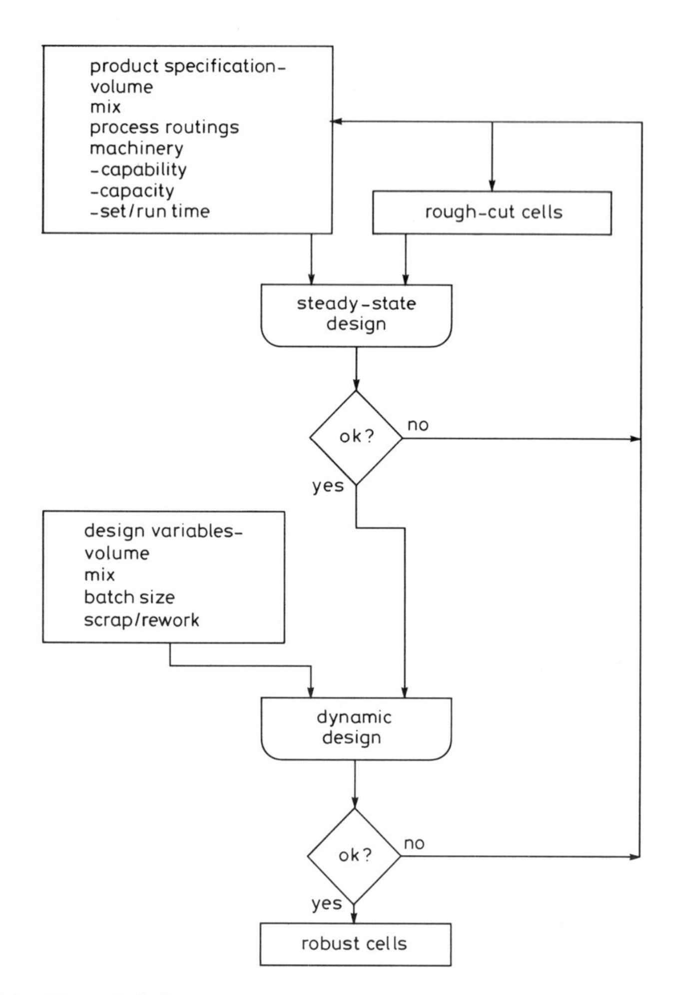

Fig. 13.6 The cell design process

13.6.4 Control systems design

The design of appropriate measures of performance and control systems, particularly for material flow, was crucial if our lead time objective of two months was to be achieved. Therefore, the taskforce spent considerable time specifying how these control systems would operate both within the cells and how they would interface with the numerous corporate systems.

In particular, the design of the material flow control system is worth describing in some detail.

From the problems with the current production control system outlined earlier it was clear that the taskforce needed to focus their efforts in two areas. Firstly, they needed to formulate a mechanism for producing a more realistic and achievable Master Production Schedule and, secondly, they needed to design localised, cell-based material flow controls to align manufacture to this schedule. The result was the creation of a top-down planning mechanism and a period flow control system, which schedules work within the cells.

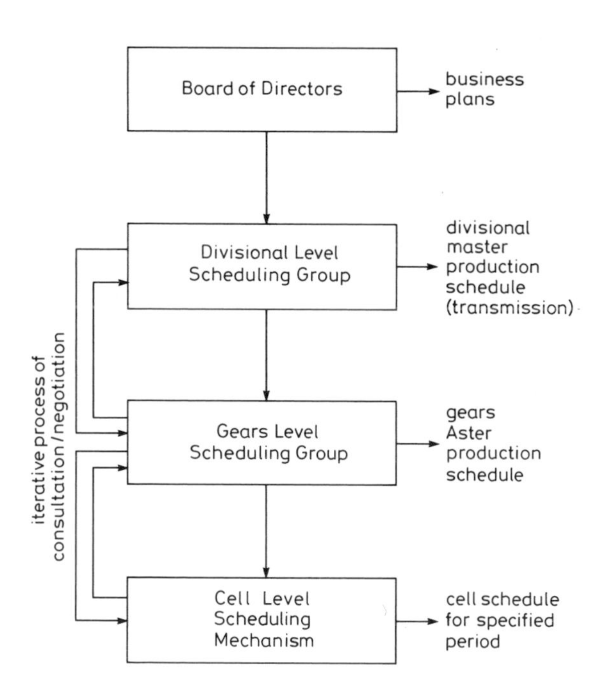

Fig. 13.7 Planning mechanism

Planning: The mechanism for producing a Master Production Schedule (see Fig. 13.7) operates at four levels. Its aim is primarily to foster communication links between the parties involved so that:

- orders can be accepted, rejected or ammended to enable levelled production and realistic, achievable schedules
- rescheduling is the exception rather than the norm so that cell-based schedules can be frozen for fixed periods. This eliminates the need for complex priority systems.

Period flow control: The erratic requirements for many of our gears makes the use of period flow control, rather than Kanban, the preferred option. With this form of material flow control, the year is divided into a number of equal periods — 26 × 2 week periods being used in the gear shop. The routing for each part number is then divided into partitions which can be carried out during this two week period.

Therefore, the start date for each batch is determined by backstaging the number of two week partitions within the routing, from the customer due date. This is similar to the conventional approach of many MRP II systems which typically backstage from the due date using one week per operation. The difference between this and period flow control is that the time horizon which the cell must concentrate on is limited to the end of the present period — a maximum of two weeks ahead.

Each part will have several due dates, one at the end of each subsequent period. The total number depends on the number of partitions within the routing. However, all parts will have the same due date for their current partition — the end of the present period. Hence, within a period all batches generally have the same priority. Work is scheduled on the machines by the cell manager using simple PC based algorithms.

Once this schedule for the two week period has been agreed upon, it is frozen. Cell targets, schedules, and consequently overall visibility, are greatly enhanced as a result of the simplified material flows and the limited time horizon. This enables the cells to review their progress to schedule constantly and take preventive action where appropriate to ensure that customer due dates are met.

13.6.5 Job design

One of the last stages in the redesign process was to create a simplified job structure that would meet three main requirements. Firstly, the new structure must encourage a team working environment. Secondly, it must enable the elimination of narrow and rigid job classifications and thirdly, it must enable the location of specialist (and previously centralised) technical and quality functions within each cell. This last requirement was seen as the key to improving product and process integrity; i.e. provision of cell-based resources to encourage cell ownership of, and accountability for, the solution to the majority of manufacturing problems.

Cell teams therefore consist of a cell manager, who provides overall direction and control, two shift leaders who act in a supervisory/leadership role, manufacturing engineers, who provide specialist and generalist technical support and manufacturers who carry out tasks previously done by operators, inspectors and manual/clerical support workers. In reducing the number of job titles within the cell to just four, we have laid the foundations for greater job flexibility. However, this will not happen overnight, but will require both commitment to the on-going discussions with the unions involved, and the continuing process of machine cross-training amongst the workforce.

13.7. The role of communications in overcoming resistance to change

The approach and methodology used throughout the project was new to Westland. The concept of a full time workforce, comprising both technical and shop floor members, using a planned and systematic approach, has proved to be a powerful factor influencing the success of the project. Equally important has been the attention paid to communicating the taskforce's ideas, progress and results to local management and the workforce.

Although initial management resistance was soon overcome, there were, and to some extent there still are, severe reservations and fears expressed by the workforce, especially over the issue of flexibility. For example, one immediate assumption was that cells and flexibility was a 'through the back door' redundancy exercise. Additionally, many skilled workers expressed fears that flexibility would bring about an erosion of their status and position of distinction over their semi-skilled and unskilled counterparts. It was also apparent that some of the narrower specialists were concerned that the widening of their roles demanded by flexible working, meant exposing their weaknesses or inexperience in other areas.

Much of the initial prejudice and anti-feeling we encountered was a result of a lack of understanding and information about the redesign process. Thus the taskforce's great challenge was to inform and persuade so that, for example, flexibility was not perceived as requiring a skilled worker to sweep the floor. We think, that for the most part, we have succeeded, but this has only been a result of the concentrated communication effort made by the taskforce. The team used all forms of communication throughout the redesign, including presentations, newsletters, open days on the shop floor and weekly meetings with shop stewards and supervisors. No communication can ever be too much, and we are carrying forward this message both into the implementation stage of this project and into further such redesigns throughout the company.

13.8 Conclusion

Although this work is by no means conclusive, it is clear that step changes in manufacturing industry performance will only be achieved by channelling the skill, knowledge and energy of every member of the workforce into the drive for improvement. This is possible only when individuals can see the actual effects of their actions, and interactions with peers and supervisors; that is, the process in which they are involved is equally visible, in its entirety.

However, there are many who have a vision for involving the workforce, and providing an appropriate framework to ensure that their involvement is productive and is fundamental. We believe we have created this framework and the results will confirm our forecasts. At the same time we have ensured that individual equipment investments will no longer be Islands of automation or efficiency, but will be justified on the basis of contribution within the system/process and overall return on assets. I submit that the benefits of organisational technology will be substantially greater than those of process technology in a batch manufacturing environment.

13.9 Acknowledgments:

Taskforce, Gearshop Supervision and Workforce

Chapter 14

Effective office support for redesigned manufacturing systems

P. Johnson
Lucas Engineering & Systems Ltd., UK

14.1 Introduction

A paper by Dr. J. Parnaby (1988*a*) highlighted the gap in performance between a typical European and a typical Japanese engineering manufacturing company using the ratios shown in Table 14.1. The clear message from the gap in performance is that Japanese companies operate their businesses in a radically different way from how they are typically carried out in the West.

How the approach Lucas has been taking to achieving international competitiveness started is described in the paper 'Lucas see the light' (1985). The continuing commitment of the Lucas group to this approach to designing businesses to be internationally competitive is demonstrated by the following three aims taken from the five published in the Lucas group's 1988 report to shareholders.

> To command a strong position in all our chosen markets, worldwide, based on leading-edge technology, total quality and high manufacturing efficiency.
>
> To achieve performance and profit levels which match and beat those of our international competitors in all our businesses.
>
> To develop, through our investment in innovation, training and better management of change, a company culture which encourages and rewards enterprise, professionalism and flexibility at every level in Lucas.

The objective of this chapter is to explain two key concepts for achieving competitive practices in the office areas, as well as manufacturing, and then give examples of the results of applying these techniques by reference to Lucas examples.

14.2 Two key principles

In this chapter, the following two key concepts will be explained. The first is the elimination of waste (Monden, 1983) and the second one is natural groups.

They have been selected for this chapter as they are concepts which are fundamental to the Lucas approach to redesigning a total business. However, they are only a small part of the total process of redesign, other elements of which have been published elsewhere (Anon, 1986; Parnaby, 1988*b*; Parnaby, 1986; Dale & Johnson, 1986; Wood, 1986).

Table 14.1 Performance comparator

Measure for an electro-mechanical engineering manufacturer	Japan	Western
Sales per employee per annum	£125 000	£50 000
Stock turnover ratio	15	5
Ratio of overhead staff to direct labour	0.5	1.5
Product cost	70%	100%
Proportion of engineers in overhead staff	60%	20%
Ratio *Engineers in product development* / Engineers in manufacturing development	1.1	10
Leadtimes in development and manufacture	50%	100%

14.2.1 Elimination of waste

The principle of eliminating waste is key to improving the effectiveness of a total business. Whilst waste is tolerated, it will continue at the same level and continue consuming resources, which adds no value to the business. Examples of business waste that should be eliminated are shown in Table 14.2.

Considering suppliers as an example, the advantages of having a small number of developed suppliers include:

- reduced overhead support costs
- reduced supplier development costs
- potential for continuity of supply
- supplier can make longer term investments.

Table 14.2 Examples of wasteful business practice

Area of company	*Business practice*	*Types of waste*	*Waste elimination*
Supplies	Multisourcing	Overhead increased to prepare requirements, evaluate tenders, manage data, manage contracts negotiate etc.	Few suppliers with long term relationships
Product introduction	Detailed design and specification of a bought out item	Overhead increased to specify, draw, communicate	Specify function/envelope and allow supplier to design
Manufacturing	Capability in all types of processes	Increase in overhead to support processes that do not offer competitive advantage at internal rates which if properly costed would be higher than a specialist subcontractor	Subcontract whole processes that do not add competitive advantage
Performance reporting	Standard hour based performance measurement	Manufacturing runs to standard hours, not customer needs; engineering runs to standard hour reduction, not system improvement	Balanced set of measures that when combined ensure good total performance

Clearly it is using the waste elimination practices shown in the last column of Table 14.2 that contributes to the performance level of Japanese companies in terms of the indirect to direct ratio shown in Table 14.1. Their performance also shows that practising such waste elimination approaches professionally does not increase costs as some might fear, but actually contributes to reducing the total product cost.

Too often senior management's immediate reaction to such examples of waste elimination is to cite the exceptions in a business that do not lend themselves to the process of waste elimination. For example, the worst supplier, the worst product, the worst manufacturing process and the other work already being carried out which will interfere. These exceptions should not be allowed to prevent the use of these techniques across a whole business. Starting and learning in those areas that will be successful provides experience and time to evaluate how to tackle the exceptions.

14.2.2 Natural groups

Organising manufacturing and office areas on a functional basis causes a tremendous amount of waste:

- moving material and information between functional areas
- inspections at functional boundaries to ensure integrity
- sophisticated control systems to track and prioritise work
- progress chasers
- storage areas
- routine paperwork.

The alternative approach is to structure both manufacturing and office areas around material and information flow respectively. Instead of the work flowing backwards and forwards between fragmented specialist areas with complex tracking and control systems it flows mainly within natural groups. Where changes of ownership are required between natural groups it is a whole work task that is exchanged with one leadtime; i.e. not many separate elements on documents all with their own individual process leadtimes. This had tremendous benefits for the leadtimes, quality and overall performance of the business. The comparison between the performance of manufacturing layouts that are functional with those that are formed into natural groups has been covered by a number of papers (Dale & Johnson, 1986; Edwards, 1988; Parnaby & Johnson 1987*a*). For the office areas the parallels will be explained here. Fig. 14.1 shows the flow of documents through a functional office area for a customer inquiry. This shows the initial inquiry causing requests for information to be sent to five other functional areas.

The leadtime to responding to the customer's inquiry is now going to be a function of the time taken to obtain all the responses. Each department

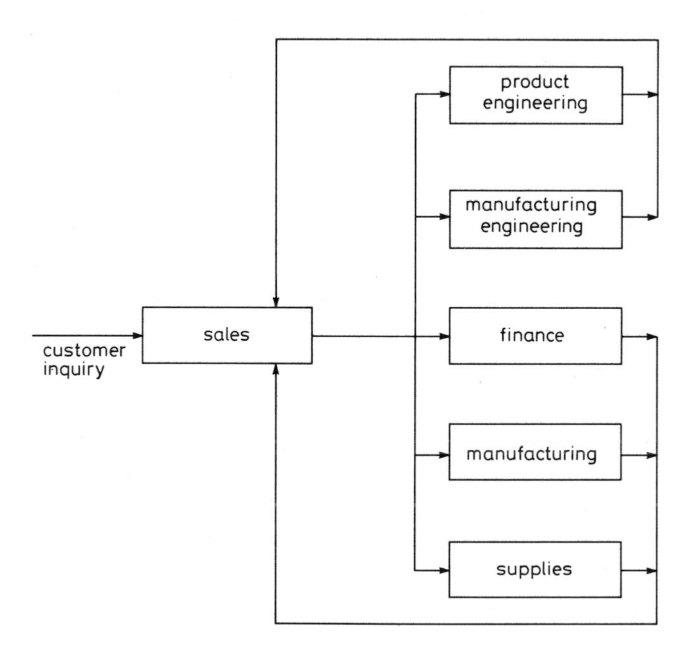

Fig. 14.1 Simulated flow of document required to prepare a response to a customer inquiry

will prioritise the task of responding to their part of the inquiry. As departmental priorities will vary according to their mix and load of work, the paperwork relating to this one particular inquiry will be at a set of levels ranging from highest to lowest priority in each department. However, the response to the customer cannot be prepared until all the responses are received. As a result, the leadtime for replying is extended to that of the longest reply, and in some cases no reply is made owing to one part of the system failing. Fig. 14.2 shows the effect of this functional layout on the number of operations, transports, delays, storages on an actual set of the documents sent through prior to redesign. If all the activities other than actual operations are classified as waste it can be seen that this document flow would have a no value added content of 74%. If the operations themselves are classified as to whether they are value added or not the percentage increases to 94%.

To change any one document requires a thorough understanding of all its uses. Given that paperwork systems grow with time and have extra complexity layered on in response to growth and problems, improving document flow in an unstructured way would be a major activity. Without changes it would also be unlikely to solve the root cause of the inefficiency; the functional layout.

▭ document	9	(plus 2 referenced)
○ operation	26	(6 useful)
□ inspection	21	
D delay	14	
▽ file	6	
△ pull from file	4	
⇩ transfer	14	

Fig. 14.2 Effect of functional layout on activities carried out on paperwork

Computerising the existing document flow is also not the solution as this will simply make the waste go round faster. The first step must be to find the natural groups that will reduce the need for paperwork, followed by a prioritisation of the activity to redesign the reduced information flow. The interrelationship between current functions must be examined by first charting their main interactions using an analysis technique such as departmental input/output analysis.

At a basic level the activities of a department can be shown in the form depicted in Fig. 14.3. Producing more detailed versions of these and then networking these departmental inputs and outputs into the form shown in Fig. 14.4. allows the next stage of locating and forming natural groups. This involves using the networked input/output diagram and examining it from the following perspectives:

- customer/supplier relationship
- necessity of activity
- potential natural groups of activity
- resource and current location
- strengths and weaknesses of creating groupings.

In this case, Fig. 14.4 illustrates 25 functional departments supplying 85 key items of information to one manufacturing cell. This also shows the need to redesign for cellular manufacture in parallel with redesigning the office areas. This will simplify the support needs of the cell to a level where a number of these tasks and pieces of information automatically become redundant; for example:

- Work to list preparation for work centres is unnecessary in a kanban environment.
- First line maintenance and engineering activities being carried out in the cell natural group will decrease the need for support.

Supplier	Input
Marketing	Forecast
Sales	Order book
Engineering	BOM
M. engineering	Route
M. engineering	Times
Manufacturing	Stock
M. engineering	Capacities
M. engineering	Manpower
Manufacturing	Exceptions
Manufacturing	Progress
Engineering	Spec. changes
Engineering	New products
Group training	Training

PRODUCTION PLANNING AND CONTROL

Output	Customer
Long-term plans	Manufacturing manager
Supply requirements	Purchasing
Manufacturing requirements	Manufacturing
Status reports	Manufacturing
Exception reports	Manufacturing
Stock data	Accounts

Fig. 14.3 Basic departmental input/output analysis

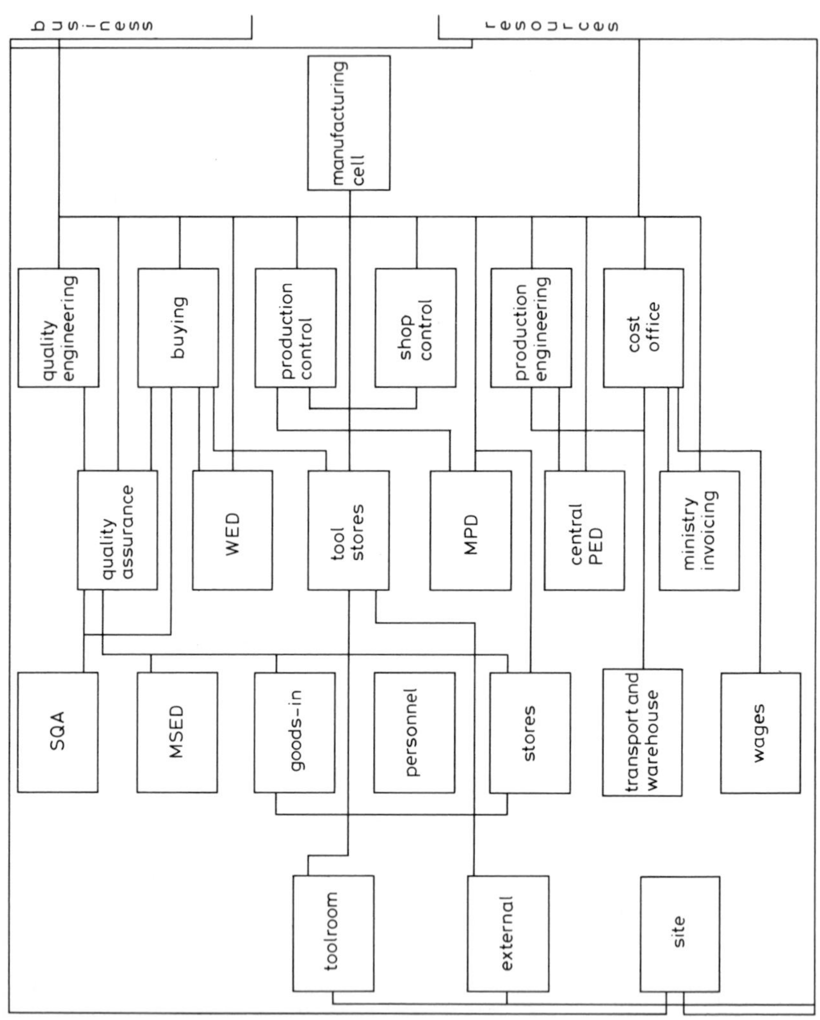

Fig. 14.4 Networked departmental analysis showing major flows

Potential natural groups are formed and then tested for suitability:

- Does it improve the reliability of the process?
- Does it reduce no value added?
- Can it be sustained?

Clearly the organisation cannot be totally one of multi-skilled natural groups but will be a combination of natural groups and functional support. The use of natural groups based around the business processes will need fewer but broader functions and jobs. The role of the functions will be to continue to act as custodians of best practice (through an audit role) and to hold resource that needs to be matrix managed between natural groups to balance fluctuating workload and skills requirements.

14.3 Results

The following are examples of the levels of improvement achieved by redesigning manufacturing and office processes around the concepts of waste elimination and natural groups:

14.3.1 Automotive factory

Product unit cell With product team

(i) Productivity doubled.
(ii) Process capability improved.
(iii) Costs of quality reduced by 50%.
(iv) Schedule adherence 100%.

14.3.2 Aerospace factory

Product unit cell

(i) Productivity improved by 63%.
(ii) Schedule adherenence 100%.
(iii) Cost of quality reduced 33%.
(iv) Throughput increased by 200%.
(v) Unit cost 65%.
(vi) Stock reduction 45%.
(vii) Leadtime reduction 50%.

14.3.3 Industrial factory

Sales order processing natural group

(i) Productivity quadrupled.
(ii) Leadtime reduced by 90%.
(iii) Reliability 100%.

14.3.4 Automotive factory

Product introduction natural group

(i) Productivity improved.
(ii) Schedule adherence 100%.
(iii) Engineering changes after introduction reduced by 75%.
(iv) Cost target achieved.
(v) Leadtime reduced by 45%.

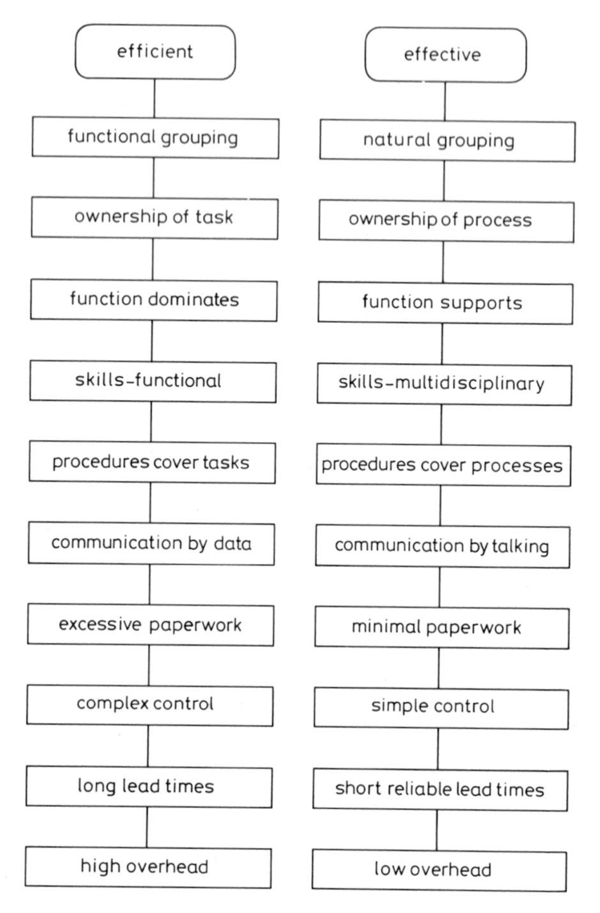

Fig. 14.5 Comparison of functional and natural groupings for office areas

14.4 Conclusion

As stated in the introduction to this chapter the concepts of waste elimination and natural groups are only two of the many behind a total systems engineering approach to redesigning a business.

The objective of redesigning a business within Lucas is to cause this to become a total quality organisation (Parnaby, 1987*b*). Total quality relates not just to the technical quality of the product or service but to the total quality of performance of every function in the organisation. However, such performance can only be achieved if the structure and systems in the business ensure that accountability and responsibility are matched. Natural groups are a key element to achieving this, and the examples quoted show that the benefits to the business are measurable and quantifiable.

A summary of the advantages of using an effective organisation design based on natural groups compared to efficient functions for office areas, is shown in Fig. 14.5.

Chapter 15

Implementing CIM: a managerial perspective

V. B. Prabhu and C. Kimble

Newcastle Business School, Newcastle Polytechnic, UK

15.1 Introduction and background

Several reports in the past few years such as those by ACARD (1983) and NEDC (1984), the House of Lords (1985) and the BIM, Ovenden (1986) have expressed great concern at the poor performance of UK manufacturing industry. Most of them have also highlighted the need for them to adopt advanced manufacturing technologies and new manufacturing techniques in order to compete and survive in the 1990s and beyond. Indeed, the importance of the crucial role of Computer Integrated Manufacture (CIM) in regenerating British manufacturing industry can be seen by the strong funding support given by our Government and the EEC on projects such as CIMAP and ESPIRIT.

CIM is in fact seen as one of the key strategies that firms should adopt in order to achieve world class manufacturing standards (Browne *et al.*, 1988). CIM, it is claimed, will improve the quality of both the design and manufacture of the product; it will reduce delivery times, increase the product variety and still maintain lower costs and smaller batch production; and reduce considerably stock and work in progress inventory. Indeed, the potential impact of CIM is regarded as being so profound and wide ranging that it has been compared with the motive power of the first industrial revolution (Slautterback, 1984). Only those companies that have the foresight and the skill to implement CIM, it is claimed, will succeed in this increasingly competitive world (Goldhar & Jelinek, 1983).

In practice, however, the successful application of CIM depends to a large extent on how the users, namely senior managers, perceive, justify, plan and implement such projects in their organisations. What are their reasons for implementing CIM and are they achieving the expected benefits of implementing the new computer technologies? In a recently completed study (Kimble & Prabhu, 1988), the experiences of nearly forty manufacturing companies in the north east of England were analysed. The manufacturing base in this part of the UK has traditionally been built around declining industries such as steel, coal mining and shipbuilding, and therefore one would expect that the need to adopt the new technologies could not be greater. A smaller subset of companies (approximately 19) from this study had clearly thought about implementing CIM and had already introduced several advanced CIM modules. This chapter presents some of the experiences of the senior managers in those organisations. In particular, it summarises some of their key reasons for implementing CIM; the problems they regarded as being important during implementation; and their perception of likely future changes in performance and working practices. Based on these observations, the chapter attempts to draw together those issues critical to the successful implementation of CIM both from the user's point of view and those external advisers recommending systems.

15.2 Some managerial experiences of CIM in NE England

Often the practical aspects of implementing CIM tend to be ignored or presented in some idealised form in many publications. To the authors' knowledge their survey provided a unique opportunity to examine the views put forward by a sample of managers who were responsible for putting CIM applications into practice. Some of the issues that concerned them are summarised below.

15.2.1 What choices of CIM are they making?

Two specific aspects are raised in this section, namely what is the view of CIM amongst our practitioners and what aspects of CIM are they implementing.

A large number of definitions of CIM can be found when reading publications on the subject. For example, Boaden and Dale (1986) identified ten different themes to published definitions of CIM. However, the executives in our survey saw CIM essentially as a means of integrating all aspects of the business through some co-ordinating 'philosophy' such as MRP or via a central database.

> 'Well my concept of it is that the various functions within the plant are computer based and essentially are all integrated . . . with as near possible a common database so that a change of data . . . is implemented in all the systems in the various departments' (Engineering Manager, Engineering Company)

or

> 'My understanding of it is a MRP II or similarly comprehensive data

system linked to shop-floor data-base collection equipment ... possibly linked to actual process control computers' (Director of Planning, Chemical Company)

This view was particularly prevalent in those industries where computer controlled production is not the central issues; instead the overall planning and co-ordination of production is seen as the major problem area. It was only in a very few cases that the executives in our survey saw CIM as a production based issue, i.e. producing goods with the minimum of human intervention, or CIM as essentially CAD – CAM.

> 'A full computer integrated manufacturing system is one that would, or could, originate from design through to some sort of CAM; that would set the machine, set batch sizes if you like and organise the tooling automatically' (Industrial Engineer, Engineering Company)

Literature searches suggest that certain clearly defined islands of computerisation and manufacturing philosophies or techniques would form part of any potential CIM system. We identified 11 such potential CIM sub-systems or CIM components for inclusion in our survey. Fig. 15.1 summarises the nature and extent of the use of these CIM subsystems in our sample. By and large, the predominance of the 'engineering' and 'general' sectors and the large companies amongst those active in CIM can be seen from Fig. 15.2 and 15.3.

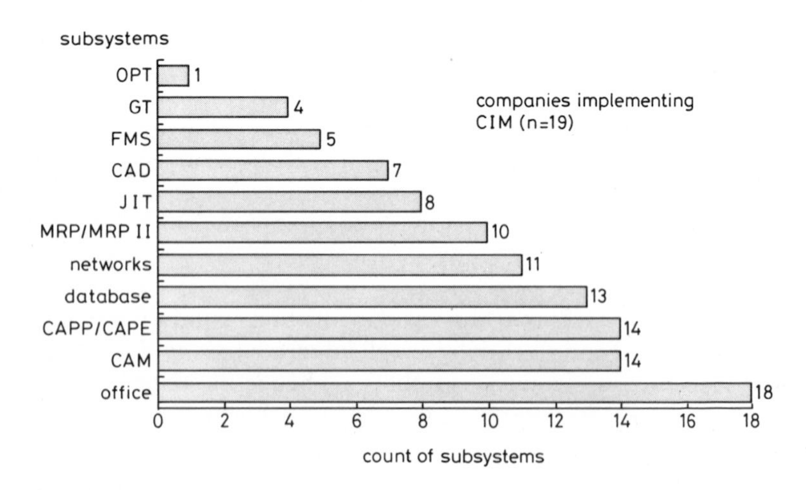

Fig. 15.1 Use of CIM modules

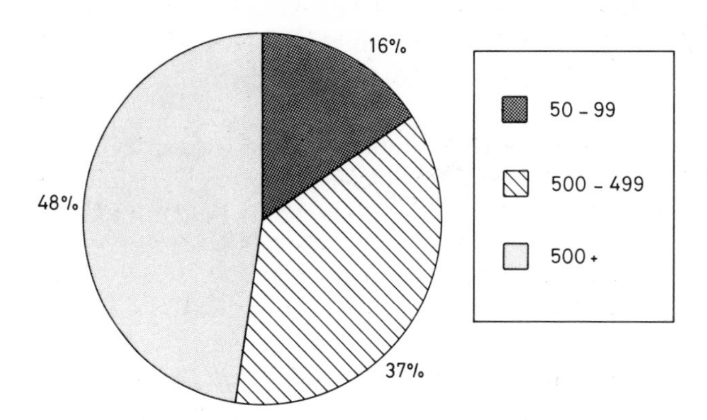

Fig. 15.2 CIM companies by size band

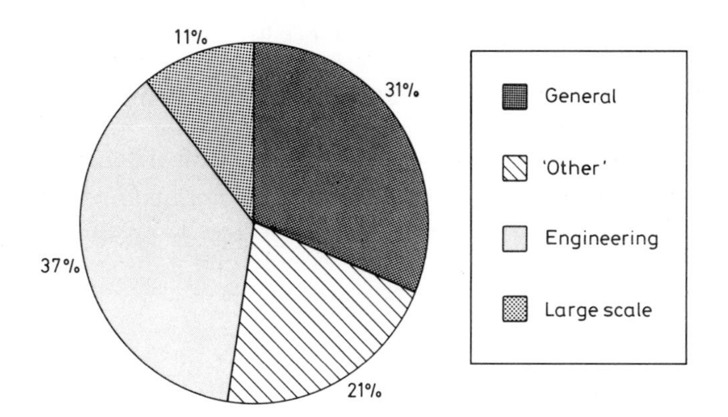

Fig. 15.3 CIM companies by sector

Some additional characteristics of our 'CIM as a target' companies are summarised below:

(i) Over half (58%) of the companies used at least six of the above subsystems. (Figure 15.1).
(ii) The companies were larger than their counterparts both in terms of the number employed and their annual turnover.
(iii) The companies manufactured a greater proportion of their output to order than the remaining companies.
(iv) A clear majority placed some manufacturing philosophy such as MRP/MRP II at the hub of a CIM system although CAD – CAM was seen by a significant minority as the focus of a CIM system.

(v) Some form of CAM and CAPP/CAPE featured strongly in CIM activities; CAPP was generally seen as a subset of MRP/MRP II.
(vi) There was a strong tendency towards the use of centralised databases although a small number of firms stated that they were moving towards distributed PC-based databases.
(vii) A relatively large proportion of JIT users were found in this group with many JIT users also using the principles of GT.

15.2.2 Why are they implementing CIM?

The need to become more competitive on an international basis has been stressed by several authors such as Hill (1985) and Ingersoll Engineers (1985*a*). In particular Schonberger (1986) has coined the phrase World Class Manufacturer as being the ultimate goal of any company that wishes to survive and compete into the year 2000 and beyond. The adoption of new technology as a means of achieving this goal is seen by all as a necessary pre-requisite.

The companies in our survey that were attempting to implement CIM appear to have done so largely because of the external pressures of highly competitive international markets which have required them to make general improvements to their levels of productivity, e.g.

> 'We operate in competitive world market, very competitive; prices are going down and down all the time . . . so if we want to remain competitive and still make a profit . . . we cannot stop improving our efficiency'

(MD, Plastics Company No. 1)

Some are hard hit by the proliferation of data needed to manage their business in a competitive and effective way.

> 'Our current data systems are stretched beyond breaking point at the moment; we don't feel that they will be able to cope with the increased data that will result in the next few years so we need to replace them with something that will cope'
> (Director of Planning, Chemical Company)

Some see CIM providing a way to overcome their own internal manufacturing problems. These problems ranged from wanting to improve control over current, sometimes manual, practices or perceived inadequacies of current systems, e.g.

> 'about three years ago . . . we found to our horror that we were only working on a job for about 5% of the time they were in the factory . . . we were looking for higher plant utilisation than you could get

out of conventional machine shops' (Operations Director, Engineering Company)

Often this external pressure results in them viewing CIM as a means of providing them with a specific competitive edge or competing in markets that would have otherwise been closed to them:

> 'basically because we see flexibility of response as our most important marketing edge.' (MD, Plastics Company No.2)

or as another firm put it

> 'a lot of our bigger competitors just don't have this system; that gives us an edge, it makes us more flexible.' (Materials Manager, Engineering Company)

CIM was clearly seen as a way of enhancing managers' control over situations and improving the volume, accessibility and accuracy of stored information.

15.2.3 What problems do they face in implementing CIM?

Essentially most of the implementation problems of CIM fall into the 'technical' category, e.g. communication between machines and database management, or into the 'non technical' category, e.g. human issues and systems management. By and large a lot can be achieved even with basic well known and established computer technology but the non technical problems covering a wide range of issues and attitudes still need to be overcome.

Our respondents who had recent experience of implementing CIM had experienced a whole range of problems although few were seen as being specifically technical in nature. Several found difficulties associated with the planning process itself. A common problem arose from trying to make a decision in a climate of constant change.

> 'things are happening so quickly and it takes so long for your plans to come to fruition (that) it is very very difficult to co-ordinate it into one master plan.' (MD, Plastics Company No.2)

Lack of information on the current 'state of the art' was often cited as a problem for those outside the engineering sector.

> 'it's making the right decision in terms of hardware and software . . . it's not obvious. Our type of industry is not like the engineering industry where you can go (and see) plenty of places where they have

> MRP II systems up and running.' (Director of Planning, Chemical Company)

Finding a basis for the financial justification of a system was seen by some executives as a problem. This was highlighted as a particular problem by satellite companies whose head offices were not in this region.

> 'I think the problems are justifying certain aspects of it from the financial point of view . . . justifying that a certain package will integrate.' (Engineering Manager, Engineering Company)

By far the most common stumbling block when attempting to implement CIM, however, was insufficient or inadequate training on how to use computer systems and a failure to take account of the need to change peoples' attitudes through a programme of education.

> 'It's a manageable problem but it's training. Training and re-training, there's a huge job to be done.' (MD, Plastics Company No.1)

or

> 'I think the whole thing hangs a lot on education, I think we probably underestimated that. There's no end to the amount of education you need.' (Operations Director, Engineering Company)

The cost and disruption of education and training was also identified as an issue by some

> 'we set up a training programme which seemed quite adequate and went comparatively smoothly, but during that training period we still had to do business . . . basically we had to manage with temporary staff.' (Engineering Manager, Engineering Company)

'CIM companies' rely heavily upon computers for their continued operation. The need to have a high level of support services to maintain the operation of computer controlled systems was seen by some as a problem area:

> 'Very high production rates mean very high loss rates when the equipment is standing. It means we have to have, on tap, some very highly skilled maintenance people. I'm afraid without them the use of sophisticated machinery is useless.' (Senior Training Manager, Engineering Company)

CIM is undoubtedly a high risk strategy and requires a corresponding commitment at the highest level. Several companies who made the highest use of CIM subsystems (at least seven) also stated that they had an MD who took an active role through the chairing of frequent and regular meetings of some form of CIM committee.

15.2.4 What changes do they see?

If one were to surmise from the literature review, that the paperless factory driven by computers and robots and basically operating with no human intervention was the idealised goal, then how does it appear in reality? What sort of changes would one expect to find in the factory of the future? Here again our north-east managers' experience throws some interesting light.

Most of our experienced CIM practitioners saw radical changes in the size and shape of their organisations and also in their working practices. Many, although not all, saw the change as being most significant for the middle managers in a CIM company.

Two main issues were highlighted: one that employees would be required to be more knowledgable, flexible and take greater responsibility for the tasks they undertook; the other that there would be far fewer of them.

> 'Well, I think the biggest change will be a more flexible type of person, I was going to say more intelligent but that might not be the right word, a more complete person who is able to come to terms with the new technology and be computer literate . . . and maybe take more responsibility than at the present time' (Engineering Manager, Engineering Company)

A director of one company summarised many of the issues surrounding the expected impact of CIM on middle managers when he compared their present role to that of 'sophisticated post boys' who process data as they pass it around:

> 'Yes I think it's going to have a dramatic change in business; my feeling is that it's going to be a bigger change at the senior-supervisory/junior-management level than elsewhere . . . I would look at it and say, where is the job for the middle manager in five years time? . . . I see the bottom level of the company reducing in numbers, yes, but I then see a very narrow neck where the middle managers used to be . . . I see it as an hour glass shape in the future, with a smaller amount at the bottom than we used to have . . . pinching in through a neck at the middle with virtually nobody (there) in the future, and opening out at the top to a wider and flatter plateau with a range of decision makers and people involved with the

> whole business . . . I see the whole shape of the company changing.' (Operations Director, Engineering Company)

Goldhar (1983) and others have labelled these decision makers 'professionals', people who have a wider view and who can take decisions on their own without the need to refer them to the next step in the hierarchy. Slautterback (1984) suggests that the climate in manufacturing industry in the future will be similar to that which attracted lawyers and doctors to their respective fields in the mid 1990s. The role of the 'decision taker' in a CIM company was recognised by one respondent who said:

> 'At the end of the day you have to rely upon your experience, your gut feel if you like . . . a decision has to be made and the quality of that decision . . . depends upon the data that is available . . . the use of the computer is giving you better data quicker . . . (however) we get down to the same thing in the end the manager has to make the decision' (Engineering Manager, Engineering Company)

Some saw similar changes for shop floor personnel. One respondent explained how previously operators:

> 'weren't interested in how the parts arrived to them, they just arrived, and if they didn't they just sat and waited until they did arrive' (Senior Training Manager, Engineering Company)

But now, through the use of shop floor terminals:

> 'the operator is controlling . . . a whole number of chain reactions . . . I would say the amount of control which is exercised now by the operators at each stage of assembly is far greater than it ever was' (Senior Training Manager, Engineering Company)

However not all the respondents in our survey agreed with this view. One MD, in answer to the question 'how do you think CIM will change working practices?', simply said:

> 'I think we will probably have a lot of bored people on our hands' (MD, Engineering Company)

15.3 Implications

The CIM experiences of our north-east managers, which have been summarised above, have been very revealing in terms of how practitioners perceive the role of CIM within the whole business context. Admittedly the views expressed are based on a relatively small sample (19 interviews); however, they represent the views of those organisations committed to implementing CIM as a goal and those who have attempted to implement that strategy in a realistic business environment. It is difficult and perhaps even optimistic to draw generalised conclusions from this limited survey; but the depth of feeling and the consistency of some of the views expressed on implementing CIM provide some powerful evidence which will be of benefit to other potential users as well as those responsible for selling and advising on the use of such systems.

15.3.1 Implications for users and practitioners

One of the main conclusions to be drawn from our sample was that the nature of CIM in the organisations we looked at was shaped by the needs and requirements of that organisation. It was clear from the interviews that, for the vast majority of them, it was competitive pressure that drove them towards CIM. From a potential CIM user's point of view therefore it is important to clearly identify those criteria in the market place which enabled them to 'win' orders, and only then to identify those 'enabling technologies' which would assist them to compete successfully. To take a very simple example if 'speed of response' to customer needs was vital for gaining the business, then all CIM efforts should be concentrated on eliminating waste in the entire business process and not just within manufacturing. In reality however, several criteria such as quality, costs, product variety, customer service levels may all impinge on the company's ability to compete successfully. It may be necessary in such circumstances to clearly identify those specific order winning criteria which are relevant for specific product groups or market sectors; nevertheless as our survey has also clearly shown, the actual CIM technology chosen for the business should be focused to achieve a given and clear strategic business objective. Often as our sample survey has also shown, simplifying the business process, integrating those simplified processes, initially manually, and then only automating, appears to be the most sensible and cost effective route for the potential CIM user to follow.

Our survey had also shown that the configuration of CIM subsystems used was diverse; there was no single given approach to the problem, nor was there any standard groupings of the modules which were used by all of the companies. Two configurations appeared to be the most common. The first had some form of co-ordinating philosophy such as MRP or JIT at its heart; the second conformed to the common view found in engineering industries namely CAD – CAM. Could both of these configurations be variants of CIM which are contingent upon the requirements of the business or are they both some intermediate stages which will eventually converge on one 'true' CIM? Irrespective of which view of CIM is appropriate, our users certainly took the approach of initially concentrating their CIM technologies on those areas of immediate concern especially where their competitive edge would be improved. The step by step approach towards implementing CIM was the norm, rather than the 'big bang' scenario when one would expect all current practices to be totally replaced in one major step by the fully 'computer-integrated' alternative. Also from the potential CIM user's point of view the implication was that only those areas of the business which required integrating via a computer ought to be addressed, rather than an indiscriminate computerisation of every aspect of the business. This would therefore lead to the view that some if not many of the CIM applications in reality would be of a mixed mode type, where only those parts of the business where the complexity justified it would be computerised and integrated, whilst the rest could operate quite effectively and efficiently with manual/human intervention only.

Most of the major concerns of our respondents during their CIM implementation were with non-technical problems. At the top of that list was the need for adequate training and education to effectively manage change. Underlying many of these problems was the pace and pressure of change; they were problems associated with managing and surviving a revolution that will affect all aspects of work. By implication the need to 'educate' all employees on the potential power of computers to manipulate and massage data in an almost endless variety of ways was seen to be of paramount importance; in particular to educate them in 'why' aspects of the technology rather than just the 'what' issues associated with training.

Another major issue of concern to our sample of CIM users was the potential impact that it had on those responsible for managing that technology. For example, our respondents expected middle management, the 'sophisticated post boys' of manufacturing industry,to coin their own phrase, to feel the greatest impact of CIM because computers can perform their tasks of manipulating and distributing data more quickly and efficiently than human beings could ever do. This in turn they felt would require more flexible decision takers, more 'whole' or rounded managers who can interpret and react to the data provided by computers. Companies embarking on CIM technologies should therefore not underestimate the need to change attitudes and thinking amongst its middle managers, and perhaps even its senior managers, if it is to achieve the full potential benefits of these new technologies. Superimposing traditional thinking and managerial practices on top of these new computer controlled technologies will certainly, according to our CIM users, not produce the desired results.

15.3.2 Implications for CIM vendors and advisers

The experiences of our CIM users also have clear implications for system vendors and those involved with providing advice on CIM implementation. As would be expected, most of our managers had dealt with both groups of people during the implementation of their own systems.

The point repeated most often by our practitioners was the need to recognise that human issues or human factors were crucial to the successful implementation of CIM. Practical experience had shown them that a failure to do so could well prove to be the road block to successful implementation for a number of companies. This did not in any way underestimate the technical problems associated with applying CIM. It was seen that most of their problems in this field could be resolved by choosing the appropriate and quite often well established technology. The issues of concern that seemed to remain unresolved and which occupied most of their time were those associated with people and their attitudes towards using the computer integrated technologies. The implication for system vendors was that technological excellence alone would not produce the desired benefits from implementing CIM; a great deal of time and effort had to be spent on making sure that the 'non-technical' user within the system was prepared and felt comfortable in using that technology.

An important element in creating that appropriate climate within an organisation was proper and adequate training and education of all concerned. This matter, as most of our respondents admitted, could have been with hindsight tackled more rigorously and widely within their companies. CIM consultants and advisory organisations have by implication a fairly major task in meeting this essential need, as often the skills necessary for developing such training programmes do not exist within the companies concerned. In some instances there could well be problems to be overcome of convincing user-managers of the need to spend that time and money for training purposes. However, in most instances the challenge is to design training programmes which will not only create the right climate within the organisation for the computer technology to flourish but which will also allow the users to participate in the design and operation of the system as a result of knowledge and skills gained from the training that they will have received.

15.4 Appendix

COMPS1 COUNT / ROW PCT / COL PCT / TAB PCT	50 – 99 0	100 – 499 1	500 and above 2	ROW TOTAL
COMPINTS	3	7	9	19
fully completed	15.8	36.8	47.4	100.0
	100.0	100.0	100.0	
	15.8	36.8	47.4	
MRP	0	6	4	10
MRP/MRPII in use	0.0	60.0	40.0	52.6
	0.0	85.7	44.4	
	0.0	31.6	21.1	
CAM	2	5	7	14
CAM/CAE in use	14.3	35.7	50.0	73.7
	66.7	71.4	77.8	
	10.5	26.3	36.8	
DBASE	1	5	7	13
Database in use	7.7	38.5	53.8	68.4
	33.3	71.4	77.8	
	5.3	26.3	36.8	

COMPS1	General 1	'Other' 2	Engineering 3	Large scale 4	ROW TOTAL
COMPINTS	6	4	7	2	19
fully completed	31.6	21.1	36.8	10.5	100.0
	100.0	100.0	100.0	100.0	
	31.6	21.1	36.8	10.5	
MRP	2	4	4	0	10
MRP/MRPII in use	20.0	40.0	40.0	0.0	52.6
	33.3	100.0	57.1	0.0	
	10.5	21.1	21.1	0.0	
CAM	3	2	7	2	14
CAM/CAE in use	21.4	14.3	50.0	14.3	73.7
	50.0	50.0	100.0	100.0	
	15.8	10.5	36.8	10.5	
DBASE	3	3	6	1	13
Database in use	23.1	23.1	46.2	7.7	68.4
	50.0	75.0	85.7	50.0	
	15.8	15.8	31.6	5.3	

NETWORK Computer network/s	2 18.2 66.7 10.5	4 36.4 57.1 21.1	5 45.5 55.6 26.3	11 57.9	4 36.4 66.7 21.1	1 9.1 25.0 5.3	6 54.5 85.7 31.6	0 0.0 0.0 0.0	11 57.9
OFFICE Computerised office	3 16.7 100.0 15.8	7 38.9 100.0 36.8	8 44.4 88.9 42.1	18 94.7	6 33.3 100.0 31.6	4 22.2 100.0 21.1	6 33.3 85.7 31.6	2 11.1 100.0 10.5	18 94.7
JITOS JIT, ZIPS or KANBAN	2 25.0 66.7 10.5	2 25.0 28.6 10.5	4 50.0 44.4 21.1	8 42.1	1 12.5 16.7 5.3	2 25.0 50.0 10.5	4 50.0 57.1 21.1	1 12.5 50.0 5.3	8 42.1
OPTOS OPT used on site	0 0.0 0.0 0.0	1 100.0 14.3 5.3	0 0.0 0.0 0.0	1 5.3	0 0.0 0.0 0.0	1 100.0 25.0 5.3	0 0.0 0.0 0.0	0 0.0 0.0 0.0	1 5.3
GTOS GT used on site	0 0.0 0.0 0.0	2 50.0 28.6 10.5	2 50.0 22.2 10.5	4 21.1	0 0.0 0.0 0.0	2 50.0 50.0 10.5	2 50.0 28.6 10.5	0 0.0 0.0 0.0	4 21.1
FMSOS FMS used in site	0 0.0 0.0 0.0	2 40.0 28.6 10.5	3 60.0 33.3 15.8	5 26.3	1 20.0 16.7 5.3	1 20.0 25.0 5.3	3 60.0 42.9 15.8	0 0.0 0.0 0.0	26.3
CIMOS Attempting to implme	3 15.8 100.0 15.8	7 36.8 100.0 36.8	9 47.4 100.0 47.4	19 100.0	6 31.6 100.0 31.6	4 21.1 100.0 21.1	7 36.8 100.0 36.8	2 10.5 100.0 10.5	19 100.0
COLUMN TOTAL	3 15.8	7 36.8	9 47.4	19 100.0	6 31.6	4 21.1	7 36.8	2 10.5	19 100.0
	Companies by size				Companies by product				

Chapter 16

CIM development: a strategic management issue

K. B. Chuah
Teesside Polytechnic, UK

16.1 Introduction

There is little doubt that much of the technology needed to implement CIM is becoming available and more affordable as the price of computer hardware and software continues to fall. However, unlike a CAD workstation, a CNC machining centre, a robot, an FMS cell or even a CAPM system, CIM should not be treated as if it were a piece of hardware or a sophisticated software which can be bought off the shelf and installed to cure all industrial ills. CIM should be viewed not just as a manufacturing issue but also as a business strategy which should be considered in the corporate planning and decision making process at the outset and not as an afterthought.

CIM is an enabling concept of functional integration. All business functions and manufacturing activities of a company are integrated to some extent with or without the use of computers. The aim of introducing computers and computer based technologies is to optimise the whole spectrum of functions and activities from processing orders to post-delivery services, from marketing research to product design, from design to manufacture.

16.2 CIM without a strategy

The various supporting technologies of CIM will undoubtedly improve manufacturing efficiencies but these efficiencies can only be turned into real benefits if the business, manufacturing and organisational issues raised by integration are tackled right from the beginning. CIM means much more than computerisation and automation. It is important to realise that the benefits of integrating functions and operations is far greater than the benefits obtained by computerisation or automation of isolated functions. CIM cannot make inefficient, over-complicated processes efficient — but if the processes are simplified and systems are integrated first, then the application of CIM concept and technologies can help to revolutionise business, management, design and manufacturing.

However, this does not mean just to automate or computerise the manual operations already in existence. Many companies have gone into isolated computerisation and automation of operations without first addressing the business and organisational issues. The result has often been the creation of the so called Islands of Automation problem. In other words, although operation efficiencies of individual functions may have improved as a result of computer application, the overall effect on the companies' operation has been small or even negative. Such lack of strategy at the outset, coupled with lack of adequate planning during the implementation stage, are perhaps the main reasons why many companies have found that investment in integrated CADCAM and CIM technologies has been an expensive exercise with much promise and potential but little realisable benefits.

In the opinion of the author, the problem is caused partly by the fact that decisions on CADCAM or CIM development have often been made only at the level of Operations and the whole issues have not first been adequately considered at Strategy level, and partly by the great pace of technological advances. In other words, decisions about investment in different areas of new technologies by different departments and functional areas of the company are sometimes being taken without full company-wide considerations.

Many such companies are not fully aware of both the technical as well as the non-technical requirements and implications of such implementations. The traditional business policy making process does not automatically have a built-in mechanism to deal with such development. As a result, CAE and CIM are being taken to mean the installation of CAD, CAM, robotics, FMS, CAPM and other computer based systems, and not as a means to achieving strategic business and manufacturing goals.

16.3 The strategic implications of CIM operations

The pace of advancement in computer based technologies in the last ten years has been such that it is no longer adequate for manufacturing companies to continue adopting the traditional approach of making incremental step changes to existing manufacturing and management systems in order to remain or become competitive. It is a well known fact that piecemeal implementation of high technology CAD, CAM, FMS and

CAPM systems in the USA and Europe has not always yielded satisfactory returns. In contrast, Japan and other fast developing countries such as Korea, Taiwan and Singapore have continued to achieve productivity and competitive gains from their investments in such technologies. The difference in results may be partly attributable to differences in social and cultural ideals, industrial structures and presence of discriminatory government policies, but, more importantly, it is partly the result of good management!

In any company, it is standard practice to treat business strategy and policy decision making, both of which are concerned with the entire business as a whole, as long termed. The aim is to provide a mission and direction for the organisation and a framework within which each department or function works.

Such strategic management process is an inherent part of most organisations in the West as well as the East. The main difference is, in addition to the other commonly acknowledged strategic decisions about finance, manufacturing capacity and facilities, product and market, vertical and horizontal integration, competition and so on, manufacturing companies in the East, Japan in particular, also treat issues like production technologies and processes, workforce, quality control and assurance, production organisation, planning and material control, sales and purchases etc. as of strategic importance. On the other hand, companies in the West tend to look at the latter areas as Operations issues. Viewing such matters as secondary issues has other unfortunate side effects. These are critically discussed by Wheelwright (1981).

One of the consequences is that managers of Western companies are unwilling or unable to make adjustments and investments whose benefits cannot be readily quantified in immediate financial returns. Development and implementation of CIM technologies is an Operations issue as well as a Strategic decision. Whatever level of automation, computerisation and integration a company is aiming for in the short term must be part and parcel of a long term strategic plan. There is no one right level of computerisation and integration for a company. A good CIM strategy is one which will help the company to achieve the level that is perceived to be right for the company's requirements.

Bad performances occur because of declining or ineffective investment (in production technology, staff training and development) and slow growth of productivity (poor management, low morale, inefficient systems and so on). Much of this is the result of detachment of strategic thinking and policy formulation at the top management level from technical decision making at the operations level and the pressure of short term gains at the expense of long term development. There is a need to re-establish the roles of engineers, designers and other technical personnel in the initiation, design, development and achievement of change. These roles have often been (or at least perceived to be) subordinated by other non-technologically sensitive groups such as accountants and financers. However, this is not to say the roles of the latter are to be ignored or dismissed.

Effective strategy can only be formulated if all aspects of the business are considered.A purely technology driven CIM development is equally unlikely to succeed. Major benefits can only be obtained from the new technologies if new organisational and operational methodologies are built into the strategic decision process of the company. Some of the main strategic questions are discussed below.

16.3.1 The strategic questions

Many companies in the West make the mistake by restricting strategic analysis to the levels of corporate, business and functional issues and regarding operations, workforce, quality, inventory etc. as 'swing factors' which can be manipulated to satisfy immediate objectives. Wheelwright (1981) pointed out that such an approach could lead to false choices being made, like cost and quality, flexibility and dependability, whether to invest (in equipments, staff development training) and so on. Companies should start by looking at how to integrate strategic thinking into the execution of operations policy.

In other words, Strategy and Operations must be mutually supportive. Operations managers at different levels must be made to understand the strategic significance of their respective contribution and, specifically, their day-to-day concern with operational details. It is with such understanding that the workforce, their supervisors and managers can be expected to work towards common goals together and consistently.

Successful development and implementation of the CIM technologies needs such an approach and a well thought out Strategic Operations policy. The aim of the corporate and business strategy is to make sure that the company is in a position to react to environmental changes and technological advances in a positive way and that it can re-direct and re-allocate its resources and investments towards improving performance and gain competitive advantage. The need for the CIM technologies, and to what extent, must be examined in this light. The Strategic Operations policy is part of this process to ensure that, firstly, there is organisational understanding and cohesion for planned changes to be effectively implemented, and, secondly, new working methods and practices are consistently kept to or maintained.

A thorough CIM strategic evaluation should aim to answer the following clearly:

(*a*) What are the corporate goals and business objectives of the company?

(*b*) Why CIM, CAE or CADCAM? What can CIM technologies offer to help achieve these business goals and objectives in terms of satisfying the functional needs of business (finance, administration, marketing, sales, purchases), product development and design, manufacturing, support services etc.?

(*c*) What are the strategic alternatives?

(*d*) What level of sophistication should the system(s) be:
  (i) for the present needs?
  (ii) for the long term requirements?

(*e*) How will the company's organisational, control and information structure be affected by CIM implementation? What changes are needed?

(*f*) How can the existing personnel be integrated into the new CIM environment? What staff development or recruitment programme is needed to use and support the CIM system?

(*g*) How can the investment be financed and justified?

(*h*) Who can put together a CIM concept, system specification and development plan specific to the company's needs? How is that plan going to be implemented? What kind of mechanism should be built into project development process to monitor and report the progress of each stage of implementation plan?

(*i*) How is the system going to be supported and maintained when implemented?

The list is by no means exhaustive. There are no doubt many more depending on the nature of the company's business and the environment or circumstances it is in. The important point is that these strategic issues have to be addressed at the outset of any development plan and not as an afterthought. CIM must be viewed as a strategy by which manufacturing companies can bring together computer based technologies and their various operations or functions to optimise the whole spectrum of business and manufacturing activities from marketing to product development, from production to post-delivery services. There is no proven path for success, but there is a methodical approach which one can adopt to avoid some of the possible pitfalls.

16.4 An approach to CIM development

As with all strategic decisions, a company must start by analysing its strategic factors, establishing objectives and resolving the order of priority. A good strategy is one which has built into its subsequent decision making mechanism and process the robustness and flexibility to respond to any subsequent change of priorities and other strategic factors. This is

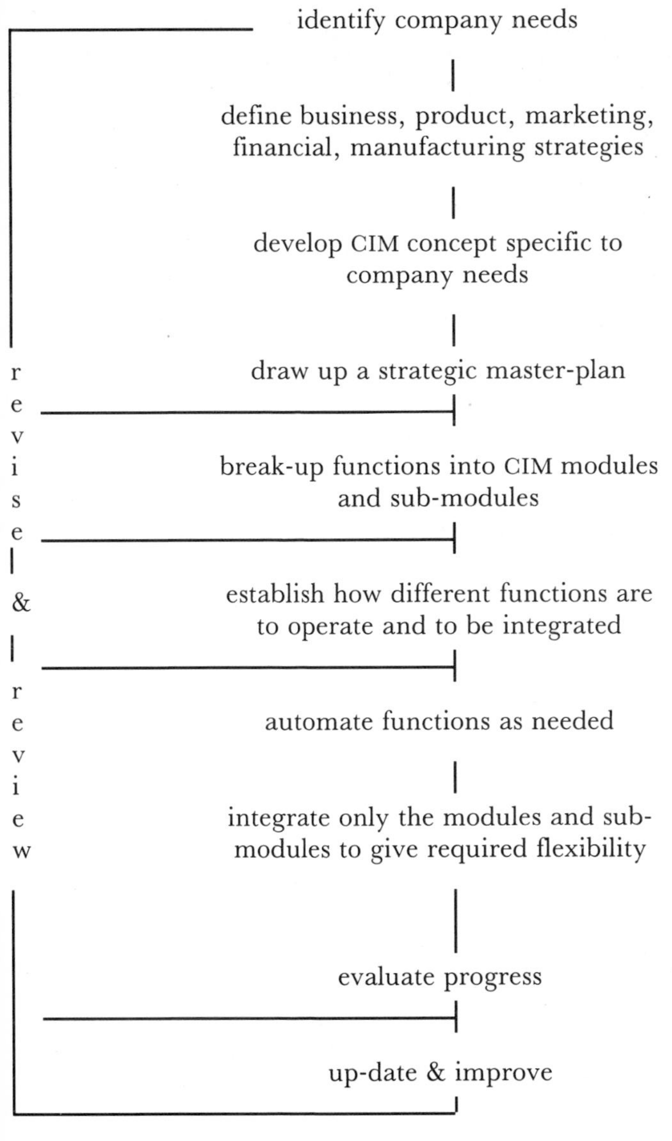

Fig. 16.1 Top-down planning and bottom-up modular implementation

particularly important in CIM strategy formulation because of the great pace the computer based technology is advancing at. What is needed is therefore a two-prong approach with careful top-down planning and systematic bottom-up implementation.

16.4.1 Top-down planning and bottom-up modular implementation

The broad outline of this methodology is shown in Fig. 16.1. Formulation of a CIM strategy specific to the company's needs, as defined by the company's business, product, manufacturing, marketing and financial objectives, is the very first stage of the top-down analysis and planning.

It should be noted here that different strategic issues can be made the focal point of a company's CIM evaluation. For instance, a company which manufactures to customers' design and requirement may wish to base its CIM activity on its CAM capability. Another which carries out highly specialised contractual machining processes may wish to base its CIM strategy on capacity and resource planning, scheduling and control. On the other hand, a company which frequently needs to redesign its products to meet changing market demands would want an integrated marketing and product development and design strategy at the core of its CIM concept. In short, there is no easy guide or quick solution. From that overview, a detailed strategic master plan is developed. It breaks down the business and manufacturing operations or functions into modules and sub-modules.

Except for companies which only require isolated application of the CIM technologies, most other implementations would be of a generic nature, e.g. an FMS cell is a sub-module of a CAM module and a CNC machine centre is an element of the FMS cell and so on. This allows the questions about systems specifications and compatibility, control and data structure etc. to be addressed at this early stage. Planning ahead reduces the problems of system interface and incompatibility later on and allows for better resource allocation and management.

Bottom-up implementation under such a master plan means that prioritised systems and sub-systems can be developed and implemented in a systematic and modular manner. It is perhaps appropriate in most situations for companies to resolve priority on the basis of 'worst first'. The advantages of adopting such a modular bottom-up approach are many; the most important of which are that there is likely to be an earlier return on investment, and that costs can be spread over a period of time. The gradual implementation also causes less abrupt changes to working practices and environment and is thus more likely to be accepted by existing workforce. Moreover, because the company has got a longer time span to move up the learning curve, there is less risk of things going disastrously wrong.

However, implementation of CIM technologies and concept is much more than installing labour saving computerised systems. It is about getting people, machines and the organisation to work together. Physical link-up of different systems is not CIM if it is not backed by the integration of people and process.

16.4.2 Integrating the organisation, the process and the people before CIM

Integration in the CIM concept means much more than integrated computerised data processing. Professor Burbridge (1988) quite rightly said that the word integration is nearly synonymous with the words simplification, combination and synthesis.

Complexities are usually the result of unchanged practices and traditions. They are wasteful and a major cause of inefficient operations. Automation and computerisation of such set-ups, without first looking at how the working of the existing systems, personnel and processes can be simplified and inter-functional activities integrated, is unlikely to improve matters — one reason why CIM has been said to mean 'computer induced madness'. Hartland-Swann (1987) suggested that the three key steps to CIM are: first simplify, then integrate, then computerise.

It is likely to be a very costly mistake if a company seeks only to automate and computerise existing activities and processes. CIM requires a thorough review and re-appraisal of traditional management procedures, business practices and manufacturing processes. Rzevski (1988) points out that this may also involve modifying organisational structures and creating a new change-oriented business and manufacturing culture before automation and computerisation. This change must be managed from the top.

16.4.3 Managing change and CIM

Hartland-Swann (1988) suggests that it is more useful to interpret CIM as 'change in methods'. Implementation of CIM is about successful management of change: change in technology; change in working method, process, environment; change in organisational structure; change in attitude, change in company culture and so on. But to challenge existing practices and to introduce drastic changes is always going to be difficult. Resisting change is an inherent part of human nature.

Installation of CIM technologies and integration of functions impose changes to traditional working methods and practices. If these changes are to be accepted by the workforce, and its co-operation and commitment obtained, the human dimension must be part of the implementation strategy. The most important key factors to successful management of change are awareness and involvement. Lowe (1987) emphasises that, to manage change successfully, there should be shared ownership of the problem (the reason for change) and solution (the new method of working) reinforced by effective reward schemes.

16.5 Conclusion

CIM implementation is an exacting task. It is a strategic decision. Its development strategy and planning should start right from the top management levels and move down to the operations and shopfloor levels. Except for perhaps the unlikely 'greenfield' development, the implementation of this CIM strategy should follow a bottom-up approach.

The potential benefits of CIM will not be realised if sophisticated computer based systems are used to solve complexities resulting from inefficient operational process and organisational structure. Management should try to first pin-point unnecessary operational and organisational complexities in the system; simplify them wherever possible; then integrate the functions where needed with the aid of computer based technologies; and not the other way round.

A CIM environment is one with its functional activities not bounded by traditional departmental boundaries. Success of CIM implementation depends not only on the right choice of systems, it needs the co-operation and commitment of the entire workforce. Only with the proper management of change can the need for change be understood, accepted and acted upon. CIM is change in methods.

Whether a company is aiming for partial or complete integration of its business and manufacturing functions, a strategically formulated CIM plan is crucial to the success of its implementation. It must be concise, clear and comprehensive, addressing not only the business, technical and organisational issues of CIM but also the people dimension.

Chapter 17
Implementing an IT strategy in a manufacturing company

H. S. Woodgate
International Computers Ltd., UK

17.1 Introduction

Effective information technology can provide an organisation with a major competitive advantage, but so many manufacturing companies fail to achieve this benefit because of the absence of an overall strategy. This chapter describes how ICL faced this issue and worked it through to become a highly successful example of IT use in manufacturing.

Many of the current problems associated with planning an IT strategy arise because of the fragmented manner in which information technology has developed. Originally data processing techniques were seen as providing individual solutions to individual problems. Some linking of separate tasks took place, but usually the whole process remained contained within the walls of the computer room and the technology had little visibility to management. Then the (cheaper) microcomputer made computing available to all and many small (non-compatible) applications emerged. Computer networks began to expand initially to service single applications from many points and later to provide the capability for remote processing on satellite computers and personal workstations.

On the factory floor, numerically controlled (NC) machine tools increased in number and some linking into 'cells' took place. These developments are aptly described as Islands of Automation.

As this evolution progressed, management became concerned about the ever increasing expenditure without an overall plan. Whilst each proposal carried its own cost/benefit anaylsis, they each gave only a small incremental improvement without any major effects on the real 'competitive edge' problems being faced by the company overall.

It was realised that what was missing was a framework, or strategy, in relation to which individual investments could be assessed. Management therefore called for thorough appraisal of the IT needs of the entire company. This was to define potential IT application areas, specify objectives and set out a 'roadmap' of how to achieve them. When this task was intitiated, management gave an overall directive that the IT investment should be targetted at increasing the long term profitability/productivity of the business.

17.2 Evolving a strategy

Before an IT Strategy could be developed, it was necessary to have a clear, and shared, understanding of the purpose of the activity to which IT was to be applied.

Manufacturing organisations are founded on the concept of financial investment growth and profit. Profit is achieved by 'adding value' to purchased materials. Value may be added by either:

- technical inventiveness
- commercial acumen.

In both categories there are many opportunities for IT intervention. Each will make different demands upon resources (money and skills) and each will make a varying contribution to the value adding process. Additionally, many of the IT processes will interact. This amalgam of issues can be summarised into a problem list as follows:

- high-risk investment decisions
- implementation of strategic changes
- inter-relationship of multiple technologies
- rapid technological change
- management information complexity.

These cultural problems form a background to strategy formulation and highlight why senior management involvement is essential. The strategy itself must include the following:

- a statement of purpose
- a set of objectives
- a list of application areas
- a recommendation on technologies
- a prioritised macro plan.

17.2.1 Statement of purpose

It was defined that the purpose of the IT Strategy is 'to increase the profitability/productivity of the business'. The simplicity and generality of this statement gave recognition to the wider role of IT in the company and moved the target away from dubious concepts such as 'reducing operating costs' and 'return on investment'.

As will be discussed in the next section, the methods of running a manufacturing company are changing rapidly and success inevitably requires large investments as well as changed work patterns. The manner in which IT can contribute to this evolution is a key issue in formulating strategy. As these changes are so important, the main components of change are outlined below before describing the IT Strategy.

17.2.2. The changing face of manufacturing

The main areas of change can be summarised as follows:

- changes in market
- changes in competition
- changes in product design
- changes in manufacturing technologies
- changes in management control methods.

New markets open up, new competitors appear. New technologies create new products which, in turn, require new production methods and new management control systems. The rates of change of these factors are also accelerating under the pressure of increased global competition. The growing awareness of the impact of these changes upon the manufacturing value-added chain, and effect upon competitiveness has given weight to the aphorism 'change' or 'perish'.

Technology has altered the manufacturing process in many ways. In electronics there has been a succession of fundamental changes in circuit design (e.g. the silicon chip), each requiring different production methodologies. Moulded plastics replace metal parts. Metal parts can be made in machining centres, or linked machine tools driven by cell controllers. Finished parts can be assembled by robots, packed by machines and distributed from automatic warehouses.

The introduction of new production concepts such as Just in Time (JIT), has revolutionised production management techniques and placed increased demands on IT systems;

All of these changes require prodigious capital investments and older methods of financial evaluation of projects are no longer adequate. Accountants are required to revise their methods as the return on these investments is dependent on securing greater sales growth and on closer assessment of the consequential higher risk. Increasingly, financial modelling is being used as a way of evaluating intangible 'value adding' changes; e.g. competitive price differentiation, perceived value of better design etc. Both capital investment requirements and risk are also increasing rapidly in step with the other changes listed earlier.

This situtation has led to two major alterations in the way manufacturing is structured, viz:

- increased specialisation in component manufacture, i.e. gaining increased volume by meeting the needs of several assemblers
- less component manufacturing by assemblers but increased emphasis upon design, assembly techniques, distribution and marketing.

A manufacturer of products now employs less direct labour, purchases more externally and seeks to add value to bought-in parts and materials by good design, choice of specification and, above all, quality. International competition is rampant in all forms of manufactured goods and the business increasingly becomes more complicated and faster moving.

17.2.3 Objectives

Objectives are the measurable criteria against which the implementation of the IT Strategy is judged. These are determined in part by the position in the market which is sought by the company. The objectives chosen were:

- faster deliveries
- keener price
- competitive product features
- quality
- variety of product
- degree of added value.

These criteria became not only the objectives of the IT Strategy, but also the key 'critical success factors' central to the Management Information System described later.

17.2.4 Integration

The achievement of these objectives is clearly dependent upon a series of company activities working in unison. The key company functions need to be inter-linked in a more responsive way than ever before. Each key company function has an IT component. An essential part of an IT Strategy is to integrate the components and represent them to management in a comprehensible way. This relationship is shown functionally in Fig. 17.1

In Fig. 17.1 the Marketing Strategy analyses the market requirement, potential volume and competitive situation. The Product Strategy designs to those requirements. These product designs in turn create downstream costs in terms of manufacturing requirements and material/component procurement requirements. The Manufacturing Strategy and Financial Strategies respond to those requirements. The Distribution Strategy provides the means of selling the product and delivering it to customers geographically.

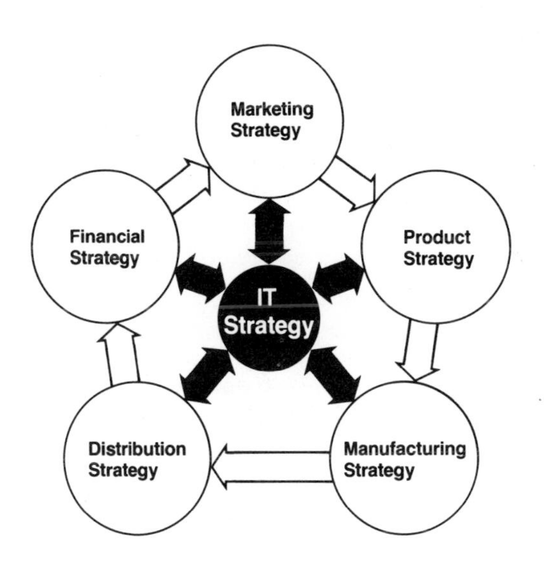

Fig. 17.1 Relationship between IT strategy and other company strategies

Policies and decisions at any point affect the efficient working of the whole. In terms of success with customers, key competitive issues are:

- speed of delivery
- ability to provide variety
- responsiveness to competition
- price.

Speed of change in design, production and delivery is at the heart of effectiveness. For example, to respond to a competitor by cutting price will reduce profit. However, if it is possible to respond with increased functionality, price can often be maintained or increased. Profit maximisation in these conditions requires that the product is designed to suit the production technology available, i.e. better integration of design and production engineering. Thus, the payoff from improved communications can be considerably more than some minor reduction in production cost.

17.2.5 The value-adding chain

Whilst each of the company functions contributes to the overall performance, the effect of improved working in each area is by no means equal.

It was stated earlier that profit is achieved by adding value of purchased materials by either technical inventiveness or commercial acumen. The components of this process can be considered as a 'value-adding chain' in which each activity makes a contribution to overall profit. The individual elements of such a chain are not necessarily sequential and they can be considered as a two-axes positioning diagram as in Fig. 17.2.

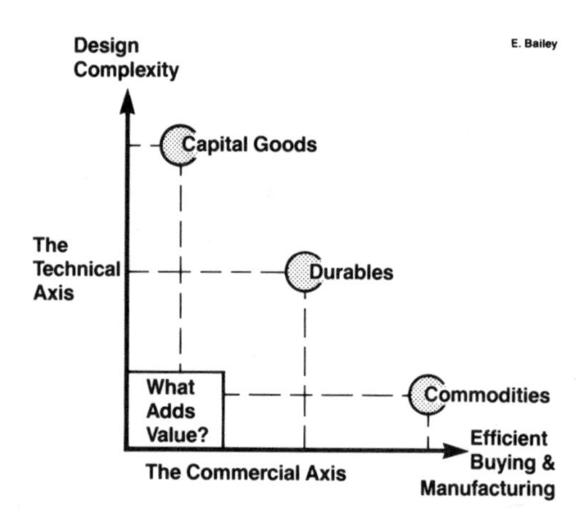

Fig. 17.2 What adds value?

In Fig. 17.2 each axis represents the group of value-adding activities corresponding to technical inventiveness and commercial acumen. Typically these will be:

Technical axis

- design
- engineering change control
- advanced manufacturing Technology (AMT)
- NC machine tools.
- automatic warehousing.

Commercial axis.

- sales administration
- production planning and control
- stock and work in progress control
- purchasing
- supplier liaison.

If a numerical value is given to each axis representing the value-adding importance of that type of activity, then a grid position can be established which indicates the driving characteristics of the company which have to be optimised to ensure success.

To illustrate the principle in Fig. 17.2 three differing industry types have been positioned as follows:

- *Capital goods*, e.g. aerospace products: contribute more added value by technical capability than by mass production.
- *Commodities manufacture*, e.g. foodstuff: simple domestic goods depend more upon bulk buying, competitive pricing etc.
- *Durables*, e.g. domestic appliances: require more equal quantities of each type of skill.

This basic diagram (sometimes known as the Bailey Box) can give guidance on a number of strategic issues, viz.:

- investment priorities
- management and personnel skills required
- manufacturing methods
- IT implementation priorities.

It is often the case that the nature of a manufacturing business will change as it grows and this basic analysis will reveal the need for restructuring. Also, if new divisions or product lines are to be added, this method will show whether different staff skills and technologies are required.

17.2.6 The product/production activity grid

To assess the impact of IT on a manufacturing business, it is first necessary to consider other value-adding variables. Many of these are inter-related, e.g.

- product type dictates the manufacturing process.
- Volume creates the need for automation.
- Complexity inevitably leads to uncertainty.

The type of grid analysis shown in Fig. 17.2 can be extended to encompass more detailed elements to assess these factors as in Fig. 17.3. Here two grids have been combined, viz.:

- product complexity
- product uncertainty

and

- production volume
- production automation.

Each quarter indicates the type of production technique that is most used with that combination of variables, and the triangular corner designates the key value-adding element. Positioning a product, or group of products, on this grid helps to clarify the priorities listed in the value-adding chain section.

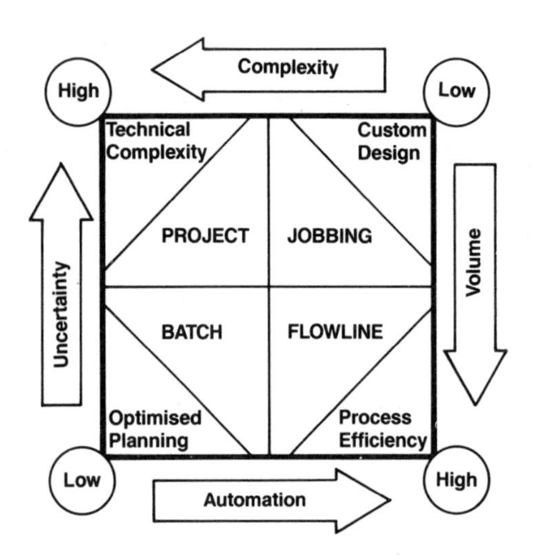

Fig. 17.3 Product/production activity grid

17.2.7 Identifying key elements

In the preceding section, IT activity was indentified by type (technical or commercial). To form a strategy, it is necessary to move into applications, technologies and integration aspects. A further analysis of these is shown diagrammatically in Fig. 17.4.

Here, key elements of each of the technical and commerical axes have been assigned numerical values. These values are a judgmental assessment (on a scale 1 — 5) of their individual importance as a contribution to the value-adding chain. The principle of comparative assessment against a common criterion (added value) is used throughout, even though there are many more than the ten elements shown in this example.

It is self-evident that a combination of the highest assigned value items represents the best investment opportunity. Thus 'priority integration links' are also added to this diagram to provide a basis upon which strategy implementation priorities can be determined.

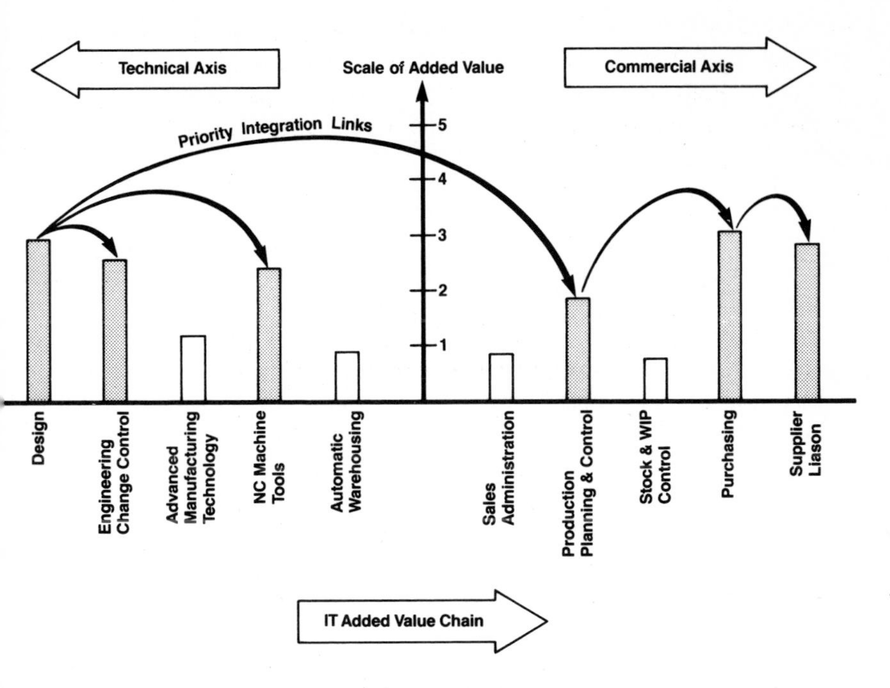

Fig. 17.4 Identifying key elements

17.3 Types of IT activity

The assessment of actual, or potential, added value is difficult, as it is necessary to translate differing functional activities into this common criterion. An attempt was made to rationalise this assessment by further analysing and categorising IT activities by type.

Data is the common element and the use to which data is put is a distinguishing factor. Three basic uses were identified, viz.:

(i) data which drives processes (NC machines, cell controllers, automatic warehouse etc.)
(ii) data which is part of the process (payroll, accounting, sales order processing, production planning etc.)
(iii) management information.

Data which drives processes usually relates to geometry, quantities or manufacturing operations. Data in this context has a precise meaning to machine tools and robots, but is not directly intelligible to humans.

Data which is part of the process is seen in most conventional data processing applications. It involves the processing of numbers to produce a calculated answer which is usually acted upon by human agency. In this category, human-readable outputs are normal, but these are only understandable by personnel trained in the application, e.g. accountants, payclerks, stock controllers, production schedulers etc.

Management information is different. Data is not information and managers require to understand the meaning of data. Unlike the two preceding categories, there are not always precise mathematical ways of processing data so that managers understand its significance. Yet understand it they must if they are to create profit generating plans and control their progress.

The characteristics of the first two classes of data are perhaps well understood, but the nature and use of management information is more problematical. This aspect of the control system is therefore now discussed in more detail.

17.4 The management information system

In the preparation of an IT Strategy, management is a particularly difficult component. Clearly, managers are part of the process in that they bring to the system:

- knowledge
- innovatory thought processes
- decision capability.

To exercise these characteristics, managers need information from the system. This creates a requirement for two further major system functions not directly in the value adding chain, viz.:

- a method of collecting, storing and updating a file of management information
- a method of presenting information to managers in a manner readily understandable.

Two unique solutions were developed to these problems which are now described.

17.4.1 The management information database

None of the proprietory types of database effectively matched the needs of the proposed management information system. The system required that the database would hold large volumes of inter-related data which could be accessed at high speed in many combinatory forms. Thus a novel kind of multi-dimensional database was designed which holds partially processed data and the relationships between that data. This system is called CUBIT as, in its three dimensional form, it holds data in three axes (like a cube). Typically the data is:

- data line items
- organisation structure
- time periods.

This arrangement is shown diagrammatically in Fig. 17.5. All key management data (order, production deliveries, stock values, financials, profit/loss) past and forecast is held on this database.

The cube database is updated regularly from active systems and provision is made for inquiries into any combination of the multiple data axis. This is a mainframe application containing gigabytes of data which is regularly accessed from over 200 terminals distributed worldwide.

17.4.2 The management information display system

Whereas most data processing systems record the past, managers need to make judgments about the future. The past may be precise but the future is not; thus a high degree of accuracy is less important. Management information is about:

- relative sizes (i.e. what is most important?)
- comparisons (i.e. are we better or worse than predicted?)
- trends (i.e. are we gaining or losing?)
- data clarification (i.e. what does this tabulation mean?)
- inter-relationships (i.e. what is the effect of this expenditure on that process?).

In this field, the precise numerical based logic of IT interacts with the wholly fuzzy domain of the human mind and a different series of IT techniques come into play. Of these, Management Graphics are the most successful and the evolving strategy now being described made extensive use of that man/machine communication medium.

The computer world and the human mind are two extremely different cultures. The analytical processes of the computer are precise, logical and consistent. The human mind is imprecise, often illogical and seldom consistent. Whereas every computer will produce the same result to a given calculation, every human interpretation of data will vary from person to person.

This dichotomy is shown diagrammatically in Fig. 17.6. When data is transferred to the human mind it is subjected to a review process which gives it meaning. The data is quantitative but the assigned meaning will generally be qualitative: a temperature becomes hot or cold, a time

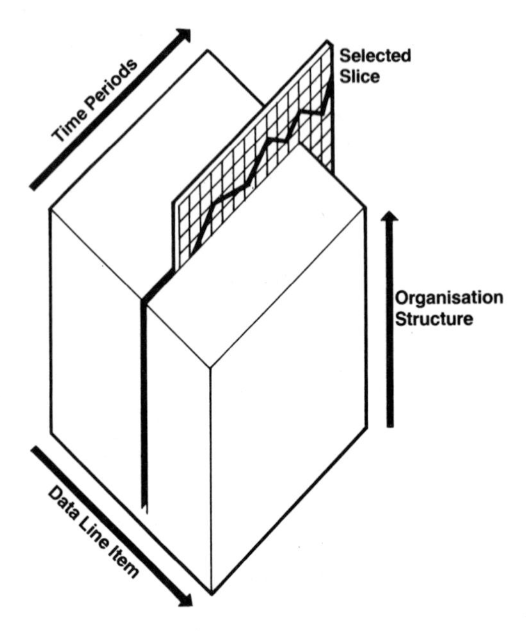

Fig. 17.5 Management information database

becomes early or late, a cost becomes high or low. The meaning which is put on the data is influenced by a multitude of previous experiences — some relevant, some irrelevant. Some of the influences are unpredictable, many are uncontrollable and all of them are variable between one individual and the next. Consistency of response seldom materialises because of the manner in which data is absorbed by the human mind.

In Fig. 17.6 this human absorption process is shown as a series of steps, viz:

- Absorb the graphical image.
- Interpret into subject matter.
- Extract meaning.
- Relate the meaning to an activity.

The most common problem encountered by users of Management Graphics is an appreciation of context, i.e. scale or relationship with other (not displayed) data. The system (when used in conjunction with the CUBIT database) overcomes this difficulty by providing a 'browsing' facility whereby the build up of data, e.g. company – division – region – area etc., can be explored graphically by interactively viewing a series of charts representing associated information.

It is argued that this method gives a close approximation of the human thought process. When a human is confronted with a new viewpoint, he/she instinctively seeks to relate the unknown to an already familiar concept. In business, this process involves a series of questions and answers which clarify the structure of the concept, and the accuracy and importance of the underlying data.

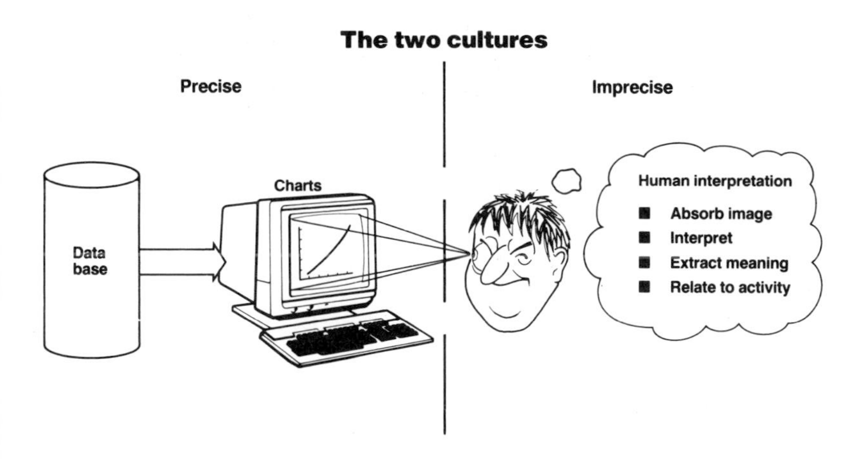

Fig. 17.6 The two cultures

This method requires a large number of charts (the Managing Director has access to over 700 charts) but does provide the ability to really understand the significance of the data.

17.4.3 Critical success factors and results achieved

At the commencement of the strategy, a series of 'Critical success factors' were set out and these are used as measures of the effectiveness of both the strategy and the company. The principal measures were on:

- profit
- costs
- quality
- service
- cycle time
- return on assets.

Improvements in all these items have occurred over the period of strategy implementation and are continuing. Unit costs have decreased and profit increased at all levels. Deliveries are faster and the cycle times for new product introductions shorter. Quality has improved dramatically. It is, of course, not possible to positively identify the proportion of improvement attributable to IT, but it is a vital component in helping the human and automation resources achieve such results.

Moreover, it is clear that the company is profitable, and is succeeding in its competitive goals. Also, there is common consent that the company could not operate for a single day without the array of IT systems now in place.

Chapter 18

Flexibility through the design of manufacturing infrastructures

S. J. Childe
Plymouth Polytechnic, UK

18.1 Introduction

The research project represents the largest single grant awarded under the ACME Computer Aided Production Management (CAPM) Initiative and is focused on the problems of implementing CAPM systems in the electronics sector. The work arises from the history of poor success in the implementation of various kinds of Advanced Manufacturing Technology. A number of authors have highlighted the problem including Voss (1985), who noted that 57% of AMT applications were seen as unsuccessful, and Ingersoll Engineers (1985), who placed the figure at 50%. These findings imply that the implementation of AMT must be approached in a different manner to conventional technologies particularly as the companies implementing AMT were generally no strangers to the implementation of machines and equipment. It is also interesting to note that many companies experiencing benefits from AMT achieved 75% of the benefits before the actual installation of the technology (Besant and Haywood, 1985).

Computer Aided Production Management is itself an example of Advanced Manufacturing Technology. The research project's working definition of CAPM includes the processing of orders from customers to the shop floor, materials provisioning systems such as MRP, WIP tracking, bill of materials control and all computerised aids to the production manager, including planning, scheduling, controlling, recording and reporting functions.

18.2 CAPM implementation

18.2.1 The single point solution

The traditional systems design approach incorporates the following stages:

(i) Definition of the current state — analysis of current procedures and problems.
(ii) Definition of the proposed system — definition of requirements which the new system must fulfil.
(iii) Specification, selection and design of system components and configuration with respect to the requirements definition.
(iv) Training of personnel and system installation.

This approach is one which models the future system upon the present and past system, taking advantage of technological developments such as faster processing, labour savings etc. and making suitable allowances to prevent the known shortcomings of the present system being manifested in the future system. This approach has been the basis for much work developing the implemenation of AMT which pays attention to issues such as user acceptability, for example Robey (1979), and the development of project management strategies, for example Tranfield and Smith (1987).

18.2.2. CAPM implementation failures

The research so far has looked at CAPM in a sample of 13 electronics companies. Failure was regarded as any shortcoming which required the CAPM system to be altered significantly from the way it was implemented, or a failure of the implementation process which resulted in the system being either very late into use or not in use as intended. From this sample comes the schematic diagram of Fig. 18.1 which shows the four typical modes of CAPM failure.

It is perhaps surprising to observe failures which resulted from an inadequate requirements definition, which led to the wrong system being implemented. Possible reasons for this, such as over selling by vendors or the installation of 'familiar' systems will not be considered here. Having correctly defined the nature of the system requirements, some companies unfortunately failed to implement a system whih fulfilled these requirements. Again, the reasons for this are not the subject of the present chapter. Another way CAPM implementations fail is by managing the project badly so that the system arrives very late, is short of necessary functionality or capacity, or is the subject of difficulties such as poor training or consultation leading to its non-use.

The most common mode of failure, however, occurs when the requirements change to the extent that the system is then short of functionality or capacity, such as when the company grows, or moves to a position

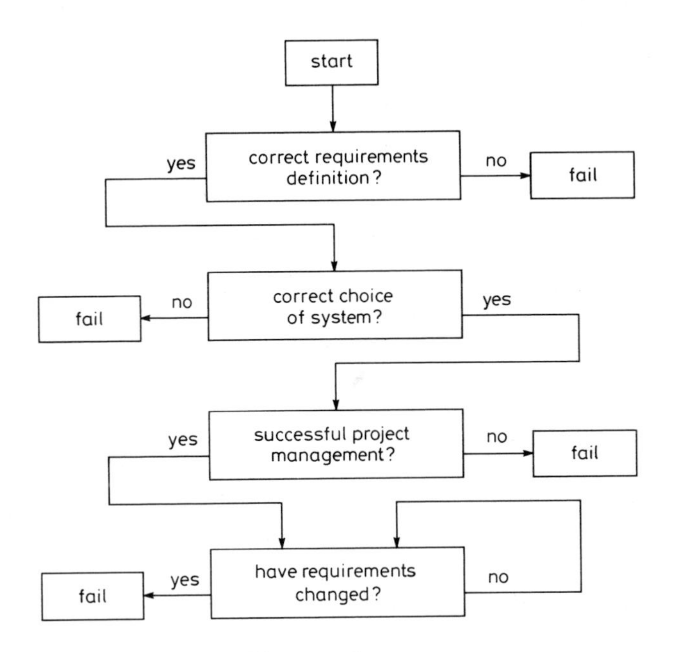

Fig. 18.1 CAM implementation failure modes

where the system is no longer appropriate; for example when manufacturing changes from MRP type batch logistics to a Kanban system. The change in requirements ultimately renders obsolete any system which has been designed and implemented in the traditional way. It is possible for this type of failure to happen even before the system has been implemented. Unfortunately, there is no easy answer to this problem beyond making 'best guess' predictions of the future. It is not easy to expect the unexpected. In our experience this is the most common failure mode, and it appears to be a characteristic of the traditional approach to system implementation, which results in 'single point solutions' which are suited to the business at a particular point in time.

18.3 The need for flexibility

18.3.1 The electronics market

The UK electronics market is one of the fastest growing in Europe, experiencing rapid change both in the marketplace and in the functionality of products. Change is common and unpredictable. For example:

- Shortening product life cycles demands more frequent introduction of new products, which require new designs, equipment, materials, and methods.
- New process technologies such as surface mount can lead to obsolescence of existing product, requiring companies to adopt new methods.
- Manufacturing volumes can change suddenly, causing changes in manufacturing strategy.

18.3.2 Some possible solutions

Some companies attempt to resist competitive pressure by carving out a niche in which there is little or no competition. For many companies this is a successful strategy but it can fail if a serious competitor moves into the niche, or if the niche is removed by a change in the market. Furthermore, this strategy leaves the company more open to the consequences of market changes because ways of coping with change will not be developed.

Under certain conditions it is feasible for companies to attempt to control the market. This strategy cannot be regarded as widely applicable.

18.3.3 How flexibility helps

Flexibility is used here to denote how well a company can respond to changed requirements, whether imposed by the needs of the market (reactive) or by the anticipation intentions of the business itself (proactive). Flexibility is often a goal of manufacturing strategy, but as Slack (1987*a*) points out it is not one that has a clear meaning. Slack was able to discern four distinct types of flexibility:

> '*Product flexibility*: the ability to introduce novel products, or to modify existing ones.
> *Mix flexibility*: the ability to change the range of products made within a given time period.
> *Volume flexibility*: the ability to change the level of aggregated output.
> *Delivery flexibility*: the ability to change planned or assumed delivery dates.'

All these types of flexibility are important to many of the companies investigated in the course of the research.

Slack noted also that managers tended to focus on the flexibility either of specific machines or of labour, rather than upon the flexibility of the manufacturing business as a whole, whilst the distinction between range and response flexibility was a source of confusion. 'Range' is the range of states which a system can achieve, such as the range of parts a machining centre can machine, whilst 'response' is the ease with which change can be made within the range. When designing a manufacturing system, it is important to understand the types of flexibility required.

Much attention has been placed upon the flexibility of machine tools, especially with the development of Flexible Manufacturing Systems (FMS) — machines which have a very fast response to change within a given range. However, it appears that flexible technology on the shop floor has to be supported by a flexible infrastructure.

18.4 The manufacturing infrastructure

The infrastructure consists of the systems which support the manufacturing activity. It has been defined by Hill (1985) as the 'controls, procedures, systems and communications combined with the attitudes, experience and skill of the people involved', and by Skinner (1985) as the 'policies, procedures and organisation by which manufacturing accomplishes its work, specifically production and inventory control systems, cost and quality control systems, work force management policies and organisational structure'. Meredith (1986) also describes the infrastructure as 'the network of non-physical support systems that enables the technical structure to operate'.

The infrastructure is critical to the manufacturing operation since manufacturing can only respond to the marketplace at the rate at which market information is translated into instructions for manufacturing in the form of orders, schedules etc., and at the rate at which new designs, equipment and materials can be provided.

It can be seen that Slack's four types of flexibility cannot be provided by structural means alone — the flexibility of the infrastructure would restrict even the most flexible manufacturing hardware by failing to provide the support required.

Hill presents an outline typology of infrastructures, describing some of the characteristics that infrastructures should possess in different types of manufacturing situation, such as batch, job, flow etc. However, this outline does not address the question of the implementation of infrastructures with particular required characteristics. Thus a more detailed understanding of the functions of the infrastructure is required.

The provision of orders, schedules, designs, equipment and materials has been mentioned earlier. It is proposed that the functions of the infrastructure are to provide the following inputs which completely support and control the manufacturing activity:

(i) designs
(ii) materials
(iii) methods information, such as manufacturing layouts, works instructions, process instructions, quality procedures, NC part programs

(iv) orders, such as work-to lists, job schedules picking lists, batch cards and travellers
(v) facilities such as the plant and equipment, machinery, fixtures, tools and gauges
(vi) trained and motivated personnel.

Thus the infrastructure provides three physical requirements (material, facilities and personnel) and three informational requirements (designs, methods and orders). The various aspects of CAPM are involved with the purchasing of materials and with the processing of orders. It is proposed that a CAPM or other implementation must aid the responsiveness of the business and must therefore form part of a flexible infrastructure.

These six functions of the infrastructure provide a framework which allows analysis and design of infrastructures in specific company situations. It provides a basis upon which decisions can be taken with regard to the amount of flexibility required in each functional area.

18.4.1 Infrastructure design

The design of manufacturing infrastructures must be based upon establishing performance parameters for the various infrastructure elements. In any area there will be limits of range and response which will limit the infrastructure's performance as well as modifiability which will affect the infrastructure's adaption to a new business scenario. The range and response criteria must be established to allow the infrastructure to operate withn a range of situations which can be regarded as likely situations in the business. Thus, as in structural terms a milk bottle manufacturer would not provide sufficient flexibility to switch production from milk bottles to printed circuit boards, in the infrastructure area a PCB manufacturer would not train designers in the area of milk bottle design. Although a wide range and a speedy response would seem ideal, the practical operationalisation is to set limits to the desired performance, within which any change could be met. The setting of these limits is thus a strategic question for the direction of the business, which is being addressed by the process methodology currently being developed by the project team.

18.5 Infrastructure functions

18.5.1 Design function

Flexibility of the design function relates to the rate at which new designs can be provided. Company A, a manufacturer of domestic electronic goods, found the increased flexibility and Just-In-Time manufacturing on the shop floor together with a continuous improvement policy created a demand for thousands of design modifications per year. The existing procedure became overloaded, and the company is now considering ways of improving this function.

Company B, a computer manufacturer, also found that increasing numbers of design changes were required, and were able to implement a computer-based design change management system to facilitate the passing of information between departments and to monitor the progress of all change requests.

18.5.2 Material function

Materials can be a severe limiting factor to flexibility. Company B finds that long procurement times, such as three months to purchase an ASIC chip from Japan, are a limit to the company's responsiveness. This has also caused quality and inventory problems owing to boards being assembled as components become available, rather than in line with the customer order schedule.

In the electronics industry the total manufacturing lead times, and therefore the responsiveness to customer demands, depend to a large extent on the procurement lead time, since the actual manufacturing lead time is relatively short. In Company C, a manufacturer of defence-related products, steps are being taken to improve the time taken to issue purchase orders and to improve supplier relationships in order to become more responsive.

Company D, a components manufacturer, has improved its ordering process so that all purchasing is managed by less than one person and a typist. This appears to provide the appropriate level of flexibility for this business.

18.5.3 Methods function

Company E, a subcontract assembler, has no area with specific responsibility for method engineering, since the business is built around one specific method only and all products are processed through the same route. This company has decided to have no flexibility of methods. If this technology becomes obsolete, the company will face difficulties in responding quickly.

Company F, a component manufacturer, requires little methods flexibility to meet day to day changes, which tend to be of mix only. However, it has assembled a highly skilled team for the development of new process equipment to create a specific market advantage.

In some companies, for example Company D, methods work is carried out by the design department.

18.5.4 Order-processing function

Most of the companies investigated are conscious of the importance of processing orders quickly. For example, Companies D and F both allow major customers to enter orders by data transfer direct into the orders database. Some other companies were considering this possibility.

Company G manufactures tailor-made electronic equipment. Customers ordering from this company are able to speak direct to a design engineer who can quickly tailor the design as required, provide a price and pass the order immediately to manufacturing.

18.5.5 Facilities

The companies investigated had made little attempt to develop flexibility in the area of provisioning tools and equipment, relying generally on external equipment manufacturers, tool manufacturers and maintenance contractors. However, this must limit the responsiveness of the companies to changes requiring new processes or tooling. The case for tool management and a review of tool management functions based mainly on metal-cutting FMS is presented by Carrie and Bititci (1988*b*).

18.5.6 Personnel

Flexible personnel approaches have been implemented by almost all the companies. This generally involves operators moving between different operations.

Company F has attempted to de-skill all direct operations and to employ operators on short term contracts to provide numerical flexibility.

Company E contracts out some operations and employs out-workers on a subcontract basis.

Personnel flexibility is treated in detail by Gustavsen (1986) and Atkinson (1984).

18.6 Conclusion

This chapter has presented the six basic functions of the production management infrastructure.

It has described the way overall manufacturing flexibility depends upon the performance of the infrastructure as well as that of the physical elements.

In a competitive and fast-changing market a CAPM or other AMT implementation can only have lasting success as part of an integrated strategy which includes the design of an appropriate flexible production management infrastructure.

Business strategy must dictate the amount and type of flexibility required in each functional area. Particularly in the context of AMT implementation, it is important to consider the infrastructure so that the AMT performance is not limited by the support functions.

Some examples of infrastructure configurations have been presented. Future work is intended, which will aim to provide an understanding of the performance of various infrastructure functions configured in different ways and to develop a processes methodology to lead companies through the implementation process.

Chapter 19

Modular systems integration for flexible manufacturing: the case of low-volume, high-variety assembly

D. Bennett, M. Oakley and S. Rajput
Aston University Business School, UK

19.1 Introduction

The concept of flexibility in manufacturing is well known (Slack, 1987*b*), and for many applications, such as parts production, flexible systems have reached an advanced stage of development (Snader, 1986). In the case of assembly, two distinct situations can be identified. The first is where automation of assembly operations themselves can be justified because only simple manipulation of parts is required or relatively high volumes are being produced. Here, fully automated, flexible, assembly lines or cells using programmable robotic devices provide a viable solution (Storjohann, 1986). The second situation is where, owing to the complexity of assembly operations and the lower volumes involved, the use of fully automated assembly techniques cannot be considered a feasible option. Competitive pressures on many firms in this position, however, demand that they still adopt the *underlying principles* of flexible manufacturing although a different system design solution is usually thought to be appropriate (Kumpe & Bolwjn, 1988).

In this chapter, a specific example of this situation within the electronics industry will be discussed and the solution developed by the firm in question, termed the Modular Assembly Cascade will be described. A research team at the Aston Business School, which has assessed the effectiveness and general applicability of this approach, is now in the process of developing and testing a methodology for firms wishing to design and implement a similar system. The DRAMA methodology (Design Routine for Adopting Modular Assembly) is in essence a set of guiding principles which addresses the question of market requirements and manufacturing strategy, organisation design, the physical design options, control and integration, work organisation, project management, implementation, justification and evaluation.

19.2 The environment of electronics manufacture

The need for flexibility in assembly operations is clearly demonstrated in the case of electronics manufacturing. In the electronics sector most UK companies face strong competition from large American multi-national corporations as well as major producers from Japan and the newly industrialised countries (NICs). The relative size or lower overhead and labour cost of these competitors usually enables them to exploit the economies of scale to such an extent that it is now virtually impossible for small to medium sized domestic enterprises to compete solely on the basis of price. The ability for products to be quickly cloned also makes it increasingly difficult to rely on supervisor design, so the approach of 'pushing' technically sophisticated products onto the market is no longer a viable strategic option in the long term.

For this reason, new strategies for production are being formulated which enable firms to compete on the basis of responsiveness, both in terms of delivery and the ability to provide precisely what the customer wants.

Such 'market driven' strategies have been adopted by a number of computer manufacturers (Knobel, 1988; Piper, 1985; Caulkin, 1987) and it is in one such organisation that research is being conducted on which this chapter is based. About five years ago the Company in question completely revised its marketing orientation following a change in senior management, and a number of 'business units' were created for selling complete 'solutions' within specific sectors rather than the previous policy of trying to sell standard products across all sectors. In turn, this new approach placed entirely new demands on the manufacturing function so that flexibility and quality became primary considerations. This change in orientation coincided with the introduction of a new range of mainframe computers which could be configured in a wide variety of ways to suit specific customer requirements. The time was therefore opportune to reconsider the Company's whole approach to production systems design, and so the idea of the Modular Assembly Cascade was first conceived.

19.3 Flexible manufacturing using the Modular Assembly Cascade

The approach to flexible manufacturing which has been developed is based on the idea of designing cells or modules for assembly which are largely autonomous, their main constraint simply being the overall dimensions of the products or subassemblies they can handle (Bennett & Forrester, 1988*a*; Nagarkar & Bennett, 1988). By adopting the 'pull principle of ordering using the Kanban technique (Shingo, 1981; Goddard, 1982) the modules manufacturing larger dimensional items draw from those producing smaller ones; hence the idea of materials 'cascading' down the various levels of assembly modules (Fig. 19.1).

The modules themselves derive their flexibility from 'generalising' the production activities and materials distribution for a range of different products. A comprehensive programme of training has also been organised to extend the range of operator skills and increase their ability to work in a less structured environment. Automation is in evidence; but

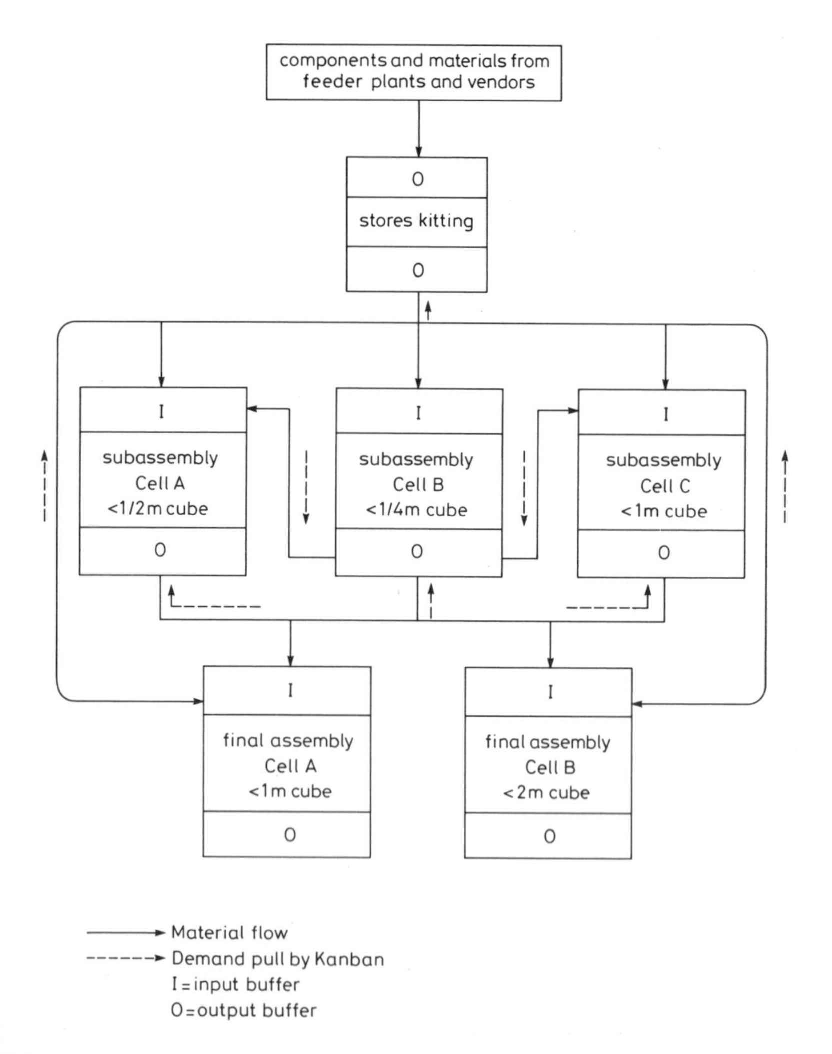

Fig. 19.1 The modular assembly cascade

rather than being applied to the assembly operations, it is used for material handling (both within and between modules) and information processing. It can be used most effectively in these areas since materials now comprise almost 95% of manufacturing cost. Thus, the system as a whole comprises computer controlled cranes, AGVs of various sizes and function, horizontal carousel and paternoster stores etc. Each cell is under the control of its own microcomputer, while the overall system is managed by a hierarchy of minicomputers under mainframe computer control employed in master scheduling. The use of the Kanban approach for controlling the flow of materials within the cascade has enabled significant reductions in inventory holding to be achieved, with the current turnover in work in progress being around 18 times compared with less than five in 1980.

19.4 Developing the design methodology for modular assembly

With the collaboration of the Company, and funding from the ACME Directorate of the Science and Engineering Research Council, a methodology for designing and implementing modular assembly is being developed. This has been preceded by a programme of research aimed at testing the viability, robustness and general applicability of the approach, coupled with an in-depth study of the actual design process within the firm. In addition to analysing the influence of the corporate strategy on manufacturing with respect to adopting a flexible manufacturing approach (Macbeth, 1985), the changing climate of organisational culture in fostering innovation and adapting to new technology is also investigated (Greenwald, 1985). The research is based on identifying the various stages involved in implementing flexible manufacturing systems (Vonderembse and Wobser, 1987), determining the extent to which the technology transforms the manfacturing operations. The methodology employed in developing and incorporating a kitting system (Conrad and Pukanic, 1986) has been analysed, and the extent to which manufacturing has been 'rationalised' in the move towards JIT (Powell, 1986) will be examined. The influence of advanced technology on the perceptions and role of the company personnel (De Pietro and Schremser, 1987) is also to be considered. The overall research method is described elsewhere (Bennett *et al.*, 1988*b*.)

The DRAMA methodology is being developed by generalising and refining the design process for the Modular Assembly Cascade (Fig. 19.2). The detailed study of the Company has considered not only its customers and competitors, but also the environment in which the business operates and the influence of corporate strategy on manufacturing. Comparisons have also been made with similar companies operating in a low volume, high variety environment.

The process of system design in the collaborating Company has been monitored and reconstructed using participative observation, an interview programme and documentary sources. Reference has also been made to equipment vendors where relevant. In moving to the generalised methodology a list of underlying assertions has been generated on which the Modular Assembly Cascade concept was based. These have then been confirmed or refuted using reference points drawn from an evaluation of system implementation and the studies of comparable companies. Conceptual models have been drawn upon, when appropriate, to assist with analysing the decision processes and to provide a framework for the design routine (Mintzberg *et al*, 1976; Doumeingts *et al*, 1986). Finally, a number of test sites have been identified for use in evaluating and refining the generalised methodology, with a view to making it generically usable in a range of product and process contexts.

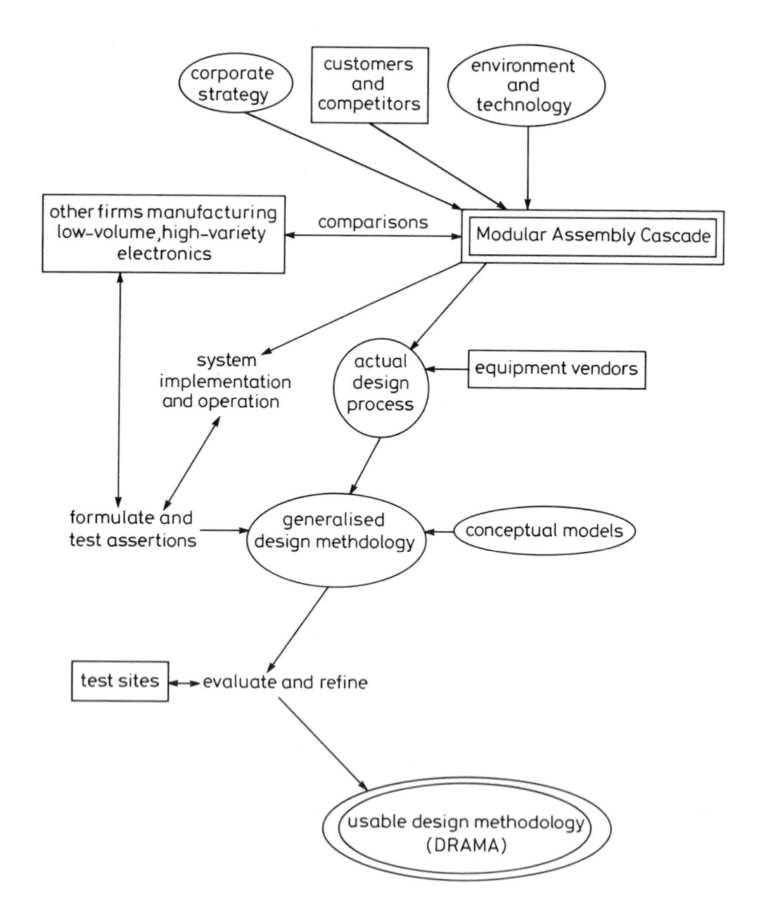

Fig. 19.2 Development of design routine for adopting modular assembly (DRAMA)

19.5 Components of the design methodology

The design methodology being developed comprises a number of components which together provide a guide for an organisation wishing to adopt the principles of the Modular Cascade for assembly operations. The methodology is, in itself, flexible in that it is based around a *philosophy* rather than offering a rigid prescription based solely on the solutions devised by the collaborating Company (Hughes, 1987). It takes various forms including narrative, flowcharts, decision trees and checklists. In a short presentation the various components cannot be described in detail, so they are simply listed here together with a brief discussion of some of the specific considerations that have been identified from the Company study.

19.5.1 Justification of new systems

The financial investment in the Modular Assembly Cascade was quantifiably justified by reference to savings in inventory holding, labour requirements and space usage. However, 'intangible' benefits from improved control and greater flexibility were also anticipated and have been achieved. In addition, the benefit to the Company of creating a 'showcase' factory was a consideration (this may be less important in other organisations).

19.5.2 Manufacturing strategy

The boundary of flexibility for each module was assessed in terms of the production requirements for a range of products, the storage and routing of materials, and the range of operator skills necessary. The modules were primarily designed around 'product dimension' parameters.

The degree of system flexibility and its ability to respond to demand was optimised by further dividing certain sub-systems into several modules, each catering for a range of product sizes. This maximised manufacturing efficiency (e.g. two modules separately cater for large and small sub-assemblies).

19.5.3 Physical system design

Facility re-layout and revision of material flow was considered on a factory wide basis. This 'total' approach, in addition to rationalising material flow through the factory, optimised space usage arising from the re-alignment of sub-systems along coherent routes of material flow, thus 'creating' substantial additional space by releasing 'redundant areas' in the old layout.

'Satellite' buffer stores were incorporated into the design of the system. The ideal philosophy underpinning JIT of operating with near zero inventory levels was considered impractical in an environment where product variety is high and production volume is low. However, a viable option was to incorporate into each module small satellite 'input' stores for kits (and 'output' stores for completed products) in order to 'smooth' the logistical complexities of material tracking and control. This provided modules with an ability to respond effectively when servicing subsequent modules within the hierarchy, and effectively reduced the main stock-holding commitment.

19.5.4 Control and integration

Although a true JIT operation involving outside suppliers was not immediately feasible, many of the benefits of operating with reduced inventory levels, and improving materials control by functioning in a JIT 'pull' mode, were deemed to be attainable by adopting the philosophy for in-house operations. This in-house JIT philosophy would then interact with MRP at the external interfaces of the business, thus more accurately matching supplier orders with long term forecasted product demands.

As expected, the 'satellite' buffers were found to assist material control between modules. Focusing materials into satellite buffer stores allows greater confidence in implementing inventory reduction programmes. The control of material flow between modules is via a minimum/maximum stock level in each buffer, and this level can be continually revised according to forecasted product demand requirements.

A 'computerised' tracking system was selected for the control of material and WIP. Accurate real-time information on the status of material flow is considered to be best achieved via a barcoding system, facilitating data integrity, allowing better production control, and enhancing strategic planning.

19.5.5 Work organisation

Module design based on 'new' concepts provided the opporutnity to re-assess operational functions within the assembly environment, in particular with regard to operator multi-skilling to enable assembly of a higher variety of products, and revising methods of job allocation and distribution. 'Quality circles' and a 'zero defects' philosophy also enabled operators to contribute to the improvement of work methods.

The kitting function was separated from the assembly function because experience had shown that the combined activities of picking and assembly by the operator involved time being wasted searching for the correct components, and delays in production owing to part shortages. Separating the two functions ensured that the assembly operator concentrated only on assembling the complete kit of components, supplied by the kitting operator, who in turn ensured that materials problems were identified and resolved prior to being distributed to assembly. Although the function of kitting may be considered to be a non-value adding activity, its costs have been found to be outweighed by the benefits of improved control and earlier identification of the problems and logistical complexities associated with a high variety of components.

19.5.6 Project management

A Working Party Team was formed to undertake detailed design of the production systems. The team included representatives from all functional areas affected by the changes. The objective was to ensure total agreement on the final solution and identify problems in the early stages of design by both system engineers and production personnel.

19.5.7 Implementation

The concept of modular design allowed the transformation of the total manufacturing operation to be implemented in phases, so minimising disruptions to production.

In a labour intensive environment concerned with assembly, successful operation and control was achieved not only by 'computerisation of production but also by the operators' ability to maintain the required disciplines of operation; e.g. ensuring kits were not supplied with shortages, maintaining data integrity etc.

A clean and tidy factory was considered highly important, so facilitating the requirements for good 'housekeeping'. This was achieved by ensuring that everything was 'in its place' by designing appropriate storage facilities within each module.

19.6 Assessment of methodology in practice

The researchers have closely monitored the system design process, implementation, and operation of the modules installed by the Company. From analysing the methodology undertaken in practice, various key issues have emerged, some of which are highlighted in the following discussion:

Unnecessary delays during the design and implementation phases can be avoided by employing a Project Leader who is also a 'champion', and who has overall authority, with the task of encouraging the participation of all team members towards a common goal. The appointment of a Project Leader with assigned responsibility for the project from initiation through to completion may provide 'continuity' in systems development. However, the complexity of a project may require the appointed leader to change at specific stages of systems development, allowing the available talents and skills resources to be optimised. In either case the Project Leaders should be formally recognised and information relating to the project channelled through them.

Capital approval may be quickly sought and approved on the basis of a very broad specification of system requirements. However, this may then give rise to pressure on the system designers to achieve detailed design and implementation within unrealistic time-scales and costs, which in turn could influence the optimality of design. Greater effort on the details of design prior to obtaining capital approval is likely to result in more realistic projections of time-scales and costs. Of course, this may create an unacceptable delay in submitting for capital approval, especially where financial conditions are subject to change. In addition, greater knowledge of the details of design might give the capital approving body more reasons for rejecting the proposed development. The success of obtaining approval may not only be based on technical issues, but also on the 'politics' within the organisation.

The adoption of 'new' operational concepts (i.e. flexibility and JIT) may not require investments in sophisticated materials handling systems. Instead, the associated benefits may be adequately gained through a reassessment of operating methods, ensuring disciplined modes of materials control, and improved methods of logging. However, the situation may be such that the traditional views and practices entrenched within the organisation can only be changed by forcing the adoption of totally new technology and facilities.

The successful implementation of major operational changes requires early recognition and accurate evaluation of the 'impact' on all parts of the organisation. Where this is not done, time-scale delays and extra costs may be incurred, and confidence in the approach and enthusiasm for the 'final objective' could decrease.

In order that production is not severely affected by hardware and software failures, 'resilience' can be designed into systems when operators are dependent on automated equipment (an example of 'resilience' is the level of stock in the satellite buffer stores). However, excessive resilience may incur extra costs because of the duplication of material and facilities. In addition, concentration on continually maximising operations may lapse, and so go on against the philosophy behind JIT.

19.7 Conclusion

The 'modular' features of the assembly system adopted by the collaborating Company have been outlined, together with an indication of the design process which was followed. Experience of this design process, including implementation and commissioning stages, is providing the basis for development of a generally applicable methodology (DRAMA) intended for use on future occasion by the Company and by other similar firms. The DRAMA methodology is being refined by testing assertions within the Company and other organisations. The objective in developing DRAMA is to provide a comprehensive set of guiding principles for achieving flexibility in assembly manufacturing via the Modular Cascade approach.

Key elements have been identified in the design of manufacturing systems adopted by the Company, and these in turn may provide the Company with the competitive edge in satisfying the requirements of a market driven business in a low-volume/high-variety environment. In particular, the benefits of improved material control associated with JIT have been achieved by initiating JIT principles for in-house manufacturing. Also, autonomous modules have been designed to cater for a family of products based on their overall size, while the incorporation of 'satellite' buffer stores allows material to be effectively controlled between the modules. In addition, kitting and assembly have been segregated, allowing operators to concentrate separately on these tasks and resolve associated problems.

Various issues for further consideration have arisen out of analysing the methodology actually followed by the Company, ranging from the requirement to formally recognise the Project Leader, to designing 'resilience' into the system. These need to be further developed and tested, and accounted for in the formulation of the generally applicable DRAMA methodology. The design of DRAMA will also be influenced by a post-implementational analysis of the extent to which the Modular Assembly Cascade has been a success in improving the Company's competitive position.

Chapter 20

CIM and the principles of business control

D. M. Love and R. A. G. Twose
Aston University, and Lucas Aerospace Ltd., UK

20.1 Introduction

The current developments in business/organisational theory, computer technology and systems are creating a catalytic industrial environment in which the adoption of a Computer Integrated Manufacturing philosophy is viable. Manufacturing organisations have in the past divided the flow of work into discrete isolated functions and co-ordinated the various activities in a serial manner. It is becoming increasingly necessary to develop a parallel and networked environment with much more interaction between the various functions, such that all aspects of the business feel part of an integrated whole and work together in a mutually supportive and integrated way to clean defined objectives. The degree of interaction and mode of integration between the various functions, and indeed the control systems within, constitute the major area of discussion within this chapter. It argues that there is a close relationship between CIM and the required levels of business control; and suggests that a particularly convenient way of modelling the integration requirements of the business is through the use of control hierarchies. The chapter discusses the effect of integration on the organisation and identifies the key elements of business integration required during the adoption of a CIM philosophy.

20.2 Discussion of CIM

20.2.1 Definitions: General discussion

There is a profusion of differing and conflicting published definitions of CIM; many concentrate on the physical technological side of CIM and reflect the parochial interests of individual authors. An analysis of CIM definitions was performed by Boaden Dale (1986); the analysis showed that the largest class of definition was the 'single facet' class and the most widely used definition was that of an integrated CAD/CAM system. This definition focuses directly on one particular element of systems technology, falls short of covering the fundamentals of CIM and typifies a narrow view of the concept.

Other authors give a slightly broader definition, by extending the scope of systems technology elements used in supporting the various functions within a business. Two such examples are illustrated by Ranky (1985) and Thompson; both authors discuss the concept of CIM as linked Islands of Automation. This view is currently shared by many authors, especially system vendors. It can be considered a simplistic and potentially rather limiting view, since setting up a link between two separate systems is merely establishing a electronic medium for data transfer, i.e. an interface.

The proliferation of CIM definitions published in recent years has contributed to the confusion surrounding this subject, which is probably one of the main contributing factors in its slow take up.

20.3 Systems integration

System Integration is much more than interfacing. It implies making systems work together as a cohesive unit, so that they can be viewed as a single entity with clear inputs, outputs and objectives. This involves a detailed examination of the systems, the business functions they support, the data required, the triggers which cause an information flow, the sign-over points (a handshake between two functions, normally performed when information crosses a functional boundary), the embracing control required, and necessitates a clear understanding of data ownership (employing the master – slave relationship) and the required speed and depth of accessibility. Fig. 20.1 attempts to illustrate this view of system integration. A network of key business processes is shown ranging from process A to process L. Process G and H are computer assisted by systems Gl and Hl, respectively. The information produced by system Gl is required by system Hl; this is shown as information flow 'X'. Process G is the owner of information 'X'; hence system Gl is the master and system Hl is the slave, that is, all changes (add, delete and update) made to information 'X' must be introduced into system Gl and then transferred to systems Hl through the interface using control parameters. Process G and H are constituent parts of a key business process which should have been properly planned and have the necessary control mechanisms in place.

Systems integration is concerned with fitting the two systems into the overall business process so that they become an integral part of that process. This would result in systems no longer being viewed as Separate

Support tools for individual activities. This also implies that triggers 1 — 6 must be natural control parameters of the overall business process.

20.4 CIM defined

Systems integration, although very important, does not represent the total view of CIM; it is but one part of its total makeup. CIM is concerned with total business integration, where all elements of the business, for example, business functions, control requirements, people, systems, the organisation (to name a few), are synthesised in a way to optimise the business performance. The author suggests that there are eight key business elements which need integrating; these are discussed in more detail later on in this chapter.

CIM is best described as a philosophy which sets a direction for the Company towards total business integration. It would be regarded as a trend, a way of life or an operating framework. CIM can, however, be expressed in a physical sense; one example of this is the systems integration model which illustrates the required systems and the degree of integration required between them.

20.5 The CIM approach

A well documented approach to CIM is to simplify and optimise prior to integration. This is one of the main messages to be drawn from the Integrated Manufacturing report written by Ingersoll Engineers (1986). Clearly it is ineffective to integrate systems which fail in any way to meet the current and future functional needs of the business (for which the systems were originally intended). It is also important that integration should only be applied to best practice. It would therefore seem pertinent to perform an appraisal of the functions, procedures, information flows and systems within the business to identify any improvements required. This redesign should be performed as an integral part of a business wide strategy.

20.6 CIM benefits

The benefits to be gained by adopting CIM have been suggested by various authors; the range of benefits are, however, subjective and may vary depending on the particular view of CIM perceived by the author. Both Jarvis (1986) and Oldham (1988) have produced a list of the benefits attainable through CIM. The authors have, however, made no attempt to distinguish between benefits from CIM integration and those which would follow from the implementation of piecemeal systems technology, e.g. CAD/CAM, CAPP and MRP etc. Both lists originate from a survey performed by the National Research Council. After further investigation, it was discovered that the list of benefits produced by the survey is in fact attributable mainly to the implementation of piecemeal systems. Many authors regard the benefits from CIM as synergistic; that is, the sum of the benefits of the parts that form an integrated whole is greater than the sum of the benefits of the individual parts. The benefits from CIM integration can therefore be regarded as the difference between the two. It is argued that the benefits from CIM should not be confused with the benefits from implementing individual systems technology elements.

To date very few companies have progressed through the implementation phase of CIM; this has meant that little experience has been gained on the actual benefits to be achieved. It is suggested that this is the main reason why the authors' extensive literature search (at the time of writing this chapter) was unable to produce a quantifiable list of benefits achievable through the adoption of CIM. Other authors, including Willis & Sullivan (1984) and Kaplan (1986) have discussed the costs and benefits associated with CIM, but only in a qualitative sense with little or no reference to actual figures. The general consensus is that CIM must be viewed in both a financial and strategic way, as some of the benefits are intangible and thus render traditional investment appraisal methods (Payback, DCFs etc) inappropriate. A key intangible benefit that is likely to follow true integration is related to gaining full control of the business. The integration of the business elements as described in this chapter would result in industrial structures and systems which support and assist the various control procedures across the business. It is suggested that there is an important relationship between CIM as a strategic integration philosophy and the procedures/mechanisms required to maintain total effective business control.

20.7 Business control

20.7.1 General discussion

The major problem in running any organisation is that of exercising control. This may be defined as 'the comparing of actual results with predetermined standards or objectives and taking corrective action where deviation occurs'. Control comprises four key elements, these elements being essential ingredients to close the control loops:

(i) *Planning*: This, the first element sets the objectives and policies both in an overall sense and on a divisional basis; it determines detailed budgets, schedules and procedures, to guide achievement of objectives. It provides also the grouping of activities and the essential staffing to cover the tasks of the undertaking, as well as the other resources and facilities essential for operation. The planning element is not just for top management but for all functions within the organisation, e.g. Production, Sales, Finance etc.

(ii) Action: This element is concerned with actually doing some work in accordance with a plan, although the work content may vary considerably depending upon its nature. The diversity is shown in the five examples given:

(*a*) To perform product research to expand the product range.
(*b*) To process an order (contract) through a business.
(*c*) To provide a method of manufacture (process plan) for a part.
(*d*) To produce a part in accordance with customer requirements and to an agreed schedule.
(*e*) To feed a tool for 'parting off'.

The examples illustrate the scope of work which is planned and actioned within a business. Many actions are inter-related and hierarchical. For example, work packet example (*e*) is part of (*d*) which in turn is part of (*c*) etc.

(iii) *Measurement and feedback*: This control element measures the current level of activity, converts it into a form to aid interpretation and feeds the information to the area or location where comparison is made.

(iv) *Comparison and evaluation*: This, the last element, interprets feedback information and compares it against the plan to identify any deviations. Any deviations are examined, and depending on the severity and the process being controlled, would result in corrective action being taken or replanning.

Control calls for control information and early dynamic corrective action at all levels of management. Deviations from plan may become apparent firstly at the shop floor level. Correction at this level is important before deviations work up through the levels of management and cause major disruptions. The demands on management are such that it is important that reports should present and interpret the *exception*, both to conserve the manager's time and to aid him or her in decision making. Indeed the adoption of procedures which promote control by exception and the use of self-correcting control should be employed.

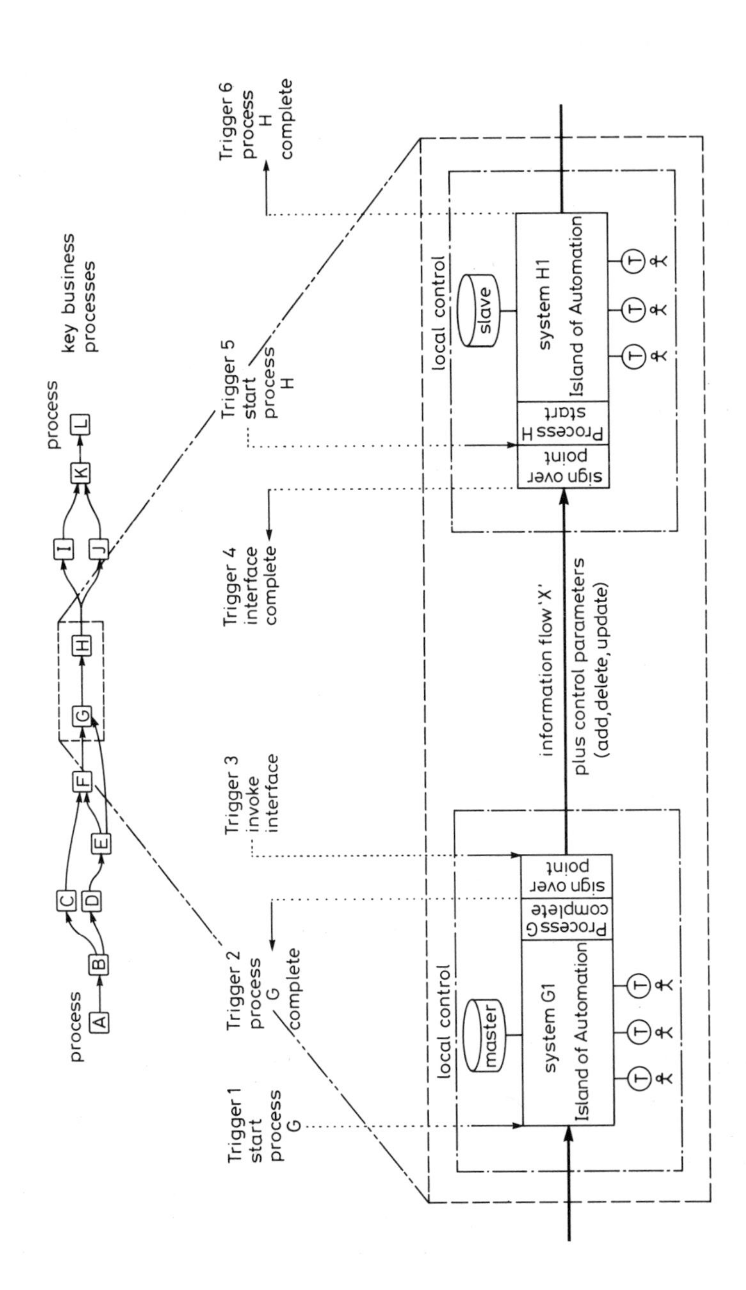

Fig. 20.1 Systems integration

20.8 Control hierarchy

In order to control a business effectively a detailed analysis should be performed to identify the control procedures and loops required to achieve the business objectives. This necessitates an examinaton of the key business processes, the performance measures which need controlling, the management structure plus their decision responsibilities, and finally the information requirements across the business.

A particularly convenient way of representing this is using a control hierarchy diagram for the business. Fig. 20.2 illustrates a simplistic view of this; five levels of control are shown, where each level has a different response time to close the control loop, requires a different style of management and demands a varying mix of decision making.

On each face of the hierarchy and in each facet of management, control activities are taking place. People at various levels and in various facets are making decisions; based on these decisions information is fed back into the overall information structure. The information flows, depicted by arrows, support the information requirements of the activities and by doing so help to close the control loops. As one can appreciate, in a real business the information flows would be rather more numerous.

The following conclusions can be inferred from this diagram:

(i) There is a distinct relationship between the structure of the organisation, the management style, the control levels/procedures and loops, the functions and the information structure.

(ii) To control a business effectively, one needs to combine the activities and control procedures to form natural groupings which simplify the information flows, support the necessary control loops, eliminate any waste and constitute an organisation which:

(*a*) Makes most effective use of people.

(*b*) Embraces the control requirements of the business in an integrated way.

(*c*) Promotes an integrated approach to problem solving and decision making where people from various parts of the business have greater and common visibility of the business as a whole and have shared values.

(iii) In such an environment it is then possible to develop an integrated approach to business control, where people have a shared vision of what is required and work together in a coherent way.

Level 2 of the control hierarchy shown in Fig. 20.2. is responsible for controlling the key business processes; it can be argued that it is also responsible for controlling the integrated of systems into the overall business process.

One such business process is the introduction of a new product which in

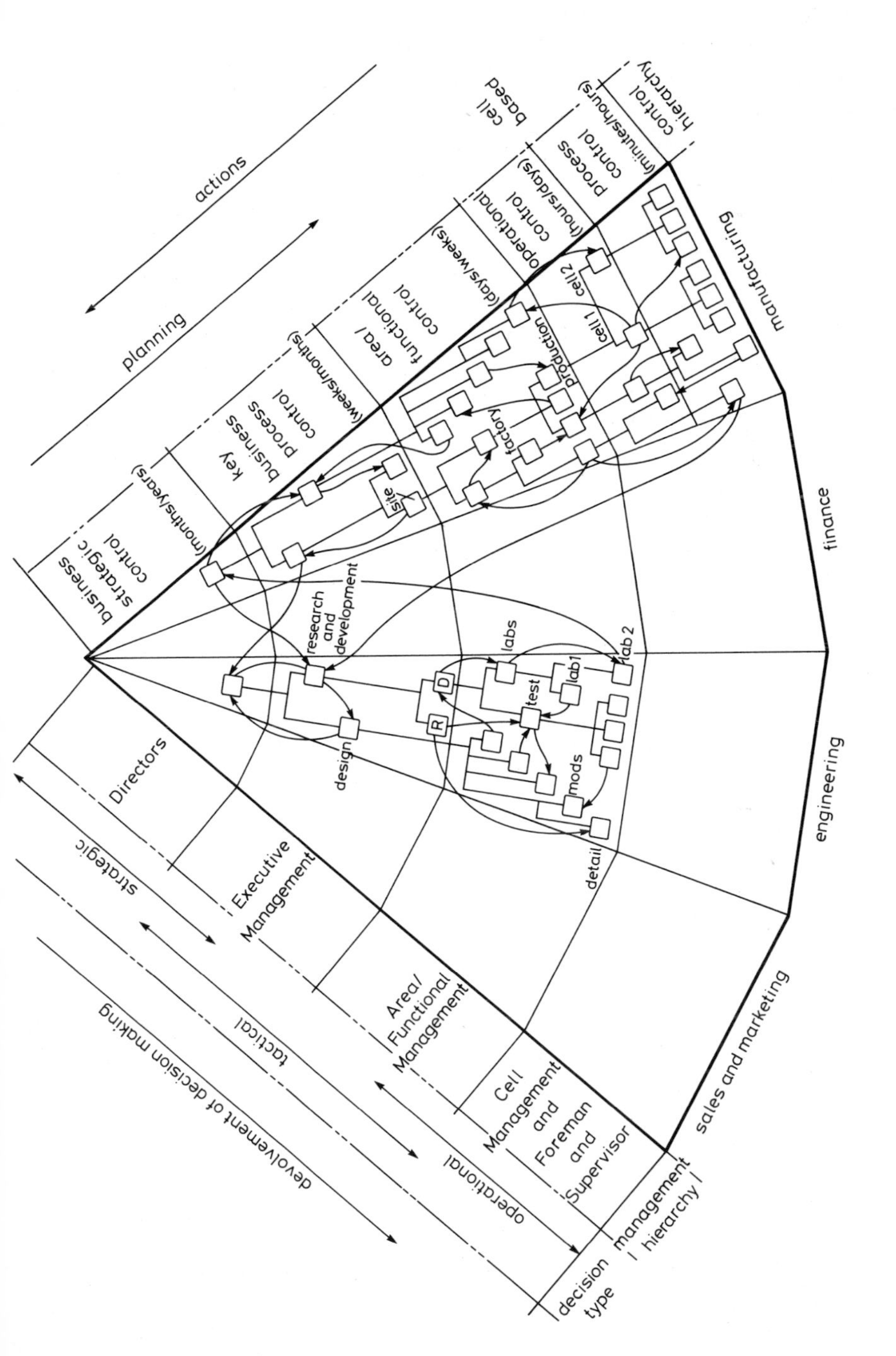

Fig. 20 Control hierarchy

many companies is controlled using a Project Control System based on Critical Path Methodology (CPM). It is suggested that in such a case the Project Control System (PCS) would be used to control the integration of systems by controlling the triggers (see Fig. 20.1, triggers 1—6). The concept of controlling systems integration in this way is illustrated in Fig. 20.3. The new product introduction process is planned using the Project Control System (PCS); a Gantt chart is produced showing the constituent processes and their start and end dates. The PCS sends out a trigger to the CAD/CAM system to start a design according to plan. Once the design is completed on CAD/CAM a trigger is automatically sent back to the PCS. This in turn executes interface software which extracts and forwards the part details to the Bill of Materials (BOM) system and the part geometry details to the CAPP system. The PCS then triggers the BOM and the CAPP system to start compiling the Bill of Materials and the process plan respectively. This chain reaction is performed until the project plan is complete. The concept of controlling work in this way has been termed Transaction Change Control.

If a control system was set up for each of the key business processes based on the Transaction Chain Control concept, it would result in great potential for redesigning the way in which businesses perform functions to respond automatically to high level inputs. The system would integrate both the actions of all functions concerned and the information flow between them using the key business activities as the basis for integration, and would thus provide visibility across functions throughout the business. Such a control system would computerise and control the inputs, reduce the number which require human attention and result in improved function control of processes and thus help control the business in an integrated way.

20.9 Integrated control and CIM

20.9.1 General discussion

A typical manufacturing company may have computer systems supporting many of its main functions. Such systems support specific elements of control, or combinations thereof, depending on the nature of the application. Indeed, it may take a number of computer systems to close a particular control loop. For example, a computer aided production management (CAPM) system would in some cases need to be linked to a shop floor data control (SFDC) system in order to close the control information loop. Much of the control previously performed by people is now being automated. Integrating these systems into the organisation and correct specification of their position within the control hierarchy is of paramount importance if the integrated control approach towards effective business control is to be followed.

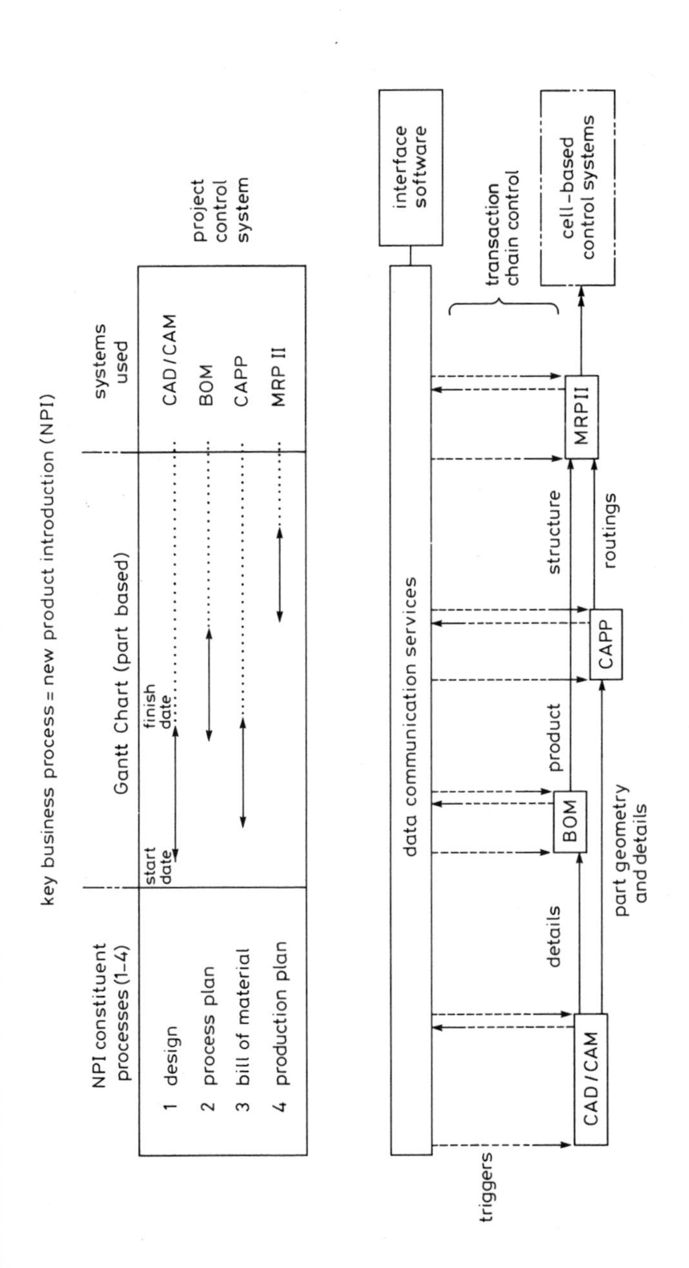

Fig. 20.3 Integration of systems into the overall business
Key business process = new product introduction (NPI)

Clearly by adopting the (CIM) integrated business philosophy significant changes will need to take place to ensure the resulting business is effective. Jones and Webb (1987) suggested that CIM changes managerial roles and the way control is exercised. In short, they argued that CIM would encourage devolvement of decision making, automate many control procedures previously carried out by man and change management roles from active involvement to passive monitoring. This they argue would have an overall affect of replacing the currently practised crisis management with strategic/tactical management.

It is the author's view that CIM would change much more than just the managerial roles and control procedures; CIM would ultimately mean changes to the business functions, the work and authority flows, the information structure and systems, the structure of the organisation, the people's roles and values and finally the culture developed by the business. This view is partially endorsed by Glen (1985) and Thurwachter (1986), both of whom discuss the impact of CIM on the organisation.

As previously stated, CIM is concerned with total business integration, where all elements of the business are synthesised in such a way as to optimise the business's performance. Clearly this would result in changes being designed into all elements of the business. The authors have identified eight key business integration elements which need detail examination.

20.10 Business integration elements

An initial review suggests that a CIM undertaking requires the integration of the following business integration elements:

(i) business strategy
(ii) business functions
(iii) control levels and requirements
(iv) people and the developed culture
(v) data/information structure
(vi) decision procedures/mechanisms
(vii) organisation structure
(viii) technology and systems.

To achieve true business integration all eight elements need to be examined in full. The business strategy sets the overall direction of the Company including strategic plans and mission statements. These plans should permeate through the Company and form an integrated network of functional plans which are executed by various functions across the business. Each function should have the necessary control mechanisms to ensure performance to plan, the four elements of control needing support to close the control loops. Each function should contribute to a business need; indeed all constituent activities should be value adding. The organisation of these functions within the business needs to be performed with due consideration to the people who perform the activities, the decisions they make and information required. It is important that the resulting organisation (functional hierarchy) makes most effective use of people and their tools and promotes a simplified information structure. The information structure should support the information requirements of all functions, control loops and decision makers.

The computer technology and systems adopted by the business should assist and in some cases automate the functions, support the information structure, aid and in some cases computerise decision making and provide various assistance in controlling the business. The culture of a Company ultimately determines the quality of its products and services. This is because the Company culture is the integrating factor for all behavioural and attitudinal patterns which prevail in the Company. Technology makes things possible but people still make things happen.

It can be seen that there is a close relationship between the eight business elements, since a change to one element would in many cases have repercussions across the others; it is argued that true integration is only achieved when all eight business elements have been addressed.

The inter-relationship and degree of integration between the eight elements needs to be examined in full, and this can be achieved using a series of models. One such model is shown in Fig. 20.4. It focuses on illustrating the ownership of systems and technology within the organisation and their positioning within the control hierarchy. Another model would illustrate the overlay of the information structure on top to show the information flows between the systems and control levels across the business.

20.11 Conclusion

The subject of CIM and its application within a manufacturing environment has been discussed for many years, but is traditionally limited to the linking of piecemeal systems technology. This chapter has sought to demonstrate that the scope of integration should be extended from the linking of a few systems to the integration of the systems into the organisation, where all elements of the business are synthesised in such a way to optimise performance. Benefits are clearly to be obtained from taking this extended view of integration over that traditionally held.

The research is continuing to investigate the detailed relationship between the key business integration elements. A number of models will

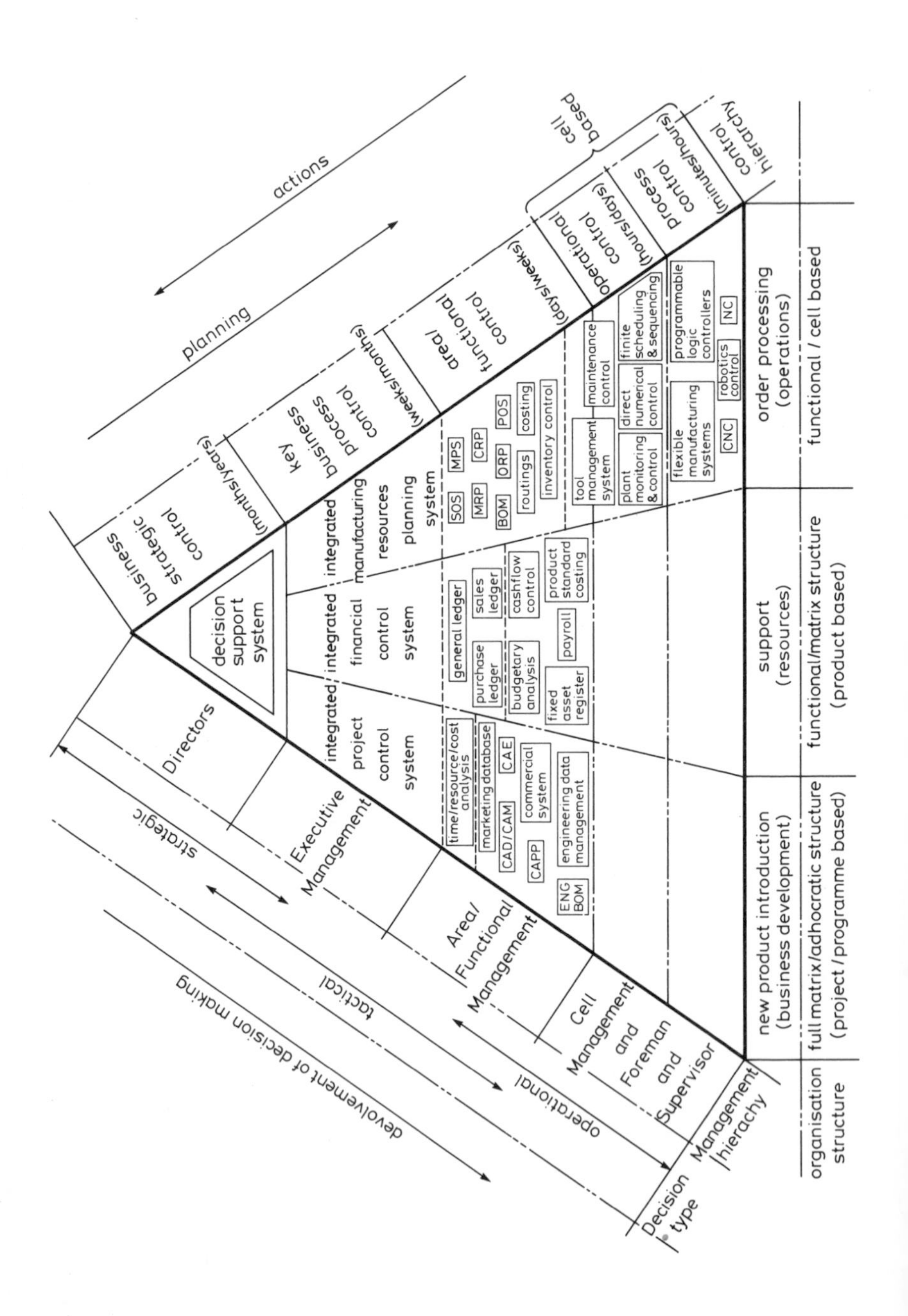

Fig. 20.4 Control hierarchy — systems and technology

be produced to illustrate the integration of these elements. These models will be based on a medium sized batch manufacturing company involved in designs, development and manufacture of technically sophisticated products.

Chapter 21

Management information: is it worth it?

C. Chapman
Wellworthy Ltd., UK

21.1 Background

The author, Production Control Manager at Wellworthy Ltd., tries to answer the question: Is the cost of producing management information value for money? He considers whether the existing system at Wellworthy could be cost justified today after seeing the benefits. He also looks at how with a small investment in the right areas large paybacks could be received.

The use of computers has revolutionaised the control of the production facility in manufacturing industry. Data is collected, processed and information produced at a speed far faster than anything imagined by the clerks of twenty years ago. Today's manufacturing managers have a mass of figures at their fingertips. But could an existing system be cost justified today?

21.2 The costs

The hardware: The Wellworthy system covers five sites in the south west of England. The basic hardware configuration provides 84.5 Mbytes of memory and 8.2 Gbytes of available disc capacity. The system is a combination of different manufacturers but is primarily based around the IBM 4341M02. The total replacement cost is £1.01 million (new for old). The average age of hardware is six years.

The software: The software is primarily based around the IBM Copec modules but these have been altered in house. The cost of the in-house modification is difficult to value; however, the cost of replacement is estimated at £535 000.

Running costs: The total running costs for system maintenance, consumables, service contracts, and employees directly costed to the system is £1.01 million per annum plus £32 000 hire charge to hire the data network.

Total costs	(£)
Investment:	1 620 000
Per annum costs:	1 042 000

It must also be remembered that these figures do not include any staff which update the mainframe as part of their day to day tasks.
Justification: Wellworthy require on capital expenditure projects to receive a payback of three years and a return on capital of at least 25%.

In order to justify this capital expenditure it would be necessary to save costs or increase earnings by:

Payback:	£1.58 million per annum
Return on capital:	£1.45 million per annum

If we take the £1.58 million as our target this is the equivalent of reducing the work force by 158 people (8%) per year over three years. Alternatively, the saving could be achieved by reducing the working capital by £1.58 million per annum (10%) for three years. Clearly these targets would be difficult to achieve and few Boards of Directors would believe these figures could be achieved by buying a computer. So how could we justify this investment?

Fig. 21.1 Three bells — I've got the payback

21.3 Actual benefits achieved

21.3.1 Reduced work force

By using the computerised systems it has reduced the number of staff producing management information and processing data. Over the six years of the system's life it is difficult to quantify the savings it has achieved as the company has changed through recession and a takeover. In order to value the savings the author has looked at the staff versus non-staff ratio six years ago and today.

Staff vs. non-staff ratio 1982 = 1:1.14
Staff vs. non-staff ratio 1988 = 1:4.75

If we assume that the company had the 1982 ratio today it would have 56 more staff than it has today (saving £56 000).

Obviously, to assume all this reduction is solely due to the introduction of a computer is an over-simplification, but we can assume that a large proportion of the saving is due to the system as there have been few other changes with such a dramatic effect.

21.3.2 Reduction in working capital

In 1982 the turnover to working capital ratio was 5.9. Today this ratio is 6.9. If we assume that this reduction is due to the introduction of the mainframe we would again be over-simplifying the problem.

However, this working capital reduction is equivalent to a saving of £1.5 million calculated as a saving of £150 000 per year.

21.3.3 Customer profile

It is impossible to quantify what the effect of running computerised systems has on the customer. It is apparent, in our case, that the customer is receiving a better service. In the competitive world of automotive components, any company which develops a competitive advantage will increase market share, and any company which does not keep step with its competitors will lose market share.

Without the computerised system employed by Wellworthy, the company would have undoubtedly fallen behind and lost a large portion of its business to its competitors.

21.3.4 Business control

A proactive manager is one who controls the business. A reactive manager is one who is controlled by the business.

Wellworthy's business is complicated. The company is split into five sites producing interactive components. It has over 5000 live part numbers and over 100 customers. Without responsive computer systems, no manager could control such a complicated business. The system network has become an integral part of controlling the business.

Interim summary (£)

System capital cost	=	£1 620 000
System running costs	=	£1 042 000
Salary savings	=	£56 000
Working capital savings	=	£150 000
Net cash flow after three years	=	£2 892 000

From this analysis we can see that at best the project lost £2.9 million in the first three years.

Clearly, on a cost payback justification the project should have been abandoned at the start. Wellworthy should have maintained its army of clerks and never invested in computers. I believe that if Wellworthy had followed this course of action it would have fallen behind and probably never escaped the recession. When attempting to justify any information systems project we cannot look purely at the bottom line. We must have 'faith' in our ideas. Information is precious and must be treated with the respect given to precious jewellery. When next trying to cost justify an information system, think of it as like trying to justify a security system—difficult to get a payback but essential to look after that expensive jewellery.

The author firmly believes that, if the money had not been spent on this system, Wellworthy would have fallen behind its competitors and a system which could not have been justified on cost could have been justified as the cost of company survival.

21.4 Conclusion

The aim of this chapter is to try to answer one question: is the cost of producing information value for money?

I have answered this question by looking at an existing system to see if it has provided value for money? By the quantifiable measures adopted it seems to have provided a net cash loss of £2.9 million over the first three years. I do, however, believe that this is not a loss but the price necessary for survival in a competitive market.

In future we have to decide on purchasing new systems equipment not primarily on cost payback but primarily on maintaining our competitive edge. Information technology investment must be made on faith in judgment rather than on a profit measure.

New systems and existing systems must resolve the problem of interfacing man and machine. Managers do not see computers as an aid and feel outputs are designed for ease of production not for ease of reading. The man/machine interface *must* be securely built for a system to be effective. A manager should turn to his VDU as often as a mechanic opens his toolbox.

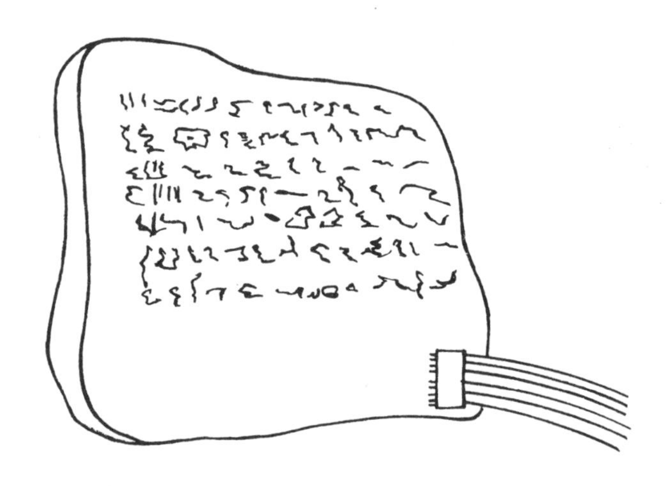

Fig. 21.2 Easy to understand the outputs

Computers today must be able to predict the future. In today's complicated business world managers require 'what if' analysis to optimise the decision making process. In the manufacturing industry of today, the costs of information are dwarfed by the enormous benefits which can be achieved. Managers must have the conviction in their own ability to reap these benefits.

21.5 Where do we go from here?

Wellworthy have purchased the IBM mainframe, the software and employ a number of people to operate the system. The company could not function without it. We can treat this as a fixed sunk cost already covered by the operation. This gives opportunity for marginal investment to give large returns.

To do this we are turning to the user for advice. Wellworthy seem to suffer from a typical IT disease; the system is designed for the ease of the system not the user. We find that the effectiveness of the computerised information processing is being hampered by the reluctance of middle managers to use it. This reluctance is based on a lack of understanding of

Fig. 21.3 Have faith

what the system can offer middle managers and a lack of understanding by system analysts of managers' needs.

To bridge that gap we are actively encouraging the managers to redesign the system to their own specification. Output configurations are being redesigned by the user. Users are being given the opportunity to interrogate the mainframe databases using personal computers. Output data sheets are now provided explaining the various output features which are and will be available, and are distributed to users.

From our investigation we found that the most commonly requested feature was to help in forecasting the future. We are planning to extend the mainframe to provide 'what if' analysis using the WASP production scheduling system and SIMAN simulation language. In this way managers can see the results of their decisions on a screen before making irreversible changes.

Chapter 22

Integrated assembly for flexible manufacture

C. Davis
John Brown Automation Ltd., UK

22.1 Introduction

John Brown Automation has been designing and manufacturing assembly and test systems for almost 25 years now and has built up a wealth of experience in automotive assembly. In early 1987 the company was successful in obtaining an order from Skoda Cars for the production of power units for a new saloon car. The system is currently being commissioned in Czechoslovakia to assemble and test both cylinder heads and engine blocks and then combine them and produce the complete engine.

The facility is significant in the way that a range of technologies have been applied in a production environment to manufacture 50 different types of engine at random. It points the way forward for the practical application of such systems in lower volume industry sectors.

This case study will describe the Skoda assembly system and highlight some of the important features that have enabled this capability.

The facility is capable of manufacturing nearly 0.25 million engines per year working at the rate of some 4.5 s per unit. The 50 variants that it is capable of producing consist of four different cylinder heads, eight different engine blocks and a large variety of different engine dress styles to suit different climatic conditions.

A distributed logic control system is used in conjunction with electronic tagging to master the complex problems of controlling and monitoring a prioritised order schedule through the system. Individual automatic stations and line control units are linked through a local area network, also facilitating the gathering of data to provide a real time Management Information System. This system is capable of displaying a variety of performance figures for the overall facility and individual statistics for each automatic station.

The facility is organised in three major parts, consisting of the engine assembly line itself, a cylinder head assembly line and an interlinking overhead robot gantry and buffer store.

22.2 Engine block line

The engine block assembly system has 64 stations and is the master part of the facility. The random order schedule can be downloaded from a central MRP type computer system or input manually at the line's master control cabinet. On receipt, the order details are logged into a prioritised order schedule which is then used to 'pull through' the different requirements against each of the order variants as they progress through the system. If necessary, this schedule can be modified and reprioritised at any time, with the exception of the engines that are already progressing through the system at that time. Even with such a complex product and over 100 pallets on the line, this represents less than one and a half hours to pull a specific variant, 'panic' engine, through the system. It is this kind of responsiveness that the lower volume, smaller company so often needs to survive in highly competitive markets.

The majority of the stations on the engine line are manually operated. The main reason is that the engine was designed in a conventional way with much of the sub-assembled peripheral equipment mounted at a wide range of angles. Expensive multi-axis robots would be necessary and would be difficult to justify against existing labour costs. However, with the distributed logic approach to the control system and freedom of workstation position in a free flow transfer system, it is easy to allow for automation of operations in a later phase. This approach is often vital for the smaller company which cannot afford to take on a complete system in one go and must plan for installation in phases.

All the major nut running operations are automated and the provision of mechanical handling equipment greatly simplifies the operator's task in a number of cases. The pallets run on a power and free roller conveyor and each pallet is fitted with a trunnion unit capable of presenting the engine to either operators or automatic stations in one of four different orientations. The line has two rework loops to which any faulty sub-assemblies are diverted for manual rectification.

All the pallets are fitted with electronic tags which are programmed at the beginning of the line according to the next variant required in the schedule. At each of the following stations this tag is then used to flag a message to the operator regarding the components to be fitted or process to be carried out for that variant. A manual flag can be set to indicate an operator identified rework or reject condition, and in this case the condition is logged into the system by the next automatic read/write head. In the cast of automatic stations, the tag calls for the program to suit the particular variant.

When the engine order starts down the line, a subsidiary order is automatically placed on the cylinder head line to provide the correct type and variant of head. By the time the engine has reached the gantry facility at station 25, the cylinder head has been completed Just In Time to be transferred and loaded onto the engine block.

In the event of a reject of either the engine block or cylinder head assembly, the buffer store comes into play and a cylinder head of appropriate type is loaded by the gantry from this store instead of the line.

After the block and head have been tightened automatically, the combined unit passes on to the main engine build phase. At station 32, cylinder head nuts and push rod cover nuts are positioned with rocker cover studs and automatically tightened. At station 38, the manifold is positioned and automatically tightened and subsequently the pre-heater, thermostat, carburettor plate mounting and ancillaries are fitted and automatically tightened. The remaining stations deal with final engine dress and assembly and a final rework loop if provided at station 60 to rectify any problems. The complete unit is then unloaded and delivered to the final car assembly line.

22.3 Cylinder head line

The cylinder head line is heavily automated compared with the engine line and has seven Bosch SCARA robots to assemble components to the head (Fig. 22.1). The head is tested for valve leaks and a novel electro-magnetic orientation unit is used to align the cylinder head studs prior to assembly.

While the robots are primarily justified by the number of different coordinates required for any one head assembly, they could easily be set up to cope with a much wider range of variants that would be the typical need of the lower volume manufacturer.

At station 1, a clean head casting is automatically loaded from one of the four infeed conveyors according to the type that has been ordered by the engine facility. The indentity of the casting is confirmed against the type called for on the pallet tag. Like the engine line, all the pallets carry electronic transponders to identify variant required and rework/reject information. At this stage, the heads are fed in and loaded on their side to facilitate valve guideway cleaning and valve assembly.

At station 2, an air blast is used to clear any foreign matter from the valve guideways and an oil mist lubricates them ready for valve insertion. An operator then loads inlet and exhaust valves and releases the pallet to station 4. This is a simple automatic station to turn the head over to present the rocker cover face ready for spring pack assembly. Grippers are used on the valve stems during the turnover sequence to ensure the valves remain *in situ*.

Fig. 22.1 One of the Bosch Scara robots picking up a stud after orientation by the EMAGO unit

Station 5 is based on the first of the SCARA robots which loads the outer spring seats to each of the eight valve stems in turn. Because the spring seats are inserted into a recess in the head it is very difficult to rectify any problems due to incorrect component size. Sensors in the gripper are therefore used to check the diameter of the component while it is in transit from the feeder to the head. Bad components can then be rejected if necessary before assembly.

A similar operation at station 6 loads the inner spring seats and a third SCARA at station 7 loads a seal on each of the stems.

Valve springs are manually loaded at station 8. While being stiff, straight sided and with closed ends, the space between the coils corresponds too closely with the wire diameter, thus providing too good an opportunity to jam to allow automated assembly. This is a typical station where allowance could be made for the introduction of automation at a later date.

The fourth robot at station 9 loads cotter and cap sub-assemblies to the springs. Conventional feed technology is used to marry the cotter halves in the cap for this operation. Once located in the cap, the sub-assembly is very stable and offers no problems during transfer by the robot to the spring pack.

At the next station, a dedicated workhead uses an outer and inner ram arrangement to compress the springs and engage the cotters on the stem automatically. While the inner ram holds the cotters in position the outer ram retracts to lock the spring pack assembly.

With the valve sub-assembly complete, the head moves on to the stud insertion stations. Following lubrication of the stud holes, three more SCARA robots insert inlet and outlet manifold studs and also thermostat housing studs.

An aspect of these stations which deserves to be highlighted is the method used to orientate the studs. Each stud is about 6 cm long with an offset unthreaded centre portion and only minor differences in the thread form at each end. To orientate these components traditionally is extremely difficult; the solution adopted came about as a result of John Brown Automation's co-operation with the Soviet Union.

The EMAGO unit used applies an AC magnetic field to the component and induces eddy current reactions. The first of two heads is located on the output flight from a conventional bowl feed unit. This head uses the field effect to find out which way round the stud is orientated. If the stud happens to be oriented correctly, the stud is allowed to fall the length of the flight to the selector positioned for the robot to collect it. If the stud is the wrong way round a second hand is excited lower down the flight which amplifies the eddy current reaction to the point where the moment of force generated is sufficient to flip the component over in mid air. Being an AC magnetic field, of course the effect can be used on ferrous or copper coated studs and the components are degaussed as they leave the field.

The robot collects the orientated stud in the head of an air powered torque unit and inserts it into the cylinder head. The flexibility of the robots at these stations allows for reprogramming to carry out each other's tasks in the event of service or malfunction. In a similar way, most of the other stations on the cylinder head line and engine line can be isolated or reorganised with a strategy to enable production to continue in the event of malfunction. This capability to react to malfunction is vital to any company, large or small, but it carries further implications for the lower volume company. It is possible for a system to be 'reconfigured' to suit particular product needs or to suit particular production requirements.

Station 15 carries out an automatic leak test of valve seats and facings. The head is temporarily transferred from the pallet into the leak test fixture where rubber seals are applied for the test. The valve face is pressurised and monitored with pressure detectors. If a leak is detected, a repeat test is carried out to check; the pallet is then coded for rework and diverted into the rework loop. Here, the pallet tag is read and a mimic diagram of the head indicates to the operator which of the valves have failed the test and also which of the studs if any failed to drive home. A maximum of three attempts at rectifications are allowed before the system automatically codes the head as reject and sends the head to the gantry unit for unload out of the system. Accepted heads also proceed to the gantry for assembly to an engine block or temporary location in the buffer store there.

22.4 Integrating gantry

A 9 m robotic gantry unit links the cylinder head line to the engine line over the top of a gangway for vehicles and pedestrians. (Fig. 22.2). The head of the gantry is therefore fitted with a special failsafe gripper system which prevents release in the event of any power loss or system failure and in addition the gangway itself is guarded.

Up to 40 heads can be marshalled by the gantry in a buffer store adjacent to the cylinder head line, stored by type of variant and held until the engine block of appropriate type calls for it. This facility allows mismatches to be sorted out in the event of engine block or cylinder head rework and also allows either of the two lines to continue production in the event of machine failure or maintenance of the other line. In normal operation of course the gantry transfers the completed cylinder head to the engine line and assemblies it to the engine block Just In Time.

A classic problem with some early systems was that all knowledge of subassembly progress and pallet position was lost in the event of system shutdown for any reason. Manual mis-direction of a pallet could also cause chaos and similarly require a lengthy process to re-input all the necessary details. The availability now of electronic tags that do not list their memory and simple positional sensors means that this is a problem of the past. For the Skoda system, the central controller can establish the status of the marshalling area and each of the two lines by reading all the transponders and sensors and creating a software model directly.

22.5 Control system

German equipment has excellent spares support in Czechoslovakia and Siemens control systems are already well known and liked. Each of the

Fig. 22.2 General view of the Skoda assembly system showing the cylinder head line in the foreground and the interlinking gantry unit in the centre of the picture

stations of both cylinder head and engine line is equipped with its own Siemens 115U PLC controller and each of the two lines is provided with a Siemens 150U master controller; all the controllers are linked by an L-local area network. Each of the robots and some of the other station equipment has its own dedicated controller and these are connected to the 115Us by serial link.

The engine line is provided with 40 Statec read/write heads for the electronic tag system and the cylinder head line with 10. Each tag can store 64 bytes of 8 bit data which is perfectly adequate when used as a flag system but keeps the cost per pallet down. The tag is powered by a sealed-for-life lithium battery with an operational life in excess of 10 years. In addition, a hand held read/write unit is provided for training and maintenance purposes to enable the data in a tag to be independently checked and reset if necessary. This, for example, enables a pallet to be reset and repeatedly sent through a particular station to check out its operation. In normal use the tag setting would validate such a repeat operation and the control system would refuse 'to assemble a component twice' with its potentially disastrous consequences.

Both the engine line and cylinder head line master controllers have colour graphic monitors which provide production managers with up to date, real time information on the line and its performance. Mimic graphic screens present the current status of the respective lines and stations and every pallet is identifiable in terms of carrying good, rework or reject products. Further information can be viewed by interrogation and management reports produced.

The local area network is used to gather and consolidate quality assurance and performance information and pass it to the respective line controllers and ultimately to the engine line master controller. Typical analyses can include units produced, rejects and by reason, cycle times and performance data on individual stations and operators. Quantitative test and inspection data can be accumulated on the products assembled and also diagnostic data on any station problems that occur.

The master control can be interrogated to show the current status of the prioritised batches that have been called for and are in production or waiting in the queue. There are also two automatic build schedules permanently resident. In the event that a cylinder head or engine is rejected, a replacement product is automatically inserted into the schedule to ensure that there is minimal delay to the priority batches. In the event that the cylinder head line does not receive new build instructions in a predetermined time, it will automatically switch to a background schedule to make heads for the buffer store (Figs. 22.3 and 22.4).

In the case of each of the robot stations, the electronic tag system can flag for the appropriate build sequence to suit a particular variant. In a situation where new options are added to the range on a regular basis it is easy to arrange for these to be added to the library in the robot controller and a new flag added into the tagging system.

The very equipment and techniques that are allowing functional flexibility in an integrated assembly system are also allowing easy expansion of the system's capability. They are also making it easier to provide the capability to successfully install a system in phase.

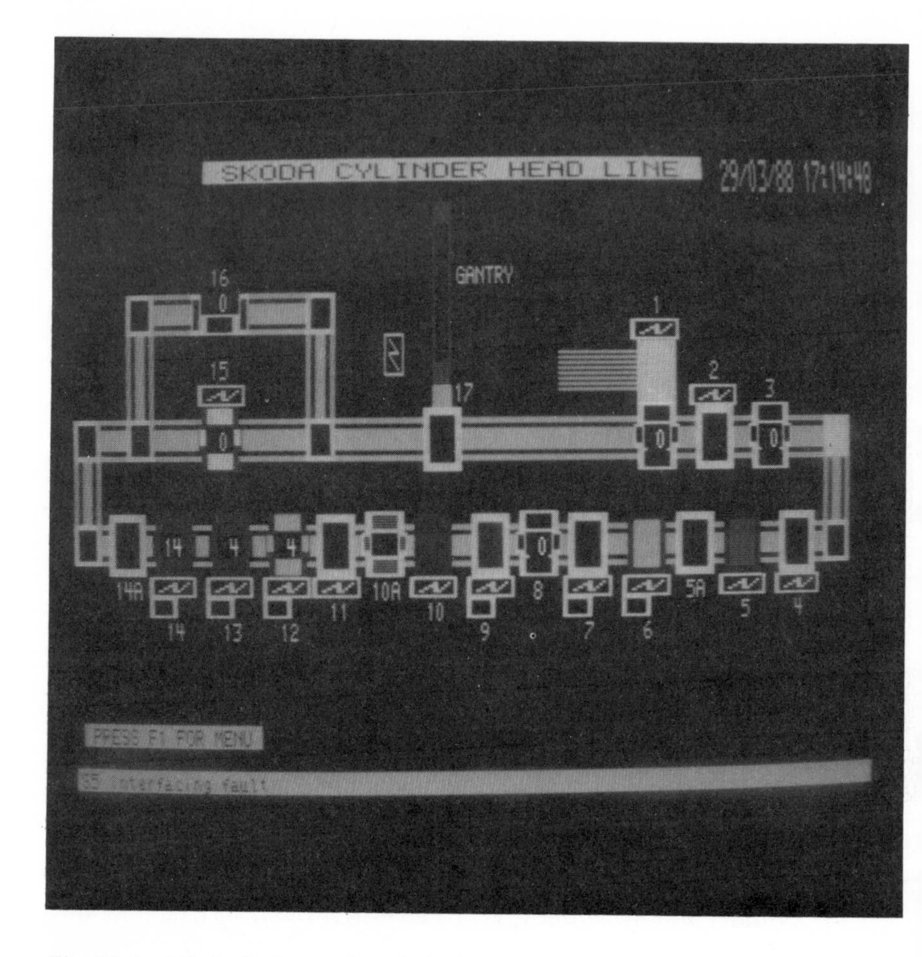

Fig. 22.3 Mimic display on the cylinder head line Master Cabinet showing status of the processes and pallets in the system

22.6 Conclusion

We now have the capability of integrating a high productivity assembly system for the factory environment that is capable of building a range of products in random sequence. Most of the technologies are in place to operate such a system on a highly responsive Just-In-Time basis throughout, pulling through components and materials to suit the incoming order schedules.

More important perhaps is the future significance to the lower volume

Fig. 22.4 Some of the 50 variants of completed engines coming off the line

producer and the smaller company. These manufacturers need this flexibility and speed of response more than the larger, higher volume organisation. We are now seeing more realistic approaches to the problem along with the capability to reduce a system into more manageable phases for such companies.

Part 3:

Controlling the quality

Chapter 23

Introduction

D. Wilcock
Sunderland Polytechnic, UK

Quality has become a major issue within manufacturing industry. It has widespread implications on product and process reliability and in the resulting maintenance requirements and in product liability. Quality has been said to originate in the design process and percolate through manufacture and into installation and operational situations. But it is more than the adoption of the appropriate technology and management procedures — it is an attitude of mind that must run right through every operation in which a company is involved. No one within the company must be the weak link in the chain. How often has industry experienced the high quality designed and manufactured product being damaged in delivery or being subject to major faults in installation?

Increasingly companies must also seek to inculcate suppliers and sub-contractors with the same attitudes towards quality. More and more there is a need for suppliers to demonstrate quality of product and in-house procedures to ensure quality of product before new business is won. The international competitive nature of business means that no company is isolated from the advances in the rest of the world and legislative standards will only add to the need to make major improvements in the quality standards of all manufacturing businesses.

The spur to much of the development in quality assurance has lain in the competition posed by Japan within the major Western markets. Most Japanese companies will point to the quality of their product as a prime reason for their success. It would be a gross over simplification to assume this is the only factor, but there is no doubt that in perception and practice the Western product, with a few notable exceptions, falls behind what is expected by modern customers — even Western ones. Most people will be familiar with anecdotes about not buying a motor car from XYZ company for a year while the first customers 'iron out the bugs' — too many people have had the bitter experience.

But where does quality fail to achieve required and acceptable modern standards? Product defects occur through faults both in design and manufacture. The design knowledge base and the tools to implement it are not deficient in Western industry. Indeed, in the application of computers to design there is a significant record of success. Within manufacture there certainly is some evidence of a lack of investment in modern equipment. However, this does not run across the whole spectrum of industry. In general the technology available for manufacture is not seriously deficient. Management is often criticised, certainly within the UK; but in other Western countries in which the quality of management is not in question quality problems still remain. Perhaps one significant area is that of full scale product development prior to a commitment to full scale manufacture. The amount of engineering that is applied to a Japanese product far exceeds that applied to Western products. This insistence of ensuring quality prior to entering the product in the market place is very significant and has perhaps not been a lesson other countries have been keen to learn. It is this wholehearted approach to ultimate service quality and all that it implies in the attitudes of both management and workforce that is the greatest lesson to be learnt in the required approach to Total Quality.

There is, of course, more cause for optimism rather than pessimism, as many Western companies are at the forefront of both technological and quality developments and those that are not are rapidly taking on board the required developments.

The topics covered in this Part cover a great range of quality problems. Hogg in Chapter 23 and Ferguson *et al.* in Chapter 24 discuss relationships between buyers and suppliers. In view of the above discussion the transfer of Japanese attitudes and procedures into a Western environment and the necessary adaption described by Hogg must be of paramount interest. Certainly there are lessons to be learned but recognition must be given to the fact that motor 'manufacturers' are large scale 'assemblers' with enormous financial power. Not all companies can, or should, attempt everything that suits a motor manufacturer.

Jones, Godwin and Gore in Chapter 26 discuss quality management systems while Gill and Holly discuss the implementation of JIT in an established company in Chapter 27. Clearly, quality is an implicit requirement of JIT and the step by step approach to its introduction within an established company provides many pointers for companies about to embark on this route. Obviously, the training, communication with, and commitment of company personnel are essential components to ensure success.

In Chapter 28, Richter and Seidel discuss the wide aspects of quality originating in design and progressing through manufacture to test. They make a strong plea for the full integration of all these activities to ensure quality is a feature that permeates all activities. Watkins and Leech in

Chapter 29 examine equipment reliability, the results of which may be used to determine the quality requirements of the said equipment. Gohari and Bititci, in Chapter 30, discuss the control of FMS systems and give a specific example of an industrial arrangement.

Elliott *et al.* (Chapter 31) Hensser and Murray-Shelley, (Chapter 32) and Catchpole and Sarhadi (Chapter 33) discuss specific examples of detecting quality defects in products. These aspects of the technology of quality control and assurance have tended to be submerged in the management of quality systems, but they are a salutary reminder that in the manufacture of products it is the technology and its proper application through fully informed and committed personnel that is equally critical.

The text provides much information and discussion of the many issues relevant to controlling quality and should give the reader much food for thought. It is certainly true that in the next few years the achievement and control of quality will be uppermost in the activities of manufacturing industry. If there is one main lesson to be learned from present experience, it is that, while the technology and management of quality is of undoubted importance, it is the enthusiastic adoption of quality as a prime goal by all the personnel within a company that is the only sure route to success.

Chapter 24

The Japanese approach to the buyer/seller relationship translated into the European environment

T. M. Hogg

Nissan Motor Manufacturing (UK) Ltd.

24.1 Introduction

Nissan Motor Manufacturing (UK) Ltd. was established as a wholly owned subsidiary of Nissan Motor Company of Japan in 1984. In February of that year NML signed an agreement with the British Government to build a new car plant in Sunderland, Tyne & Wear.

Our first employee was hired in October 1984; he was the Personnel Director. The plant was built in 62 weeks and handed over at the end of 1985. During that year we sent the first 160 or so of our new employees to Japan for training. The original reason for this was straightforward: we did not have a plant in which to train them and the part of the world where we are based has no motor industry tradition; it is historically shipbuilding, mining and heavy engineering.

Originally it was envisaged that the agreement with the British Government would be implemented in two stages. Phase one involved the construction of a £50 million plant for body assembly, paint and final assembly of 24 000 units per year. Under the Phase I programme Nissan Bluebird cars were assembled from body panels and parts shipped mainly from Japan; this was an important point since initially there was very little purchasing involvement.

Production commenced in June 1986, when we produced 5000 cars with 500 employees. During 1987 this increased to 28 700 with 750 people. In 1988, we produced approximately 56 700 vehicles with a head count of 1500 people.

In September 1986, two years earlier than originally expected, we instituted the Phase II expansion programme. This will take us to a planned 100 000 cars annually by 1991. This £340 million expansion programme will mean expanding our existing body assembly, paint and final assembly facilities and adding a press shop, injection moulding shop and engine machining and assembly.

In line with the significantly expanded production facilities, local content of the Nissan Bluebird exceeded 70% during 1988 when exports to Continental Europe began. This is two years ahead of the agreement signed with the British Government and will enable the Bluebird to be designated a British car. By 1991, local content will have increased to 80%, we will be producing 100 000 vehicles annually and employing 2000 people. Well before completion of our Phase II programme we have announced a Phase III expansion plan which includes a second model which will be the replacement for the Micra and will take our annual vehicle build to 200 000 per year, employing 3500 people and spending approximately £500 million per year with local suppliers.

24.2 A Japanese approach to buyer/seller relationships

It is against this background that I want to talk about the Japanese approach to buyer/seller relationships in a European environment. My general theme is how we are going about our purchasing business differently from a typical European motor industry. I should stress *motor industry* because other industries can be, and indeed are, different in their approach to purchasing.

The approach we have adopted at NMUK is one based on the Japanese approach suitably modified to cope with our environment. The main areas I will be covering are the customer/supplier relationships; supplier performance and purchasing — the Nissan approach.

24.3 The customer/supplier relationship

Firstly customer/supplier relationships: how they differ between the United Kingdom and Japan and the responsibilities that these relationships entail for both parties will be discussed.

24.3.1 Japan

- mutual dependence
- long-term commitments
- shared benefits.

In Japan's history of supply there is a basic concept of mutual dependence: the customer is as dependent on the supplier as the supplier is on the customer. They see it not as a totally dependent business, more of an integrated business: the supplier and the assembler or customer planning together so that one cannot survive without the other. This frequently extends into shared ownership or infant companies being spawned by the parent organisation.

This mutual dependence leads to a fundamental issue — the long term commitment. Both the supplier and the customer have this not necessarily specific contract, but a clear understanding of a long term commitment to do business with each other. If you are a customer you will go and find further business for your suppliers. If you have a supplier who makes one type of component and you are about to eliminate that component, you will normally try to find him some other business. For example, if certain parts of the wiring have been substituted by something else, you will normally try to encourage your supplier to provide that substitute so that he will keep a business relationship going. You are committed to keeping him in business. Similarly, a supplier is committed to keep a customer in good business; that means a supplier will do things in a way that they will not normally do in Europe. They will do things that might in the short term be extremely difficult for their business in order to support their customer. They will undertake responsibilities for their customer that would not normally be considered here in Europe.

A third part of the Japanese system is this concept of shared benefits. This means they see the total profitability, quality improvement, all aspect of the business, as something to be shared by both, not negotiated one from the other or obtained mainly by one or the other; but a very open relationship about what is involved for both parties and a willingness to share it. The sharing may well be different at any given time, for whatever benefit is involved, but over time it is seen as a very equal partnership without any particular power on one side or the other. If you compare that with the UK, the historical pattern is completely different.

24.3.2 United Kingdom

- arms length
- multiple sourcing
- retendering
- contract return review.

The traditional relationship between the supplier and the customer is seen as an arm's length relationship. It is not good to get too close. It has even, as in the past, been considered unethical to get too close. You must keep at a distance and you must never become too dependent on the supplier and you do not want the supplier to become too dependent on you.

With that goes an approach to multiple sourcing. It is common practice not to have just one supplier for one component, but maybe two, sometimes even three. By and large this is an approach which says you cannot afford to give any other business that much power over your own business; you cannot afford to put a supplier in a position where he is your monopoly supplier for a type of component. This is now changing, not necessarily across the board, but it is worthwhile saying that these Japanese aspects are gradually being adopted by certainly the more progressive companies in British industry and to an extent the European ones also, because of the benefits being seen. Our claim at present is that we are further down that road already than any other European businesses because we started that way and do not have the historical encumbances to shake off.

One other big difference is re-tendering, and to explain that a little let us take an example of a supplier supplying one of the plastic mouldings for a car. He has been supplying you with that part for a particular model for three years. You introduce either a change to that part or you have a new car coming in or you just have some significant economic change; for example, more sales less sales. You go back to this supplier and his competitors and you ask them to re-tender. You give them specification documents and drawings and you say to them please tender against that specification for that part. That is completely different from the Japanese view of long term relationships. If you are asking the supplier and his competitors to re-tender, you are automatically suggesting to him that, from that point on, the relationship you had is over until it is restarted. Even if he does get the business you have weakened the business relationship you had with him because you told him on the day you requested him to re-tender that he might not have the business any more. We are talking about principles: if you look at Japanese industry and UK industry these are fundamental differences.

What that leads to in the UK is what I call contract return review. Because the supplier knows he might lose the business then he is forced as a businessman to look at each contract on an individual basis. Each time you approach him for a component he has got to look at that piece of business and say 'can I make that business, on its own, pay. If I am going to produce this part, it is going to be good for three years; is that going to be a paying business. How much do I need to charge to do that?' It is a detailed review he must go through each time he is asked to make a business proposition. As an example, I recently talked to a supplier of injection mouldings who in discussing this general issue make a point to me that in the preceding year he had been out seeking new business, and on an existing £18m. business base he had acquired an extra £6m. worth of new business. But when he had actually come to look back a year later his turnover had increased to £20m. not £24m. because he had lost £4m. worth of his established business. It had gone to somebody else; therefore he had put a lot of time, effort and resource into trying to grow his business and instead of growing the 33% he had expected, it had grown by about 10%.

All that effort essentially for no gain. If you look at that more globally, if you step back for a minute from that single picture, you only have a limited number of suppliers of certain types of component and you only have a limited number of customers. What that reflects is people going around a roundabout. It is customer A saying this year I am doing business with supplier C and D, next year I will do business with D and E and the year after F and G and then I will be back to C and D again and I will have finished the circle. You have other customers doing the same and there is a lot of effort by both business parties to no real positive gain. All they are doing is chasing round a circle; that effort is not directly gaining anything, it is just moving stuff around wasting management resources and being relected in the cost of the part or the service being provided which automatically increases the cost of your finished product. This results in:

24.3.3 Supplier objectives: UK

- search for new business
- single contract review
- investment and pricing (RUI).

Take that philosophy and what does that mean for a UK supplier (forget the customer for a minute). It means he has to put a lot of time in searching for new business even if he is not after substantial growth. If his objective is gradual growth he still has to put in enormous resource because he has to be prepared to lose some of his business by the nature of this process. Secondly he has to put a lot of effort into a single contract review. He has to look at his investment, try and maintain that a minimum, because he might only need that investment for three or four years maximum.

He has then to look at his pricing and make sure that it recovers his investment within that period, which brings UK businesses to the concept of a satisfactory rate of return for investment. It is worthwhile to note that if you talk to most UK motor industry suppliers about what they except as return on investment, it is considerably higher than that of most Japanese business for a particular industry. What comes with that is a philosophical difference that is harder to see but is there. UK businesses when they present themselves to you for initial discussion will tell you that they are actively managed on the financial front and will have a clear idea of what a satisfactory rate of return is for their business. Most Japanese suppliers would find it very difficult to answer that question.

24.3.4 Supplier objectives: Japan

- steady growth
- long-time scale
- support of customer.

Japanese suppliers do not even have this concept in the traditional way; they want to be profitable and to grow, but this is usually expressed the other way round. They will usually say we want steady growth and for that we need profitability. They do not have the same concept of return on investment, the concept is not as critical to them. That is not because they are bad businessmen, it is because their view of the business is different.

A Japanese supplier will see the growth of his business over a long time frame. This maybe from obtaining new customers, but will be primarily be either through development of a new product so that he has something new to offer, or that he can use more of, or that has higher added value, or alternatively through growth of his customer's business.

Therefore, a supplier to the motor industry will do his best to support the growth of the assembler's business, because that is his way to growth. He will do this by assuming a wide range of responsibilities. The supplier in the UK, in the West generally, will tend to think much more about business growth by new contracts with new customers. They tend to see the way to grow business differently. Again, that is a reflection of this concept of long term relationship and shared business compared with arms length stand along business. Because the UK supplier operates within that environment he has to look at it that way; when a Japanese supplier operates within the different environment he is able to see things differently, and in a way all those things take pressure off a supplier, but in return for that removal of pressure there are certain responsibilities that the supplier must accept.

Because of a long term relationship, the Japanese supplier can concentrate on increasing his investment; he is looking over a long term, he is not looking at a three year contract. He is notionally looking at it as long as the equipment will last, because if he does not produce this part, he will produce the next one; and he knows that he will have the business for that next part, so he will be prepared to invest in better facilities. Facilities in this context is actual production facility rather than tooling. In addition to this, because he has security of business and he knows he does not have to put together packages and go in search of new business and make proposals and evaluate new drawings, he is able to better utilise his manufacturing engineering professional strength by directing his attention to improving productivity instead of searching for new business. These are two positive things that can come out of a long term relationship with a Japanese supplier, resulting in a focusing of attention on cost reduction. British suppliers, on the other hand, direct their attention to minimising investment because they have got to be prepared not to need that investment after the life of a particular part, and secondly they have to look to finding new business to replace the business they might lose, indeed anticipate losing, and also to evaluating ways to sell to a new customer. The attention is being directed a different way.

24.3.5 Supplier responsibilities

- quality
- continuous improvement
- development and testing
- productivity improvement.

Continuing the customer/supplier relationships theme, I mentioned that with commitments goes responsibilities. The first of these responsibilities is to maintain a consistent and acceptable quality level. We are not talking of quality in the sense of a Rolls Royce compared with a mini, we are talking of quality in the sense of making the right component to the right production quality, of achieving the required standards, and making sure the component is produced to specification and it is produced with a zero defect rate.

It is clearly a supplier responsibility to achieve that, to improve it and to police it. It is never a customer responsibility. Customers should not have large inspection departments, they should not have incoming inspection reports, they should work to the principle that they have delegated this responsibility to suppliers who know the standard and are prepared to accept responsibility for consistently achieving it: regular routine achievement of quality targets means not having rejections. It also means quality through the warranty stage because we believe a supplier has to be responsible for the warranty of his components right through the life of our vehicle; so he has to have enough control of his process to be happy to warrant that part for 100 000 miles or three years, because that is our vehicle warranty.

Most Nissan suppliers in Japan are now operating to rejection rates, defect rates, measured in parts of a million. We are still talking in the West of percentage points; if we are lucky, we might achieve less than 1%. By contrast, the Japanese are orders of magnitude ahead in terms of their ability to control production quality.

In addition to the responsibility for quality, it is the ongoing responsibility of the supplier for continuous improvement. Because we are not going out for re-quotes every year does not mean we abdicate responsibility, nor do we expect our suppliers to abdicate responsibility for trying to improve productivity and cost; so they have to be looking at continuous improvement in the products. The supplier feels responsible for making improvements to his product, to his process, to his material usage, to show up in terms of quality and cost. That is not forced on him by contract review, it is his bringing to the table what he offers for that long term relationship — continuous improvement.

Another aspect of supplier relationship is a very strong requirement on the supplier to undertake development and testing work; again it stems from the basic relationship. If you are saying to your supplier, long term you are going to make my door trim pads, I do not want to buy these from anybody else, then as a customer you are also saying to yourself that means I do not have to commit myself to producing lots of drawings of these things to send out every three years to get people to quote. On the same basis I can trust my supplier to do that; he is the expert. So suppliers are given much more responsibility to develop their part of the product, and therefore for testing it in its early stages. I am not talking about production testing I am talking about new product testing. You find in Japan, for example, big vehicle assemblers have far less component testing facilities than in Europe; that is because they do not do it, suppliers do it and each supplier has the equipment required to do new product testing on his component, not the assember. It is a very different pattern here in Europe.

Finally, one of the supplier's responsibilities is continuous improvement. Japanese suppliers expect and are expected to improve productivity year after year on any component. They are expected to share the benefits of that·improved productivity with their customers. They have at times startlingly high targets for that.

We are aware of Japanese supplier improvement rates consistently in the high single figures per annum, e.g. 9%. Sometimes, for example over the recent 18 months, we have experienced 15—30% per annum improvements in productivity. Sometimes, some years, on some components, these improvements are not startling. What is true is that they are consistent; there is always improvement. We know because we buy from Japan and we do get component prices coming down year after year. What tends to happen in Japan is that the customer and the supplier talk each year about how much the price can come down, whereas in Europe the supplier is generally talking to the customer about how much the price is to go up. That is what we mean by commitment to productivity improvement. My belief is that it is not because they are better individuals, it is because in knowing that they have a long term business, they have the ability to concentrate over time on improving investment and improving productivity, instead of minimising investment and maximising the search for new business.

24.3.6 Nissan Motor Manufacturing (UK) approach

- principles of supply relationship
- supplier selection
- 'Nissan way'.

Having covered some of the basic differences we as a company have identified between Japanese and European suppliers, I would like to move on and explain the specific approach Nissan UK has taken in establishing a wholly new supply base, one that has taken us to over 70% local content by 1988 and 80% local content by 1991. To do this I would like to review three specific areas: firstly principles of supply relationships, secondly supplier selection and thirdly review what we term the Nissan Way.

24.3.7 Principles of NMUK/supplier relationship

- long-term relationship
- single sourcing
- shared benefits
- mutual dependence.

Firstly let us look at the principles of the supplier relationship. You will see we have adopted many of the Japanese ones, modifying them as appropriate to suit our particular requirements. We go for the long term relationship. We tell our suppliers up front, right at the start of any business with us, that we expect to stay doing business with them. What we say to a supplier is you will not have to re-quote, we will not put your part out to tender on the next model or in three years time; if you are making that part for us you can expect always to be making that part for us. All you have to do, supplier, is to stick to our way of doing business; that means adopting the supplier responsibilities we talked about earlier. If your quality is right, if you improve your productivity, as long as you do those things you get to keep the business.

We add to that single sourcing for the very reasons we spoke about earlier. We do not believe that we can say to a supplier, and ask him to believe us, that you have that long term relationship, without our demonstrating that by 'putting our money where our mouth is' and saying you are the sole supplier for the part.

We believe in the concepts of shared benefits. We expect our suppliers to have profitable businesses. We do not expect people to do business with us for the sake of it, we also expect to share our benefits with them and vice versa. Therefore when they come up with cost reduction ideas, which we expect them to, we will not expect to acquire all those cost reductions ourselves, but neither do we expect any secrets; we expect to sit down openly and talk about how we share those improvements.

Many suppliers are not conceptually happy with this idea. They are not yet, in many cases, fully convinced that it is the truth. They are not totally sure they believe us; however, they are all prepared to go along with us in principle and we have actually started to turn some of into practice. We have achieved agreement with suppliers about what they are going to do on productivity, what it is going to do for their price, where it is going to go.

We now have suppliers coming along to us with ideas for improvement. We are agreeing to many of these and are prepared to consider much smaller improvements than most motor companies would be interested in. Not because they in themselves are of enormous benefit but because we have a belief that over time all those little things add up and do make them more significant, and also because it is a necessary gesture of faith on our part to demonstrate that we are prepared to do it. It is part of building that credibility with our suppliers and creating the mutual dependence.

We have for example one supplier who, because they have obtained our business and because they wanted to do it our way, have effectively roped off a part of the inside of their plant and said this is for Nissan. Within this clearly defined area they have been able to make significant changes in employee attitudes, in plant and equipment, and they have obtained agreements about flexibility and ways of working that they have not been able to obtain elsewhere in the plant. Now clearly they will hope to expand that opportunity once they have shown people it is worthwhile; once they have shown people there is nothing to be afraid of, then they will adopt these working practices through the remainder of their plant so in that sense some of the benefits we bring will spin off to their customers, because it will make that supplier more efficient.

What is proving more difficult at this stage, if we are very honest, is getting suppliers to the point where they are able to undertake the product development and testing role. Many are simply not geared up for it and we are the only customer presently that asks them to do it. In many of those cases we have sometimes had to directly fund from Nissan the investment in developing and testing facility from them. What is true is that this task is easier in Japan because in Japan there is usually some form of link between major supplier and major customer. Sometimes there is a shareholding by the supplier in the customer and vice versa. Also very often there is a common major shareholder; for example, one or other of the Japanese banks will be a major shareholder in both businesses and that leads to a much greater understanding.

What is much less common in Japan than here is a supplier supplying more than one customer. You tend to get suppliers who supply to Nissan or to Honda or to Toyota, but rarely to all three. It is only the big, really big, majors that are suppliers to all three, people like Hitachi. We believe that the approach can still work when the supplier works for more than one customer, and we are quite happy with our suppliers also being suppliers to others.

24.4 Supplier selection factors

24.4.1 Quality

- quality philosophy and policy
- agreed objectives
- SPC
- improvement programme
- zero rejects.

Supplier selection is probably the most single important part of the purchasing procurement role. Because once you have said you are going to stay with somebody and work with them to improve, then the important part is to have chosen the right partner. Quality first; this should be relatively straightforward: does he produce to specification. However, to have confidence in a supplier's capabilities it is necessary to go somewhat deeper.

He should have a quality philosophy that he can explain to you. Not just because we like to see these as good things but because if he has not already thought through and developed a philosophy it is a reflection of his approach to quality improvement. Secondly, is he able to develop objectives that give year on year improvement. We do not want somebody who says my objective on quality is to do this. He should already have sight of where that is going to lead him in the future, how to keep the quality trend improving. Thirdly, because it is an important tool in the quality improvement, does he understand statistical process control. Is he approaching his production process in an analytical engineering fashion, or is he still back in the days of 'throw more people at it'?

Does he have an improvement programme: I do not mean a suggestions programme, I mean an improvement programme for housekeeping, for safety, for productivity, for quality? He does not have to have all of these but has he got that approach to his business? How well developed is his supplier involvement programme? It is important to remember that we do not deal with anybody who does not in turn have some suppliers. Has he thought of working with them to improve their delivery quality to him? Because if he does not he will undoubtedly have problems in fulfilling both his and our objectives.

And finally is he a supplier that really sees he has got to achieve zero reject? It is a commonly used phrase nowadays. I mentioned earlier to you that there are now Japanese suppliers that genuinely deliver parts per million of failures and they are still trying to improve. We are not interested in somebody who just says I want zero rejects, we want to see somebody who appears to be turning his organisation towards it. It will take a few years, that is fine, it will take us a few years.

24.4.2 Product development

- investment in facility and expertise
- design in components
- investigate 'systems effect'.

We look at the suppliers' ability for further development: what investment have they got in facilities, what investment have they got in the expertise and people?

Do they already do design on components or do they buy their design of components and tooling? Are they used to working from absolute detail or can they be asked to provide something, a component or an assembly, to do a job rather than provide something from a drawing on a piece of paper?

I call this 'systems effect'. I do not think there is a proper name. But if we consider a supplier of a certain component there is often another single or related group of components that together with that major one make up a system. We like to look at a supplier and to say: 'does he understand the system?' We talk to him about the whole system instead of about the component within a system because that makes our life easier and in the end makes his life easier. It makes for better management because you are managing fewer things.

A good example of the systems effect is brakes. You can buy brakes where the only people who see a braking system is the motor vehicle assembler. There are calipers here and tubes there and something else here and something else there, and none of the separate suppliers has any concept of the total brake system. Alternatively, you can also buy brakes by going to one of the big companies, because a few of them do have the expertise to manufacture a complete system, and saying we need brake systems for this motor car. And there are all sorts of stages in between these two extremes.

24.4.3 Production capability

- production engineering
- automation
- inventories — especially WIP
- productivity improvements — faster than economics.

When looking at production it is important to look at engineering capability, it is surprising how many British companies do not have real production engineers, but they are fundamental to a manufacturing business. We look at their level of automation, but do not insist on it. However, we are aware that well chosen automation is important for consistent quality achievement and it is important for maintaining productivity levels.

We always show a healthy interest in inventories, especially WIP. Many motor industry suppliers carry weeks of stock of finished product, not necessarily because they want to but because it has been a requirement in the past from their customers. Our particular interest is how much is within the production process. If it is work in progress, unfinished stock part way through the manufacturing process, why is it there and what is that telling us about the ability of the supplier to adequately manage his business? And lastly we are interested in a supplier who genuinely believes his role in life is to improve productivity faster than economics, if he sees an inflation rate of 5% on labour then he should be aiming for an improvement of 6% on labour productivity as a minimum.

24.4.4 Pricing

- produce the minimum
- pull scheduling
- eliminate inventories
- eliminate handling.

In order to maintain minimum pricing we look at the way suppliers manage their business, especially the way they manage in production control terms. The first task is to find people who recognise that their job is to produce what they have to and no more. Whenever you walk through a production plant and see somebody who wants to pour things into a black hole at the end of it just to maintain a stock of goods, then his production management system has got some fundamental flaws in it.

Secondly, we want people to understand the concept of pulls scheduling. Ideally we want to move to the position of having a part produced as you get the order for it, not in terms of anticipation of that order. Cascading this down the production system ensures that the WIP inventory is kept to an absolute minimum, which in turn minimises the amount of material movement and handling.

It is still true that if you compare the average British company or European component plant with a number of the better Japanese ones, you would see effectively no inventory at all in a Japanese plant because a part is always in motion somewhere; it is hardly ever waiting. And handling has been refined to such a degree that it is an integral part of the process rather than an actual operation.

24.5 Purchasing—the Nissan way

- supplier selection
- supplier liaison and support
- support of long-term relationships
- ‘negative pricing’ review.

The first and most important task of the Purchasing Department is supplier selection. The traditional and rather crude method of supplier selection is to write to a number of known potential suppliers, then bargain with the most likely ones to get the lowest price or best package deal.

The next two points deal with supporting the supplier. Practical support is only possible if you are close to your supplier, not close in a geographic sense but aware of his needs on an ongoing basis; this is a continuous process and large customers can frequently provide invaluable support to their suppliers in many guises, further developing the dependency aspect of their relationship.

Long term support is equally important if somewhat different. Both supplier and customer require to be creative if a good supplier is to be encouraged to grow into other fields; this may be necessary for a number of reasons, for example, due to technology change. Even good suppliers will not necessarily do the same job or the same amount of business over an extended time frame; it is the responsibility of purchasing to recognise this and identify other work opportunities, ensuring that over time, whilst relative workloads may change, suppliers can continue a successful and beneficial relationship, and not be discarded because we have just cause.

Finally, purchasing have responsibility for pursuing a policy of negative pricing; they are responsible for teaching the concept of negative pricing and working with suppliers to achieve a continuous cost reduction effort. The concept of purchasing, being pro-active in discussing price reviews well in advance of implementation, is somewhat different from the traditional adversarial negotiations that traditionally take place. This process works well in Japan; we currently buy a number of parts from Japan and benefit from their continuous cost reduction achievements, and a number of parts show price reductions year on year.

24.6 The differences summarised

Western

- adversarial
- retender
- multiple sources.

Japanese

- mutual dependence
- long-term business
- supplier development.

Finally to summarise what we see as the differences: the Western approach is based on an adversarial relationship requiring re-tendering and multiple sources with the customer always protecting his ability to be independent of his supplier and the supplier having to divert his resources to seeking new business rather than improving current business.

The Japanese approach is fundamentally different, with the customer taking the view that he and his supplier need each other equally and that they must have a long term business relationship together and much of the future development work will become the responsibility of the supplier.

Chapter 25

New tools of analysis in buyer/supplier relationships

N. F. Ferguson, L. F. Baxter, D. K. Macbeth and G. C. Neil

Glasgow Business School, UK

25.1 Introduction

This chapter describes work being done at the Glasgow Business School on the subject of managing suppliers in an AMT environment, as discussed by Macbeth *et al.* (1988).

AMT (Advanced Manufacturing Technology) is another one of these acronyms which is open to interpretation. In the light of substantial changes taking place in manufacturing philosophy based on the introduction of Just in Time techniques (Schonberger, 1982; 1986) in UK industry, our work has focused on the input of this AMT on the overall supply chain.

Just in Time (JIT) encompasses four key characteristics, i.e. perfect quality, making the material flow, waste elimination and people involvement. It pursues these concepts through a number of well-known approaches which hitherto have not been engaged in Western industry.

Most of these approaches are based on sound industrial engineering principles, e.g.:

- Set-up time reduction allowing economical small batch production.
- High quality levels preclude the need for safety stocks and re-works.
- Improved material flows reduce work in process and non-added value operations.
- Involving all personnel in problem solving and decision making makes use of a substantially untapped resource.

The changes resulting from these JIT techniques being applied in one company have significant ramifications for the suppliers to that company. In essence, a whole new set of demands from the customer going JIT means a complete re-appraisal of business operation for the suppliers.

It was this reaction to changes taking place that formed the study field for our project.

25.2 Research process

From the start the study has been guided by industrial collaborators who form a management steering committee for the project. Our team has worked with five major buyers, and around 16 of their suppliers, to develop case studies of existing practice in buyer/supplier relationships. The information gathered from these companies has subsequently been enhanced and refined by the use of a survey covering a larger cross-section of the electronics and mechanical engineering sectors. The data collected from this research has allowed the classification of existing relationships and given a useful insight into how companies have to change their philosophies and operations in pursuit of world class manufacturing excellence.

25.3 JIT and the supply chain

Our study of buyer/supplier relationships in an increasingly JIT environment has allowed ready determination of the theory that companies going JIT will move towards closer relationships with fewer suppliers:

> 'the firms . . . that were most effective were so because of their ability to gain competitive advantage based on establishing important relationships with suppliers . . . relationships which enabled their companies in a fundamental way, to compete effectively' (Shapiro)

One view which has been expressed is that there is a distinction to be made between JIT in the company and JIT in the supply chain to the company. This extends to the belief that the JIT material procurement by a company should be one of the last areas in overall JIT implementation to be tackled.

This suggests that JIT can be 'plugged in' progressively, whereas we would suggest that the taking on board of a new philosophy impacts on so many manufacturing businesses and requires modifications in so many directions that, unless attention is given in parallel to improving supply chain, the internal advantages of JIT will not be fully realised.

Inherent in the waste elimination goal of JIT are the ultimate targets of *perfect quality, instantaneous delivery* and *minimal cost.* It is widely recognised that these are the three main areas on which manufacturers have to perform on. More specifically, quality is the foundation on which success in the other manufacturing criteria is built. For 'advanced' manufacturing companies very good Quality is now often taken as 'given' and Delivery aspects are coming under closer scrutiny. Cost then becomes an outcome of correct decisions and performance on the other two, and can be controlled via these factors. Hence, cost is now being defined in new ways, e.g. total acquisition cost (Neil *et al.*, 1988), and the term encompasses measurement in many hitherto unthought of realms.

25.4 A means to improvement

The basis for our 'deliverable' is shown in Fig 25.1. Stages 1 and 2 exist side by side as the means by which a company can 'locate' itself with respect to the best or ideal manufacturing performance. The 'ideal' could be JIT ideals, although some of these targets are unhelpful, being Utopian in scope, and only serve as an indicator of ideal direction. Our current research can provide a means of determining the 'best' or 'ideal' practice, but ultimately the optimum approach is to establish the best practice via a rolling benchmarking programme, which takes in national manufacturing performance and determines the top notch companies specific to the industrial contest.

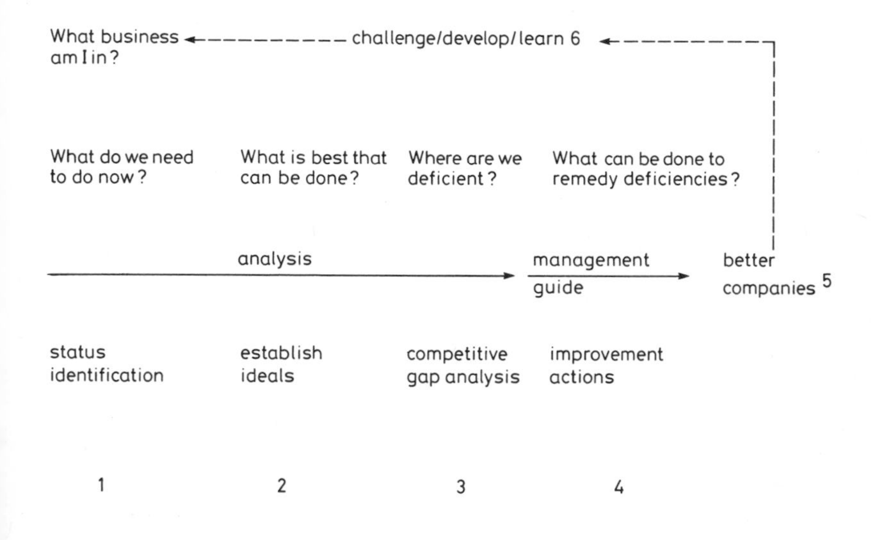

Fig 25.1 Improvement model

Stage 3 is the point where the specific identification of deficiencies in present performance for a company takes place. This pinpoints weaknesses in a firm's operations and lends itself to be the basis of information to which improvement initiatives can be applied (Stage 4). These paths to improvement can therefore be specifically targeted to reduce the competitive gaps highlighted by Stage 3.

Any successful outcome in this process should result in a more efficient and competitive operation (Stage 5), but, of course, in accordance with JIT philosophy the next stage (Stage 6) should be the initiating of a whole new cycle of self-examination and improvement.

25.5 The positioning tool

The means by which Stages 1 and 2, identified above, are progressed in the Glasgow Business School concept is called the Positioning Tool. The Positioning Tool models a company's operation as being composed of three key Dimensions — Capability, Requirements and Performance. 'Capability' can be described as the company's raw ability to meet demands placed on it. 'Requirements' are the demands placed on the company by its customers or by the company on its suppliers and/or internally on itself. 'Performance' relates to how well the company has converted its raw Capability into actions to meet the Requirements placed upon it.

The analysis of a company's relative operational efficiency in each of these three Dimensions is enabled by the subdivision of each Dimension into three relevant Factors, which in turn are influenced by a variety of Topics impinging on them. Finally, the Topics are more completely described by being derived from a number of Questions, which are grouped into nine main sections. Fig. 25.2 shows the sub-division of the three Dimensions into Factors.

By means of scoring answers to individual Questions, strengths and weaknesses in Topic areas become apparent. These Topic indicators contribute to measurements for individual Factors, and hence a pattern of a given relationship, such as shown in Fig. 25.2, can be built up. As all the scoring is referenced to an ideal of some sort, as discussed previously, it becomes easy to identify the relative gaps in competitive ability. However, the scoring used is not an absolute measure, rather it serves as a problem area indicator.

For example, referring to Fig 25.2, a deficiency in the 'People' Factor within the 'Capability' Dimension is highlighted. This could be a result of a variety of situations within the company; poor training set-up, minimal workforce involvement in decision making, limited employee job flexibility etc. In each case, the root cause can be traced back through the

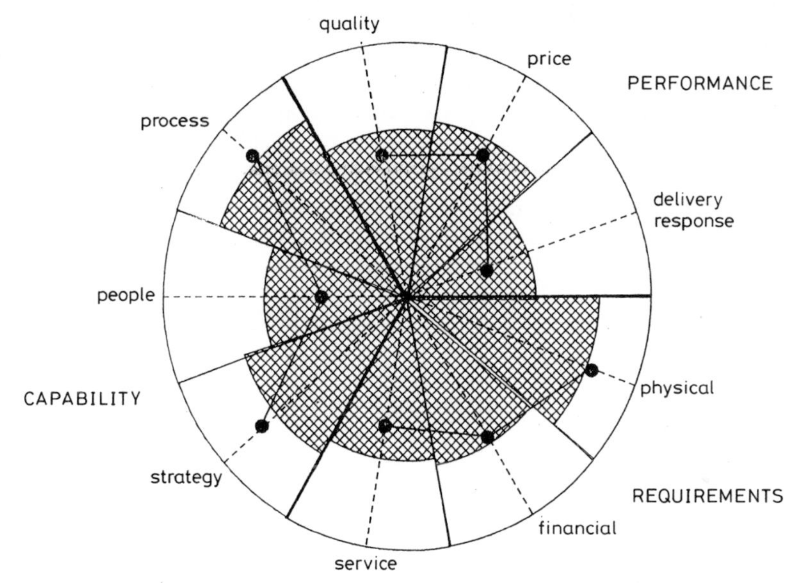

Fig 25.2 Positioning tool

Factors and Topics to the original Questions. It is then possible to highlight action points and indicate management choices.

At present, the Positioning Tool is created following off-site analysis of answers given to the nine section questionnaire administered by the research team at a specific company.

We would hope that future versions of our concept could be self-administering, with a variety of application options; e.g. we would envisage the Tool being useful for pursuing improvements within an existing Buyer/Supplier relationship, providing a means of self-education for a supplier in relation to its competitors, and also for allowing a supplier company to determine the types of operational requirements necessary to break into a particular supply chain. The last example could be of major significance in the area of import substitution.

25.6 Improvement paths

In pursuing our research on Buyer/Supplier relationships, we have identified examples of improvement paths which have been successful in closing some of the competitive gaps.

In association with our major industrial collaborators, these improvement paths will be further developed to enable the choice of a number of routes to the desired relationship state. We see these routes as being composed of analysis, evaluation and communication between participating companies, highlighting 'priority treatment' areas and being specific to the undoubtably unique business relationship existing, rather than being a prescribed universal panacea. The management guide, which is the major deliverable of our project, will assist this process and will encourage the approach of continuous improvement, rigorously expanded by JIT followers, to the variety of strategic and operational issues which confront management on a day-to-day basis. Of course, as current 'best practice' will not necessarily be 'best practice' for tomorrow, the analysis and improvement devices employed must be continuously developed and refined.

The requirement to instil a culture of continuous improvement in companies has been made all too evident through our work, where we have recognised a reasonable base of JIT knowledge, but a failure to convert that awareness into understanding and action within the company, especially with regard to supplier companies. Our other findings to date are detailed elsewhere (Baxter *et al.* 1988; Ferguson *et al.*, 1988; Macbeth *et al.*, 1988).

25.7 Further development

Originally, the deliverables from our project were conceived as being purely paper based, but, as the work has progressed, it has become apparent that tasks such as data gathering and analysis, and the Positioning Tool score and graphic generation, benefit greatly from computerised techniques. Also, aspects of the management guide might also lend themselves to the advisor mode of knowledge based systems. This is a further avenue to the project which will be examined in due course, but obviously we must develop practical paper solutions to fit the requirements of our industrial collaborators initially and then the general industrial environment on which our work is based.

The use of benchmarking to establish current best practice could be an important technique to provide the basis of information for our development and analysis tools. We would envisage that large scale benchmarking would require co-ordinated effort from a number of parties, both in the set-up and maintenance of such a process. With increasing interest becoming apparent within the industrial community for means to measure aspects of business hitherto untouched, such a benchmarking exercise might have some commercial viability.

25.8 Conclusion

Just in Time manufacturing is not just another 'quick fix' solution to immediate problems arising in industry. The more famous elements of JIT may not be appropriate in the short term for certain industrial operations. However, the underlying principles of JIT, such as training, organisational change, waste elimination and continuous ethics, can reap immediate benefits for any company, and require only minimal investment.

The real problem for most organisations lies not in the scope for improvement, but in the initiation and indoctrination of JIT philosophy.

We have developed the Positioning Tool to aid the prioritising of improvement activities and, more specifically, to point the way towards developing more integrated supply chains. Much work is still required in refining our Positioning Tool and we look forward to working with interested parties in the further development of our concept.

25.9 Acknowledgments

The authors gratefully acknowledge that the funding for the research project from which this chapter is drawn — 'Managing suppliers in an AMT environment' — is provided by the ACME Directorate, Science and Engineering Research Council, Swindon, UK; Grant Reference No. GR/E/12337.

Chapter 26

Applying the Five-Step Method to modelling quality management systems

C. M. Jones, A. N. Godwin and M. Gore

University of Warwick, and Coventry Polytechnic, UK

26.1 Introduction

This chapter focuses on the application of a new method — the Five Step Method — to model quality management systems within supply chains in manufacturing industry. The problem it addresses is that we currently do not have integrated information systems to support decision making in manufacturing that consider suppliers' and customers' activities adequately or that show the interdependency of business decisions relating to quality and logistics, as examples.

It is useful to be able to model the supply chain management system and its various sub-systems including quality: (i) to improve understanding of the decision making processes involved, (ii) to highlight problem areas and (iii) to be able to simulate the effect of possible decisions. This can lead to better design of systems to support business activities or to better business decisions. The Five-Step Method has been designed to enable all three of these — steps one to three address points (i) and (ii) and steps four and five, point (iii). This chapter concentrates on the application of steps one to three of the method and therefore looks at improving understanding and highlighting problems in existing systems. The work is part of a research project co-funded by the SERC and Lucas Automotive to use a systems approach to investigate supply chains. The case study relates to the quality management system in Lucas Automotive.

Throughout the chapter the quality management system is seen from Lucas's perspective; customers are, in the main, original equipment manufacturers (OEMs) of motor vehicles, such as Ford and Volvo. Suppliers are suppliers of raw materials and components to Lucas.

26.2 Available methods to model manufacturing systems

Manufacturing is a complex environment to model; there are a wide variety of structures within it. A great deal of existing work shows the need for methods to help design manufacturing systems. Parnaby (1986) stressed the need for a structured and methodical approach to redesigning manufacturing systems, but it is difficult for a single person to have the necessary knowledge and expertise that will allow a comprehensive description of many of the recurring problems in the industry. Feigenbaum (1983) emphasises the necessity of an effective system to integrate all aspects of quality.

Model building in manufacturing systems is evolving in a similar way to commercial systems; new methods adopt good ideas from existing ones. However, manufacturing system methods are not adapted to the environment sufficiently. Three example methodologies that have been used for the description of manufacturing systems illustrate many of the difficulties. The ICAM (Integration of Computers in Aircraft Manufacture) methodology discussed by Bravoco *et al.* (1985) argues strongly that three notations — IDEF0, 1 and 2, (IDEF is an abbreviated form of ICAM Definition) — are required to give adequate support for the description/design process. IDEF0 supports the modelling of functions, IDEF1 supports data modelling and IDEF2 supports modelling over time, or simulation. However, there are still some gaps not filled by the IDEF set, as highlighted by Godwin (1989), and as tools they are very generalised. The early stages of modelling and the so-called soft systems aspects are not properly accommodated by IDEF. The (Checkland, 1984) Soft System Methodology is centred on these aspects but is not explicit about how to incorporate other aspects. A brief paper showing some simple links between the Checkland method and IDEF0 has been published by Ming Wang (1988) but there is little detail and no reported actual combined usage. A second example is the GRAI methodology discussed by Maloubier *et al.* (1984) (see Chapter 8); it provides a bias towards the manufacturing production process but inhibits inclusion of commercial aspects of the system and inter-organisational links. A third example considered is Input – Output Analysis; the main fault found with this is its lack of specifics related to manufacturing (the opposite problem to the GRAI method). It has advantages in that it is easy to produce broad initial descriptions, but the development from these is not well defined.'

What is needed is a method that is relevant to manufacturing, which allows soft as well as hard issues to be considered to support system description and modelling. The Five Step Method was developed in response to this need.

26.3 The Five-Step Method

The Five-Step Method is a set of methods and tools to assist the business manager and the systems analyst in building models of existing systems and simulating changes. The method starts from the ICAM notations but adds extra notations and procedural links between them. A prior step to cope with the influence of soft systems has been included. Details specific to manufacturing have been added to the use of IDEF. An outline description of all five steps of the method is given, but only the first three steps are shown in the example. The main concern of this chapter is the application rather than the details of the methodology.

The first of the five steps establishes what is called in systems analysis language (SAL) terms described by Ross (1977) the purpose, viewpoint and vantage for the model. (IDEF0, the ICAM derivative of SAL, only includes purpose and viewpoint in a minor position.) The technique used for step one is brainstorming, associated in Checkland's methodology with the production of a rich picture, though this is not a requirement of this method. However, a rich picture may be useful in the brainstorming as a basis for discussion. CATWOE (the details of which are given in the Lucas case study) analysis can be used in this step to assist understanding of the problem.

In the second step, a set of IDEF0 diagrams identify the main functional areas for attention. The purpose and viewpoint of the model defined in step one dictate the level of detail of the functional model built in step two; it is recommended that step one is refined during the construction of the functional model. Having selected the viewpoint, the model can be looked at from different perspectives or vantages. Skeleton IDEF0 hierarchies for significant vantages are provided as a starting point to users of step two. The example quality model shows how useful these templates are.

Step three analyses the activities from step two in more detail. Part one of this analysis is carried out in the context of nine generic headings; planning and installation, operational objectives and plans, input management, resource management, operation, output management, monitoring, evaluation and maintenance. The second part of step three is the building of an appropriate data model for those activities that have been analysed. This data model can be expressed in IDEF1 form although the method does not insist on this.

In step four, the static model obtained in step three is quantified and the dynamics of the data model are introduced. This is done by determining frequencies and resource usage for actions and formulating standard sequences of actions. The data collected in this step allows hierarchical simulations to be constructed. Each single level simulation can be expressed in terms of an IDEF2 diagram. The combined IDEF0 – IDEF2 model can be translated into SIMAN (a simulation package). The method of data collection for this step is like the sequence of questions used in the first part of step three.

Step five tries to assess the efficiency and effectiveness of the performance of the activities. A further round of questions with roughly the same sequence is used to gather this data. The main difference between step four, the simulation step, and step five is the consideration of how decisions are made and their quality.

26.4 Lucas Automotive: a case study

The objectives of the study in Lucas are to investigate quality and logistics related activities and decisions, to model them to improve understanding and highlight problems, and to simulate effects of decisions. It is hoped to provide Lucas with a user led method for modelling that enables improved business decision making and improved system design. The study as a whole is interested in the integration of quality and logistics decisions in the supply chain; this chapter considers some of the issues relating to quality discovered in the investigations. The case is based in one operating site of Lucas Automotive that manufactures braking systems. It illustrates and evaluates the application of the first three steps of the Five Step Method to the quality management system.

26.4.1 Step One: identification of purpose and viewpoint

Application of step one of the method involved using brainstorming and CATWOE analysis. The initial brainstorming to broadly define the problem was conducted by academics and Lucas managers over a long period of time as it was part of the pre-proposal stage for a funded research project. However, managers in industry using the Five-Step Method could brainstorm in their own company relatively quickly. The results of the brainstorming defined the areas for investigation as being the general structure of the supply chain system, and within the supply chain the systems to manage the flow of quality information, the flow of orders (the logistics system) and the physical flow of materials. Examples of Checkland rich pictures (Checkland, 1984) were shown to Lucas but the perceived triviality of the symbology (for example, stick men and eyeballs) caused adverse reactions to the technique. Therefore the principles of rich picture analysis were used without the diagrams.

The CATWOE analysis for the Quality Flow system was derived from investigations within Lucas.

After this, the purpose and viewpoint of the model was stated as being 'to analyse how customers' desires for quality are achieved by the manufacturer and its suppliers within the supply chain.' It was found to be useful to make a list of things that the model would and would not consider in order to clarify the viewpoint:

Table 26.1 CATWOE analysis

Category	Participants/entities
Clients	Customers (Original Equipment Manufacturers — OEMs)
Actors	Suppliers, Purchasing, Engineering, Production, Quality, Sales and Marketing
Transformation	Materials are transformed into braking systems to customer quality requirements
World view	Success is viewed by general market reputation and by specific customer measures of quality performance and capability
Owners	Quality Manager, General Manager
Environment	The main constraint is imposed by customer (OEM) quality requirements

Will consider: Managing supplier quality (quality performance monitoring, incoming goods inspection, detailed material specifications, supplier inspection systems); managing manufacturing quality (process design, quality performance monitoring, in house inspection, customer audits); managing customer quality (customer desired quality, product design, quality performance monitoring, in house inspection, customer inspection); communication and co-ordination of quality information up and down the supply chain.

Will not consider: The number cost of quality. This viewpoint drew the boundary around the system that was considered, i.e the quality management system. It was decided to investigate particular vantages, or perspectives of different people of the quality management system, by classifying activities in three ways; namely a functional classification that considered departmental divisions, a hierarchical classification that considered strategic, tactical and operational activities and a theme oriented classification — this project was concerned with quality and logistics; so themes of quality, orders and physical material flow were considered.

Choice of the above classification criteria could have been made at each level of functional decomposition. For example, if the model was focused on management reporting structures, the primary classification choice could be hierarchical. If, then, reporting between business departments was of interest the secondary choice could be functional. By combining these classifications, an overall vantage or perspective of the business was gained. With a choice of three classifications there were 3! or six permutations of possible vantage. Two vantages were selected for comparison; the first considered quality, then functional issues, then management hierarchy; the second looked at quality, then management hierarchy, then functional issues. Because the purpose and viewpoint of the model focused on quality in the supply chains, the theme classification was chosen as the primary choice in both cases. Tannock and Wort (1988) state that analysing quality as a separate sub-system is preferable to modelling the entire manufacturing system to see where quality fits in, if the focus of the analysis is to be on quality; both methods are allowable within IDEF0 rules.

26.4.2 Step Two: basic functional modelling

The purpose, viewpoint and vantages within viewpoint identified in step one lead to a different set of decomposed IDEF0 diagrams for each hierarchy; the first set is shown in Table 26.2. It was found useful to express each set in the format used in manufacturing for indented bills of materials.

Table 26.2 IDEF0 diagrams for hierarchy 1

IDEF0 activity	IDEF0 name	Appendix
A0	Supply chain management	
.A1	Quality flow	1
..A11	Customer quality	2
...A111	Customer quality policy	
...A112	Manage customer quality	
...A113	Operate customer quality	3
....A1131	Inspect/test manufactured goods	
....A1132	Deal with customer concerns	
.....A11321	Deal with minor customer concerns	
.....A11322	Deal with serious customer concerns	
....A1133	Implement corrective action	
..A12	Manufacturing quality	
..A13	Supplier quality	
.A2	Order flow	
.A3	Materials flow	

26.4.3 Step Three: data collection

The description under the nine headings in step three was obtained using a standard questionnaire. The example function chosen was ‘operate customer quality' which analysed activities, within Lucas and customers related to quality, from completion of manufacturing at Lucas through to acceptance of Lucas's products by its customers. The results can be summarised as follows:

Initialisation: The quality manager and other senior management set up a system to deal with customer concerns. Customers set up the inspection system.

Operational objectives The quality manager supervises quality improvement. Records of parameters, such as number of returns, are used to measure improvements.

Input: Quality manager, sales or manufacturing receive customer returns and complaints. Manufactured goods received by customers are queued, awaiting inspection or testing.

Resource: The quality manager organises meetings to discuss and prioritise customer concerns. This determines the effort put into organising and communicating internally and with the customer. The manpower for the inspection process in this example is provided by the customer.

Operation: There are three activities associated with this function: inspection of manufactured goods; analysing customer concerns; implementation of changes in response to customer concerns.

Output: The activities produce inspected goods; notes to customers on the actions being taken.

Monitoring: Inspection logs are kept by customers and there are records of the meetings related to customer concerns.

Evaluation: The quality manager is responsible for evaluating how the two internal activities are carried out. The customer must evaluate the efficiency and effectiveness of the inspection.

Maintenance: The quality manager ensures that manpower is properly trained and the lines of communication to the departments involved in customer concerns meetings are efficient. The customer will maintain inspection equipment and manpower.

Abbreviated results of the second substep for the example activity are given below. The data model is written out in non-diagrammatic form, but the results could be expressed with modification as an IDEF1 model with diagrams:

Entities: The following are entities within the data model: PRODUCT_IDEA; DESIGN (product_idea); PRODUCT_SET(design);

CUSTOMER; CUSTOMER__INSPECTION__SCHEME(customer, design); QUALITY__ASPECT(customer, design); CUSTOMER – __MEASURE(customer__inspection__scheme); CUSTOMER__STANDARD(customer__measure); ACTION__MEETING; DECISION(discussed__at); ORGANISATIONAL__STRUCTURE; COMPONENT__OF__STRUCTURE(organisational__structure) PARTICIPANT(component__of__structure);

Relationships: These include: FEEDBACK__REPORT(customer, product__set); DISCUSSED__AT(feedback__report,action__meeting); TAKE__PART(action__meeting,participant);

Properties and Rules. These are as below: CUSTOMER__RETURN__RATE(customer__standard,design); STATUS__CONCERN(decision); CUSTOMER__RESULT(product__set,customer__measure) CUSTOMER__ASSESSMENT(product__set,customer__standard) CUSTOMER__ASSESSED__STATUS(product__set,customer—__inspection__scheme);

The data model will be used as the basis for simulation in step four of the method, but for the purpose of this chapter the findings from the first three steps are considered.

26.5 Discussion of the case study results

26.5.1 The method

The application of the Five-Step Method to the quality management system in Lucas Automotive's supply chain demonstrates the benefits of the pre-planing steps before data collection to improve understanding of the system to be modelled. Performing this as a stand-alone exercise could be useful to managers to help them analyse the activities that their departments are involved in, and how they fit in with other departments and suppliers' and customers' activities. Using CATWOE as a tool in step one and IDEF0 in step two forces consideration of what constrains and enables activities as well as inputs and outputs. Crossfield *et al.* (1988) report similar benefits from using IDEFc, a variation on the IDEF tools, to map quality management systems.

The method was shown to enable a more selective approach to systems description than total system mapping, an exercise which often companies give up part way through because of the lack of focus on business problems. Because the example focuses on quality, the chapter does not show the full scope of the method to assist integrated model building. Effects of quality activities on logistics activities and vice versa are therefore not explored.

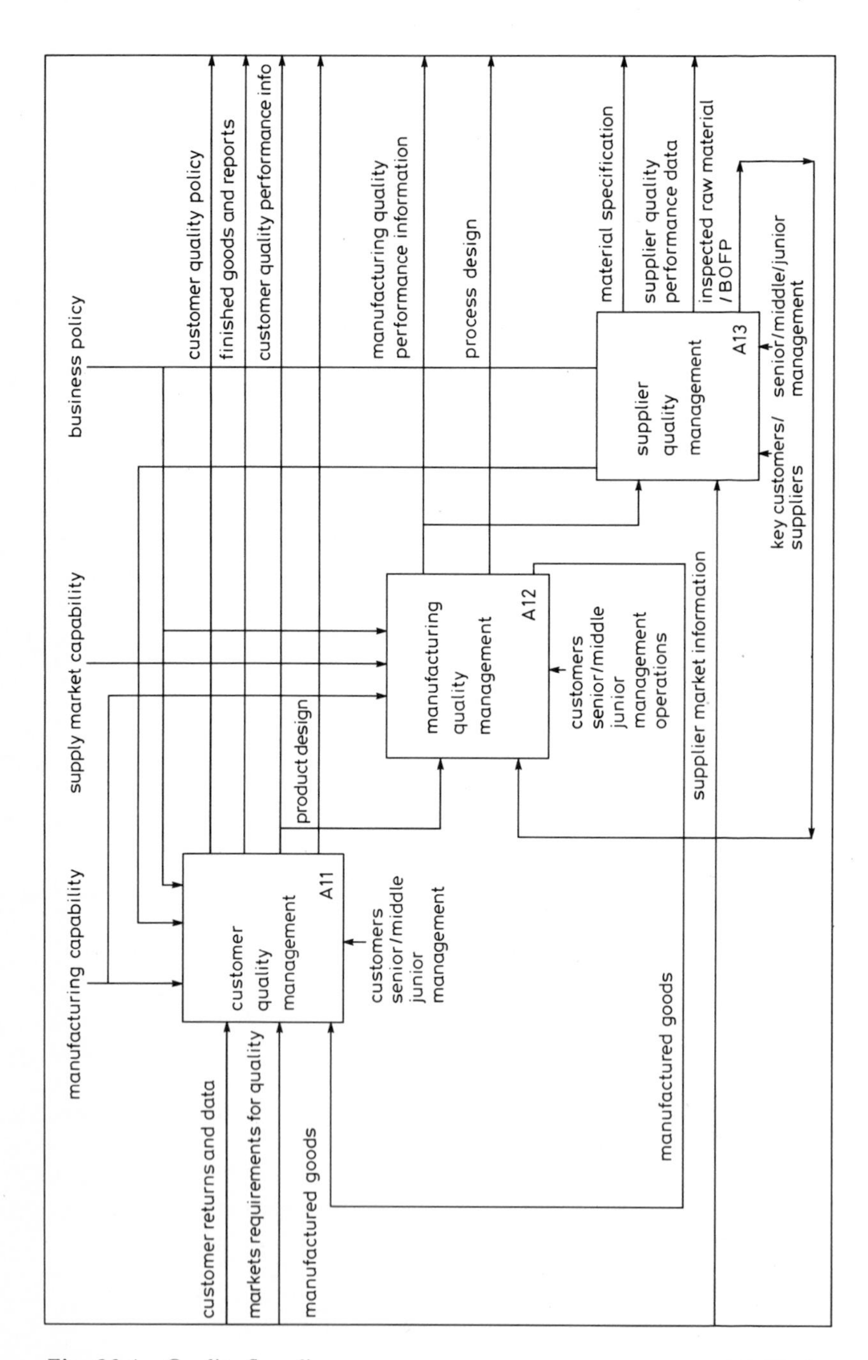

Fig. 26.1 Quality flow diagram

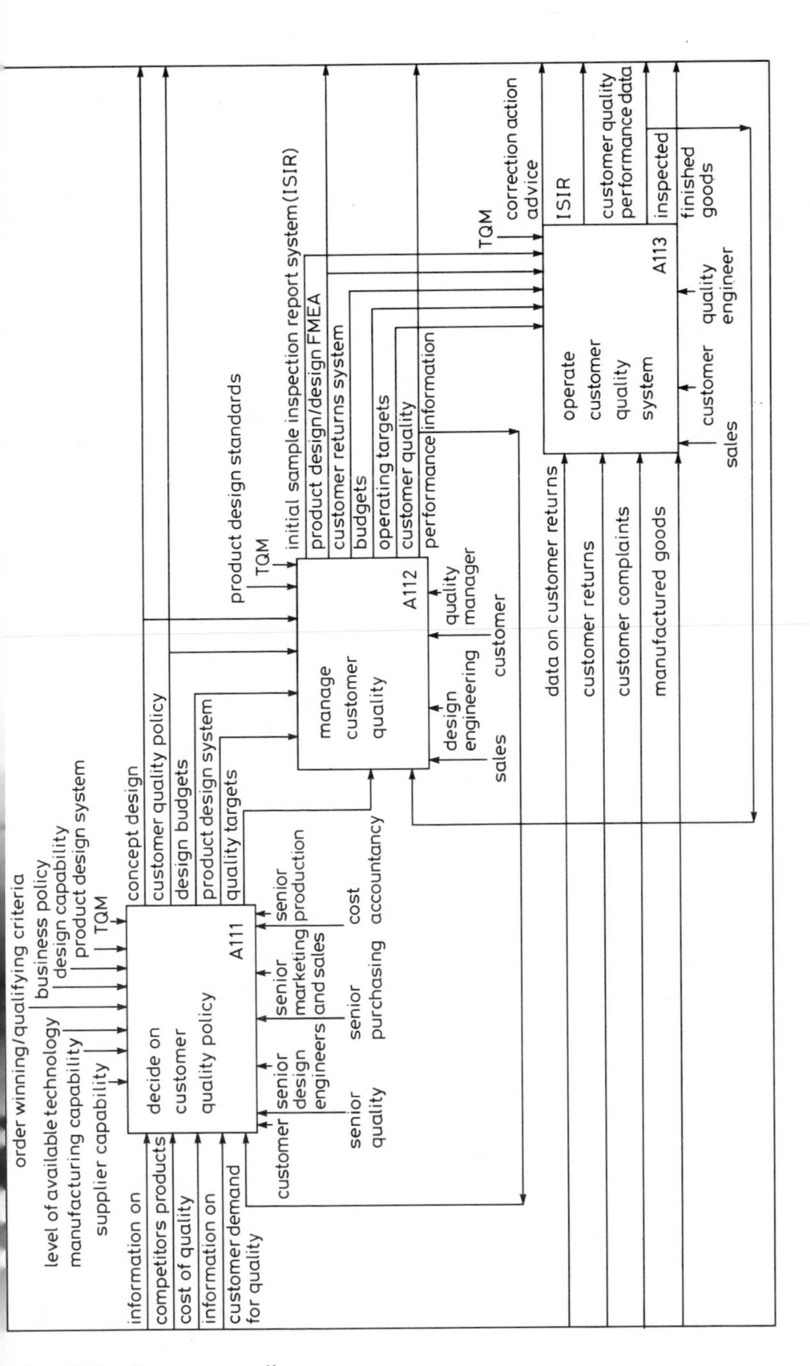

Fig. 26.2 Customer quality

The example draws out the complexities of managing quality in functional organisations and highlights the effects of vertical reporting structures on the quality management system.

26.5.2 Findings from Steps One and Two

The final version of Step One provided a framework within which quality in supply chains could be considered as containing a functional view, a management hierarchical view and a quality theme view. Comparison of the Quality Flow IDEF0 diagrams from two different vantages (Figs. 26.1 and 26.4) highlighted the effect of vantage selection. Fig. 26.1 shows relatively little interaction between functional areas. Feedback is in the form of materials flowing through the supply chain. The Quality Flow Diagram in Fig. 26.4 shows the outputs of strategy constraining middle management, who in turn provide and manage systems that constrain operations. Feedback is in the form of data and information on quality performance. Staveley and Dale (1987) describe how in many organisations this feedback may be informal owing to the difficulties in obtaining meaningful information from links further down the supply chain.

The observations from the comparisons are not unexpected; they confirm the reporting structure in functionally organised businesses as being upwards within a function rather than across functions. Cross functional reporting happens mostly at senior management levels, so performance measuring data and information traffic will flow vertically not horizontally. Galbraith (1971) and Barrar (1988) highlight that product oriented organisations should steer systems integration to reflect the product influence. Matrix management is viewed as an intermediate structural form between functional and product oriented structures.

In the case study, when quality went wrong, i.e. when a customer returned goods or registered concern about a product, quality was focused in on by a multi-functional team who identified the cause and implemented corrective action (Fig. 26.3). The Quality Manager integrated all the relevant business functions to solve specific problems, which was difficult because so many departments were involved.

In many of the EDEF0 diagrams, mechanisms included people outside the organisational boundary of the company. A further point to note about mechanisms for quality related activities is that not only do they span departmental boundaries but they also involve many people at all levels of responsibility.

26.5.3 Findings from Step Three

Combining the data model given with other models from other perspectives would provide a data model that would be stable with respect

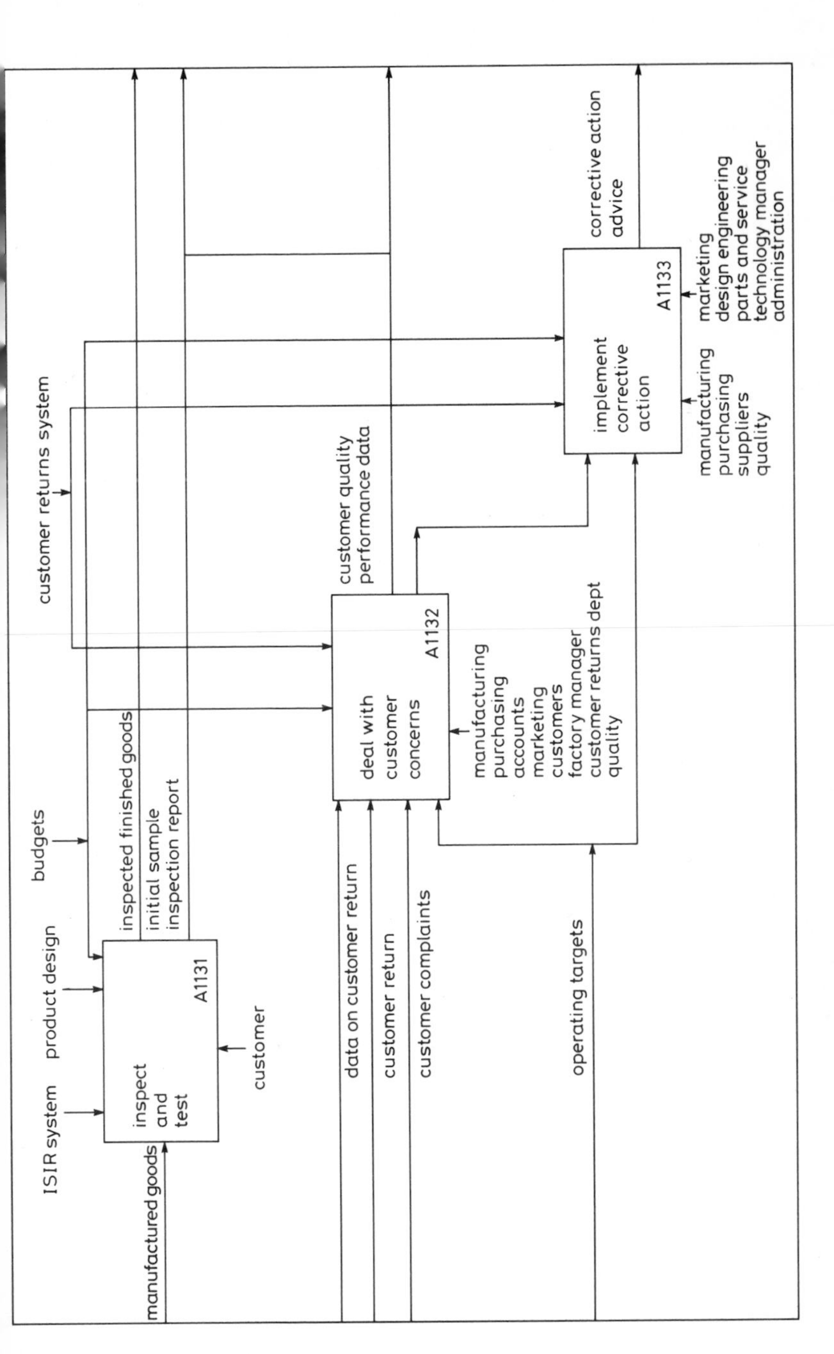

Fig. 26.3 Customer quality system

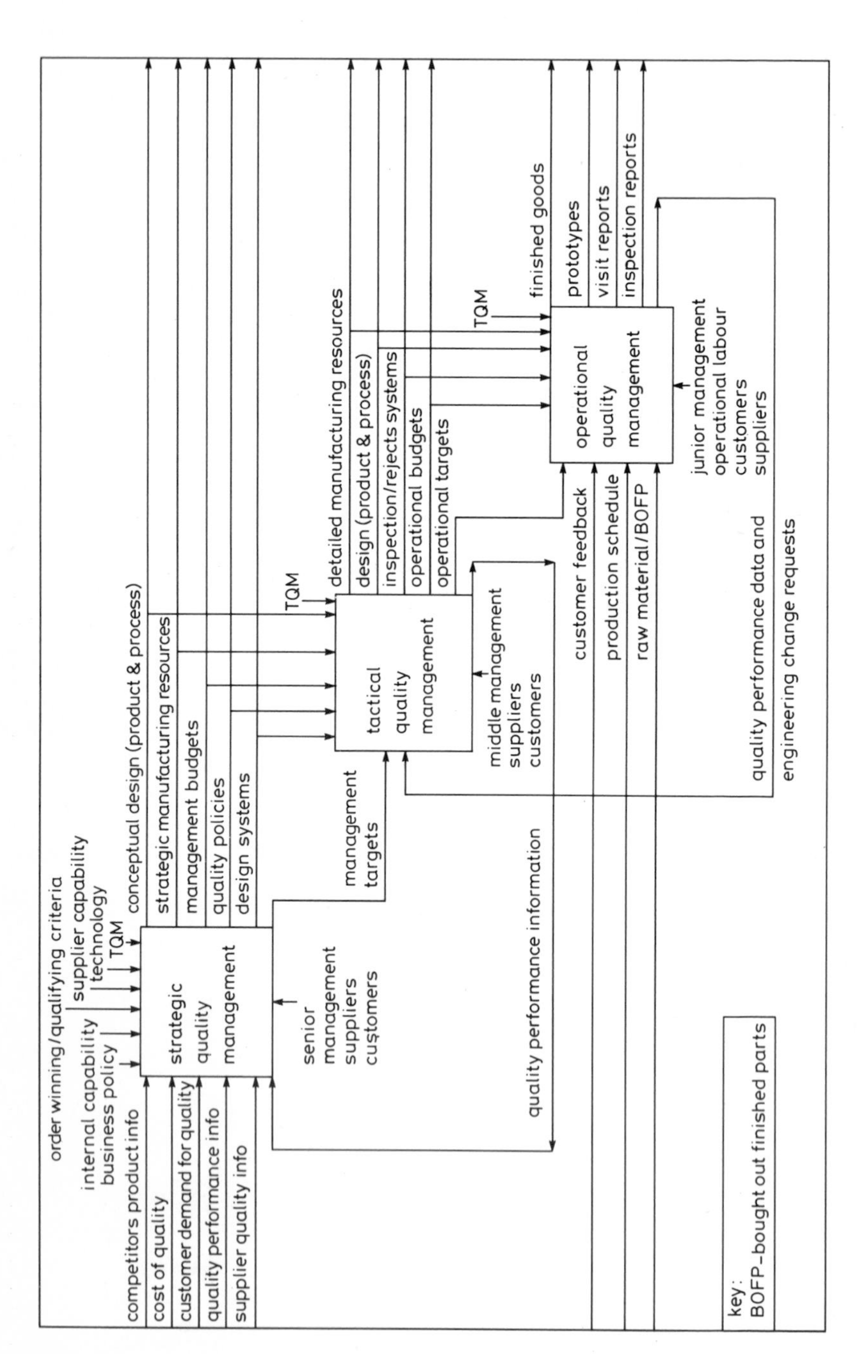

Fig. 26.4 Quality flow

to quite a wide range of changes in the organisation. There is considerable work remaining before implementation, but much of the structure of the database for the information system has been established. The two large areas of remaining work are: (i) file design; (ii) design of the data capture processes and ensuring integrity.

26.6 Conclusion

The findings from the case study showed that applying the first three steps of the Five Step Method to a sub-system such as quality management can yield sensible and useful results. It would now be useful to apply the remaining two steps to such an example and to apply the full method to more complex, integrated systems.

26.7 Acknowledgments

The authors wish to acknowledge the IMMS research group at Coventry Polytechnic, of which the authors are team members.

Chapter 27

Implementing JIT

R. Gill and C. Holly

Middlesex Polytechnic, and ITT Jabsco, UK

27.1 Introduction

One of the major attractions of Just In Time (JIT) manufacturing is its simplicity. Everyone knows JIT is a good idea — in someone else's factory. There is always some resistance, usually middle management rooted, expressed in reasons as to why it will not work in their own factory. The goal of JIT is the manufacture and delivery of the right product at the required time, in the correct quantity, and of assured quality. Demand for components, subassemblies and finished goods therefore 'pulls' goods through the manufacturing system.

The Japanese have perfected JIT systems in industry and numerous manufacturers have adopted the methods, along with the 'Kanban' system developed at Toyota. The origin of JIT philosophy has been accredited to various sources; the Japanese shipbuilding industry (Schonberger, 1984) and the construction of the Empire State Building (Lawerabcem, 1987) have been amongst those regularly quoted. In JIT manufacturing, inventories are kept to a minimum and planning is short term. Therefore, it is necessary to adopt a number of parallel objectives, including zero defects, minimum set-up times, minimum handling, zero breakdowns, short lead times, and a batch size of one. Manufacturing hold-ups are the arch enemy: faulty output is eliminated by Total Quality Management (TQM) while breakdowns are avoided by implementing a programme of preventive maintenance. JIT places those work centres which feed one another in close proximity, improving visual contact and reducing inter-operation handling. Parts are processed one at a time at each work centre and move right through the manufacturing process.

ITT Jabsco Ltd. (Jabsco) are one of the world's leading manufacturers of pumps, from flexible impeller types, whose invention dates back to 1940

(Fig. 27.1), to more recent ones which include lobe and drum varieties. Production of the flexible impeller pump commenced in 1941 in the United States and in 1956 in England. The ITT Corporation took over Jabsco in 1968 and at present pump manufacture is based at three locations employing about 500 people. The English base at Hoddesdon employs 160 and exports 75% of its output (Table 27.1). The flexible impeller, lobe and drum pumps are capable of handling a wide range of products in the food, dairy, chemical, pharmaceutical and engineering industries; the choice of pump type depends on the medium to be transported.

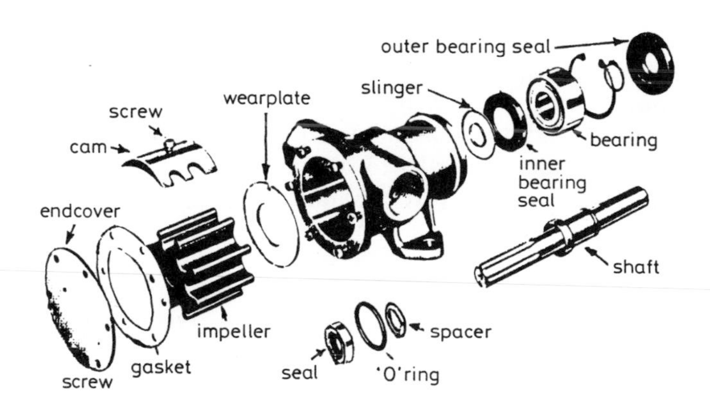

Fig. 27.1 Flexible impeller pump

Table 27.1 markets/industries served, percentage of sales

Boat OEM	3
Marine engine OEM	10
Marine distributors	42
Fuel transfer	3
Food, dairy, beverage, pharmaceuticals	17
Industrial & other	25

In any country, for manufacturing to stay competitive and ensure survival, change is necessary. Markets are dynamic and continuously changing with the addition of new products and the emergence of new suppliers. Furthermore, the customer's demands are increasing as exposure to a wider choice continues. In order to respond to these factors, manufacturing must be dynamic and prepared to face up to the change so as to meet the requirements of the market. The steady (and sometimes rapid) decline of British manufacturing industry has been observed universally when faced with the strong overseas competition. This decline has fundamentally taken place because industry did not (and in some instances still does not) recognise the need for change.

Although Jabsco had a strong market position, the management was quick to recognise the need for change in order to retain and increase their markets. However, the ultimate reason for change was the need to meet the financial targets of ITT. As a unit, Jabsco wanted simultaneously to reduce delivery times and manufacturing costs. In the not-too-distant past Jabsco tried to accomplish this by:

- reducing machining times
- pressuring vendors to reduce costs by encouraging competition amongst them
- implementing sophisticated management techniques such as efficiency, productivity, MRP etc.
- instilling fear and division amongst the workforce
- holding large stocks in order to respond to customer demands
- employing Just-in-Case (JIC) manufacturing.

In acknowledging this, Jabsco recognised the need for radical change in its manufacturing strategy. The management decided to implement JIT as part of a five year plan to eliminate the typical 95% downtime which a product has during its factory life (Fig. 27.2).

27.2 Planned change

Implementing JIT manufacturing involves major changes to existing thinking and current manufacturing practices. Thus, to ensure that the flow of products to the customer is not interrupted any changes had to be planned in a systematic manner. The primary aim was to reduce waste in any form (Fig. 27.3) and the objectives were:

- 10 day order turn around
- to reduce work in process
- to minimise machine downtime
- no rework
- to eliminate scrap
- to minimise changeovers/settings
- to eliminate material shortages
- to eliminate non-standard operations.

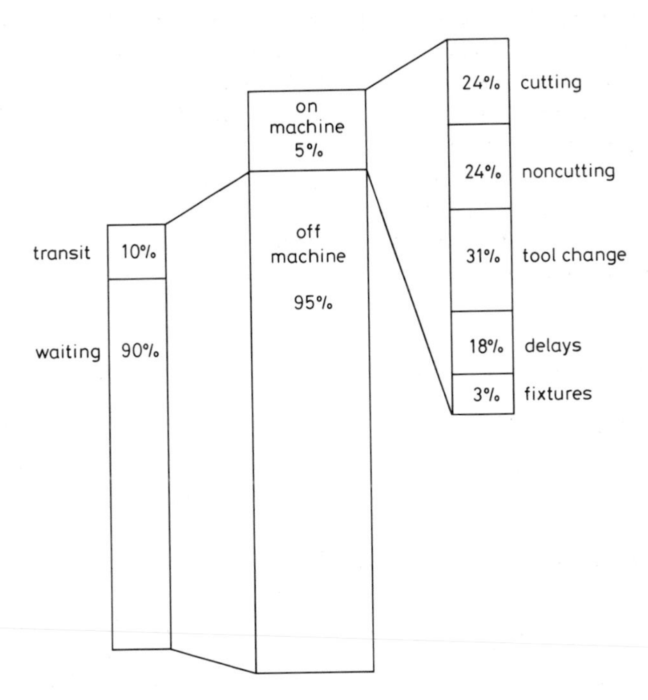

Fig 27.2 production time breakdown

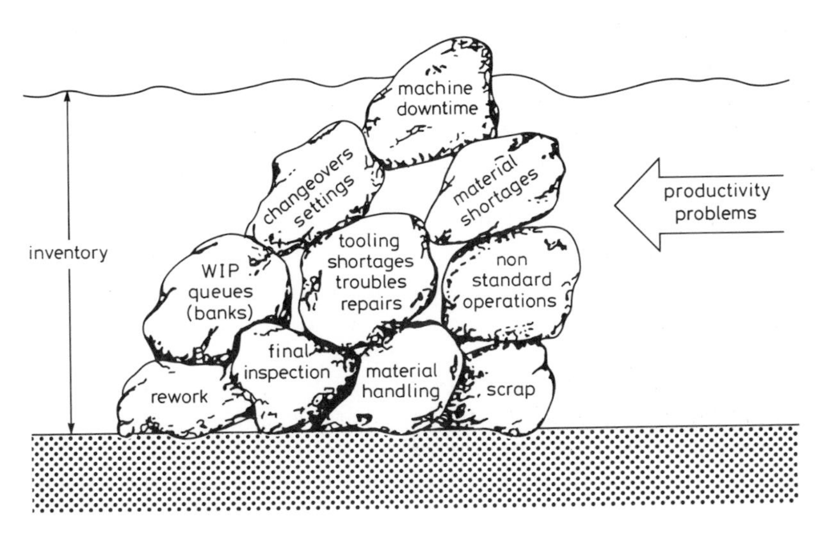

Fig. 27.3 Causes of waste

These objectives were to be achieved by:

- setting up a steering committee to oversee and implement JIT
- in-depth training of all persons directly involved with JIT (e.g. production control)
- providing JIT and TQM familiarisation programmes for all current and new employees
- training all persons in Statistical Process Control (SPC)
- setting up a planned maintenance programme
- setting up teams as and when necessary to implement JIT in various areas
- informing all that JIT is mandatory not optional
- introducing JIT and TQM to vendors
- reducing the number of suppliers and introducing co-makership
- constantly questioning why things were done in a certain manner
- adopting realistic accounting methods, using time based systems rather than cost based systems
- constant performance measurement to ensure improvement on a continual basis.

27.3 Current status

Senior and middle management as well as supervisors have all been trained in JIT and TQM. The employee familiarisation programme is continuing. The company has managed to get the message to its workforce and managers that high quality performance is best achieved by increased employee involvement and the use of processes which consistently produce parts to full specifications. Both the machinery and the operators determine the capability of each process to avoid sub-standard or scrap components, while the process control systems detect and feed back any running deviations in a manner which does not detract from either the flow rate or the quality standards.

Jabsco uses a control methodology which studies the capability of each machine or process to produce parts, combining them with statistical process control (SPC) of the finished dimensions. Having established this initial data with a large sample of parts, operating control is maintained by the use of a charting system (Fig. 27.4).

With the continuation of the extensive training programme, the steering committee was set up and the company felt it was ready to install the first JIT manufacturing cell. The crucial decision had to be made to select a component for the venture, and the choice of component was the flexible impeller pump body.

Manufacturing engineers under continuous pressure have long been

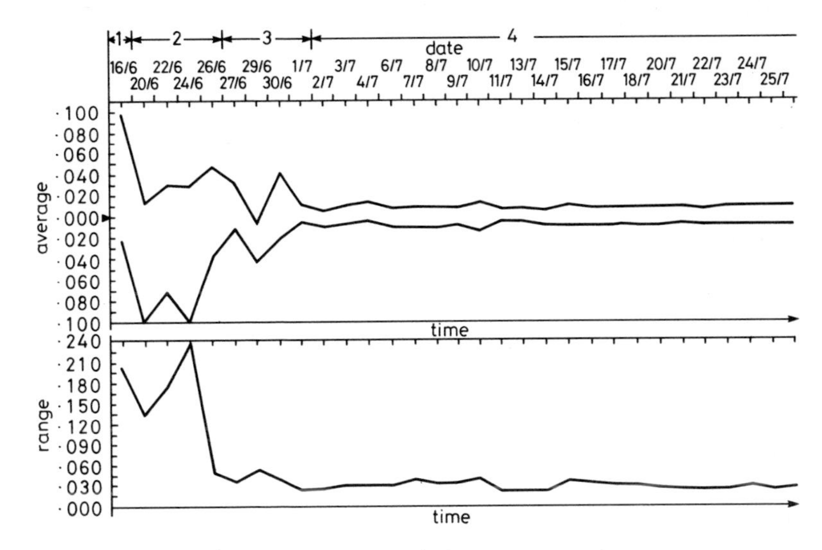

Fig. 27.4 An example of the use of statistical process control

SPC Toilet piston rod

The graph illustrates how statistical process control has been used to control the diameters of two retaining ring grooves on the piston rod for the new marine toilet.

1 Shows the results of a random sample taken from production and measured by members of a statistical process control training session.

2 Illustrates the continuous quality improvement over four working days which was made possible by major changes to the set-up. (Note: the drawing limits were also increased.)

3 This improvement has continued, failures on assembly have been eliminated and the product now meets customer expectations.

4 The control limits were set around nominal and the improvement has continued so that we now use only 40% of the original drawing limits.

aware of the need to reduce processing times, of which machining time is usually the smallest proportion (Fig. 27.2). Therefore, investment in faster machines would be counter-productive although time-saving features such as pallet loading and automatic tool changers have made significant reductions in total processing times in many industries.

Jabsco has adopted all these features over the past few years and if the installation of the body cell was on a greenfield site, the installation of the latest machining equipment could culminate in the reduction in processing times. However, like most manufacturing organisations, Jabsco had to start with existing plant and an expenditure budget which did not allow for new machines.

Using JIT philosophy, Jabsco decided to select a single component which was plagued with bottlenecks between machining operations and built them into their first cell (the impeller pump body). With machine operators who had been trained in all aspects of JIT philosophy, new procedures and the flexibility to conduct any operation necessary in the cell to maintain flow, the results have been amazing. The body cell set-up times have been reduced from five hours to one hour, and the pump body, which formerly required a 30-day processing cycle, is now ready for the next stage in one day. In addition to the dramatic reduction in processing time, the quality has been improved by 50% (of a product which was of a good quality even before employing the body cell).

With these significant gains, the company was encouraged to conduct further reductions in set-up times on other components, and has achieved 35% savings on some machining centres.

Other benefits noticed since the introduction of JIT (body cell) are:

- Instant demand for response from management. If a problem occurs it has to be resolved on the spot, whereas previously the problem would be side-stepped until time could be spared.
- Creates a direct line of communication. Production control load the cell direct rather than going through shop supervision.
- The whereabouts of the job are always known. Previously the job had to be 'hunted', but now it does not go out of the cell.
- Missed operations are no longer possible.
- Increased worker participation and satisfaction. Encourages individuals to respond to teamwork within the cell and take pride in the manufacture of right first time components.
- Has accelerated the need for standardised drawing requirements. In order to reduce set up times the cell team investigates each component going into the cell with the view of standardising all parts.

27.4 The future

The 'think big start small' tactics have provided Jabsco with the ideal launching pad for implementing the JIT philosophy primarily owing to the remarkable success of the body cell. The long term aim is not to take any action in the company until an order is received when the body (and, shortly, other components) can be ready for assembly in time to meet the customers' delivery dates even for single items ordered. One major problem has been the time taken by vendors to deliver castings, forgings and many other items in accordance with the strict schedules necessary for JIT.

To overcome these difficulties, Jabsco is now insisting that all suppliers make a firm commitment to the introduction of the appropriate quality systems (BS 5750:ISO 9000), SPC and JIT. It is not expected that the suppliers will take on board these changes immediately, but will introduce them gradually with assistance from Jabsco in training and other areas.

27.5 Acknowledgments

The authors would like to thank ITT Jabsco for giving others the opportunity to share their experiences.

Chapter 28

Integration of design and assembly planning: quality aspects

M. Richter and U. A. Seidel

Fraunhofer-Institute for Industrial Engineering (IA) Stuttgart, FRG

28.1 Introduction

What is product quality? Andreasen *et al.* (1987) put it as follows: 'The quality of a product is the customer's perception of and evaluation of the properties of the product . . .'. But he continues to show that this quality cannot be recognised until the product enters its market. Therefore, we prefer a more general definition: 'A product has to meet its specifications'.

In many companies the sales department specify the requirements regarding the functionality and the quality of a product. But unfortunately they are often not well informed about the production possibilities of their company. Already we arrive at the first problem: the specifications. The task of the designer, then, is to plan the production according to these specifications. He makes the basic determinations for production by creating or selecting technical solutions for the product functionality and quality requirements. So he is already better informed about the manufacturing possibilities than the sales people.

But the man who knows most about the current shop floor facilities, apart from the production people themselves, is the last one in the planning process — the manufacturing/assembly/production planner, whose task it is to set up production equipment and processes in order to meet desired production rates, cost limits and product quality requirements.

Finally, the production people have to produce components and ready-for-sale products corresponding to the specifications and requirements fixed in the planning phase. Of course they have a strong influence on the product quality, but they are not able to correct wrong decisions, and it is hard work to compensate for bad planning. If there are problems in production you can of course manage your product quality by test processes and mechanisms. There is no argument that tests are necessary, because neither planning nor system is error-free; but tests cost a lot of money and do not add any value to the product.

What consequences are implied by all this? Product quality is first of all a planning problem, then a production and finally a test problem, as illustrated in Fig. 28.1. And it is sensible to invest time and money into the planning work. By starting and running production under better planned conditions, your investment will receive a more rapid return and reach the break even point sooner, as shown in Fig. 28.2.

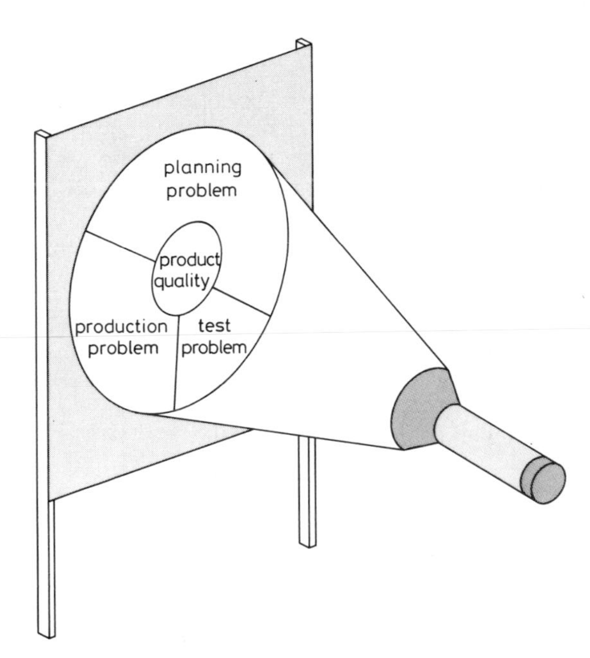

Fig. 28.1 Product and production quality — problem areas

28.2 The test problem

As mentioned above, tests are not productive (if we do not look at test documentation), but none the less are quite expensive. Test equipment is sometimes more expensive than production equipment and test time is often longer than production time. But, on the other hand, it is dangerous to run production without test processes.

In general, tests are more effective the earlier they are applied, because errors become more expensive the later they are detected, as shown in Fig. 28.3. That means that, first, planning work should be checked. During the

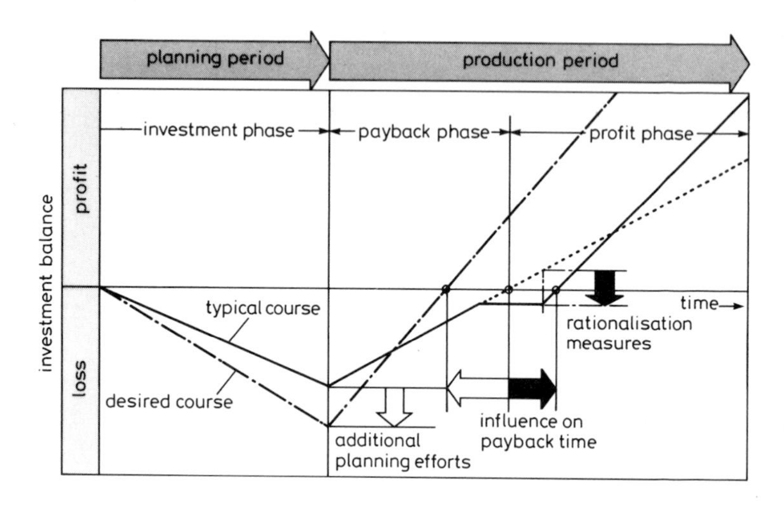

Fig. 28.2 Planning efforts and investment balance

production run the check of production equipment and processes is at least as important as the testing of product components and sale ready products.

So tests and test strategies have to be planned and integrated in production. It should be guaranteed that test results are immediately available. If products are tested one week after their production and errors are detected then, you can only choose between repair or rejects without the possibility of influencing production.

28.3 The production problem

First of all, suitable production facilities have to be set up in order to meet production rates, cost limits and quality requirements; and here we are back to planning the work. But your production equipment should be driven as efficiently as possible. Production facilities include technical equipment and, even more important, manpower resources. As far as quality and the technical aspects of new equipment are concerned, it is very important to start with test production (pilot lots) in order to detect serious problems. With regard to manpower, it is very important to have well skilled and trained workers and employees. This means that training should have happened before production starts, but also should continue during production.

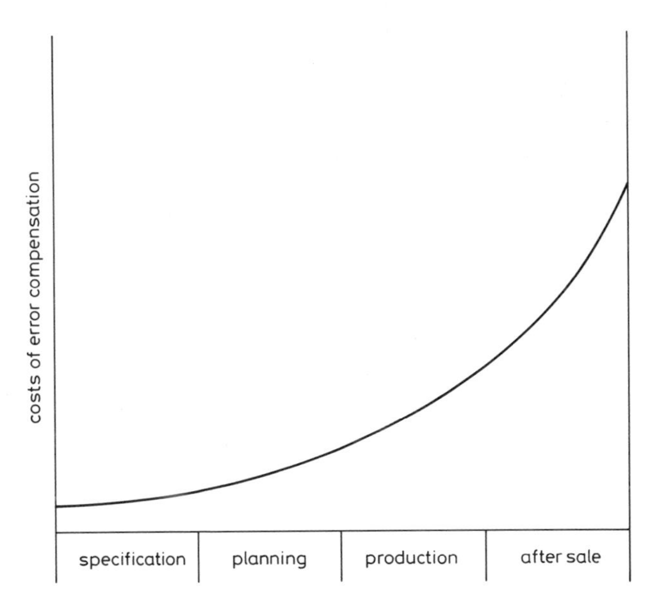

Fig. 28.3 Production and production quality — error compensation costs

28.4 The planning problem

The quality of planning results substantially depends on planning know-how and the information available. There is a lot of know-how and information in each department — so we would only have to map the required planning know-how with correct and sufficient information. That is easier said than done. There are problems of co-operation and information transfer. Departments tend to do 'closed door work' of the black box type: input — nobody knows — output.

This is one of the main reasons why our Research Institute started an international project under the EUREKA/FAMOS initiative, named Integrated Design and Assembly Planning (IDAP).

28.5 Integrated design and assembly planning

Planning work is getting more and more complex. On the other hand, planning time should become shorter. To meet these demands the typical planning procedure of today has to be changed. Planning procedure in European industry is usually sequential and departmental. Tayloristic concepts have been transferred from the direct layout to the planning work. But the essence of each Design For Assembly (DFA/DFM) methodology must be the integration of product design and process planning into one common activity (Stoll, 1988).

It is not possible to concentrate all product and production know-how of a company within one planning department. So we have to build up planning teams depending on the planning task, and support them with actual and reliable information produced with the help of integrated planning tools. Within IDAP such organisational concepts and an integrated planning system are developed for assembly applications. The philosophy of this project is shown in Fig. 28.4. The co-operation of FRG and UK research institutes and industrial companies promises to be successful.

28.6 Objectives of IDAP

The overall objective of IDAP is to design, develop and install an integrated design and assembly planning system. This system is based on the concept that product planing (i.e. design) and assembly planning need to be integrated parts of the whole CIM process, thus providing the basis for CIM realisation (Scheer, 1988). This integration has a potentially high impact on the methods and economics of design and assembly planning as well as on assembly itself. The potential for significant advances through a unified approach is high.

This concept is to lead to a system which enables designers and engineers to improve substantially 'right first time' execution of designs and assembly planning with less cost and in less time. Success should possibly be provable in terms of tangible quantitative measures, like:

- reduction in product and assembly planning time
- reduction in production lead times
- reduction in unit production costs
- reduction in product changeover time and costs
- enhanced quality assurance and product liability.

Besides the technical aspect of creating an adequate computer system support, the definition and establishment of adequate organisational structures and a supporting and highly motivating work design are of vital importance.

28.7 Designer support system

The main objective here is to design, develop and install a new generation of intelligent CAD workstations, which, besides the traditional CAD

EUREKA-FAMOS Σ! – IDAP	Philosophy	

Integration of product planning (development and design) and assembly planning (process and systems) by development of methods and computer-aided tools and new organisational concepts.

Today:

- successive planning of product and assembly
- isolated tools, some of them computer-aided
- fixed organisation structures, separated departments

Problems:

- long planning time
- many iterations in planning process
- assembly problems

IDAP:

- integrated computer-aided planning tools
- parallel processing of product and assembly planning
- new organisational concepts

Benefits:

- short planning time
- immediate feed-back of assembly planning to product design
- increasing quality of planning results
- common data basis, CIM integration possible

Fig. 28.4 IDAP philosophy

functions, will allow computer supported in-process creation, analysis and evaluation of designs. We call this system a Designer Support System (DSS). The system has to be highly integrated into the Engineering Support System and compatible with Open Systems Architecture principles. The requirements of the Engineering Support System have to be integrated into the product modeller to achieve a closer link between both support systems.

This is in contrast to most of the available CAD systems which only support the drawing and, to a limited extent, drafting process. Beyond the capabilities of well-known Assemblability Evaluation Methods (e.g.

Boothroyd & Dewhurst (1983) or Hitachi's AEM (Myakawa and Toshijiro, 1986)) which are generally carried out on completed designs, IDAP will provide procedures which will guide the designer through even the early conceptual design stages, e.g. drafting (Redford & Swift, 1980).

First attempts will be made to supply the designer with real tools for generating. The technical problems of assembly must systematically be included in the design process. It is therefore necessary to identify the points in the design process where the technical aspects and requirements of the assembly process can influence the product design. Generally, manufacturing facilities will influence each product design, but it is supposed to be left to IT techniques to provide a set of more formalised tools for taking these constraints into account. Studies will be made of how to evaluate design alternatives in terms of absolute cost instead of a less transparent and less comparable rating in some given point system.

The new system will depend heavily on a product modeller which will handle all product-oriented information, dealing with elements (e.g. components, geometry data), relations (e.g. functional relations, tolerances, assembly sequence), rules (e.g. standards) and integrity rules (Shah & Rogers, 1988; Ullman and Dietterick, 1987). This is also vital for the integration of engineering functions like process planning and tool design. In fact, assembly structure, feeding and joining are based on component relations. The model employs a hierarchical concept.

28.8 Engineering support system

The main objective is to design, develop and install a new type of Engineering Support System (ESS) which will, besides traditional CAP and CAPP functions, allow computer supported creation, analysis and evaluation of flexible assembly systems. The system has to be highly integrated into the Designer Support System and compatible with Open Systems Architecture principles.

This part of the IDAP system will support both:

- assembly process planning (cf. Amme *et al.*, 1986)
- assembly systems planning.

Support will be given to the following tasks:

- analysis and preparation of product information
- generating of assembly plans on different levels of abstraction
- mapping of product, technology, feeding, joining and assembly sequence requirements to technical solutions (assembly building blocks)

- developing and description of assembly system structure
- simulation of the production process
- technology and/or utilisation driven optimisation of task assignment in a given assembly system (including system balancing)
- generation of work plans and programs for NC system components
- production programme planning.

The computer support can take place with varying intensity depending on the planning task requirements. Emphasis should be laid on generative functions, which provide proposals and eventually degrade fail-safe to dialogue with the planner. The system should provide support at different method levels, if applicable.

The engineering methods will depend on an assembly process model, a production programme model and a production system model. Assembly plans are structured sets of assembly activities. All models will be object-orientated models and will employ a hierarchical concept.

28.9 Organisational embedding and work design

The new technologies and methodologies are regarded as having considerable impact on organisational structures, procedures and work design. This is true not only for the shop floor but also for the planning departments. In particular, we concentrate on three topics:

- *Human factors*: To create a system which is adaptable to the application environment, the user and his abilities. Guidelines on how to design satisfying work content.
- *Educational*: To provide adequate training and seminar material for the teaching of the IDAP planning philosophy and methodology.
- *Organisational/management*: To provide guidelines on how to design and build up supportive organisational structures and management procedures. One vital element will be the integration of design and engineering people to form 'Product and Production Planning Teams'.

Success could be indicated by sociometric and economic indices; e.g. work productivity, quality of the planning results or enhanced skills and qualifications of the planners.

28.10 Definition phase

FAMOS is a group of EUREKA projects which aim directly at the specific area of flexible automated assembly. IDAP is an agreed EUREKA/

FAMOS project (EU 289). The Definition Phase of IDAP has the following main objectives:

- conceptual framework and system requirement catalogue
- planning projects as first test of 'IDAP' philosophy
- definition of the objectives and detailed work plan of the main phase
- identification of potential synergy effects with respect to the results of other ESPRIT and EUREKA projects
- build up of contacts with interested European companies or institutes
- contacting IT vendors.

28.11 Conclusion

When we started IDAP we did not emphasise the quality aspects. But we know from our co-operative projects with industry that many quality problems arise from the lack of planning.

Integration in production, e.g. thinking in terms of product families and manufacturing centres, calls for integration in planning. So we expect the IDAP system and organisation to provide benefits to product and production quality by improving planning quality. One final aspect should be considered: in contrast to Taylor's time, today's employees are usually well skilled. Their motivation will rise by solving integrated tasks in teamwork, and highly motivated people produce better quality.

Chapter 29

The continuous analysis of equipment reliability

A. J. Watkins and D. J. Leech

University College of Swansea, UK

29.1 Introduction

The owner and operator of a bank of equipment need to consider the lifetime distribution of individual items of equipment if they wish:

(*a*) to discuss whether specified performance standards are being attained
(*b*) to determine overall availability of equipment, and
(*c*) to determine spares-holding and maintenance policies.

The bank or population of equipment can be viewed as the result of a birth (as items enter service) and death (as items fail) process. Consequently, the lifetime distribution of equipment may be estimated with this population still in a transient state, but with some information on arbitrarily censored lives available. We consider some aspects of, and procedures for, the continuous monitoring and analysis of such reliability data.

Our discussion centres on a convenient and widely used class of lifetime distribution, the two parameter Weibull distribution with cumulative distribution function

$$F(z; \beta, \theta) = 1 - \exp(-(z/\theta)^{\beta}), \ z > 0 \qquad (29.1)$$

for positive parameters β and θ, and we initially assume that lives of items of equipment follow this distribution for fixed and unknown parameter values. To begin to assess equipment reliability, we first need to estimate these parameter values from our observed data, and in the next section we consider the data we assume to be available to us.

29.2 Data available

We observe the population from its origin, and consider the data available after a certain time period has elapsed. We assume that, from a total of N (>0) items entered into service in this time period, we have recorded m (>0) failures with times to failure

$$f_1, f_2, .. f_m,$$

so that, at the end of the time period, $n = N - m$ items are still in service, with lifetimes

$$c_1 c_2, .., c_n,$$

which we now regard as *censored* observations.

Example
The operator of a fleet of vehicles begins to use a new sub-assembly, with an advertised average lifetime of three years. In the first year of operation $N = 60$ new sub-assemblies are fitted, and at the end of this first year $m = 6$ have failed, and $n = 54$ are still working with (censored) usages.

We note that, in general, the restriction m >0 ensures that we know some times of failure. Thus we must wait until at least one item has failed in service, and the larger m is, the more accurate our parameter estimates are. However, the benefits of allowing m to increase may have to be balanced by the need to challenge, within a guarantee period, the specification of lifetime distribution of items provided by the manufacturer or supplier. Hence, it should also be noted that, when the cutoff period is small relative to the average lifetime of equipment (as above), then the failure data comprises early failures only, and, to obtain a truer picture of lifetime distribution, this data must be augmented by information on censored lives. In terms of our example, the times to failure are useful only when we bear in mind that the majority of sub-assemblies have yet to fail.

We now wish to estimate β and θ in eqn. 29.1, on the basis of such information. If we then wish to use these estimates in assessing equipment reliability — perhaps in terms of objectives (*a*)—(*c*) above — we should also gauge the precision in these estimates. Kalbfleisch (1979) discusses a likelihood analysis, and the role of relative likelihood in this analysis; in the next Section we discuss how to obtain the maximum likelihood estimators of β and θ.

We finally note that, for the birth and death process described above, we do not have either a type I or type II censoring regime, but instead observe data subject to arbitrary censoring. Thus, it is entirely possible for there to be censored lives both smaller and greater than all recorded times to failure.

29.3 Location of the maximum likelihood estimators

From eqn. 29.1, the log-likelihood of our data is, apart from an additive constant,

$$L(\beta,\theta) = m \log(\beta\theta^{-\beta}) + (\beta - 1)s_e - \theta^{-\beta}s(\beta) \qquad (29.2)$$

where

$$s_e = \sum_{i=1}^{m} \log(f_i)$$

and

$$s(\beta) = \sum_{i=1}^{m} f_i^{\beta} + \sum_{j=1}^{n} c_j^{\beta}$$

We can attempt to maximise eqn. 29.2 with respect to β and θ jointly, but, as is directly observed by Kalbfleisch, (1979), the value of θ which maximises L for fixed β is

$$\theta(\beta) = (s(\beta)/m)^{1/\beta} \qquad (29.3)$$

and if we now substitute eqn. 29.3 into eqn. 29.2 we obtain a *profile* log-likelihood, which, apart from an additive constant, is

$$Lp(\beta) = m \log(\beta/s(\beta)) + (\beta - 1)\, s_e \qquad (29.4)$$

This is a univariate function of β, and can be displayed rather more easily than eqn. 29.2

From such a diagram we can locate $\hat{\beta}$, the value of β which maximises eqn. 29.4, and which is thus our maximum likelihood estimator of β. Once $\hat{\beta}$ is known, $\hat{\theta}$ is, from eqn. 29.3,

$$(s(\hat{\beta})/m)^{1/\hat{\beta}}$$

Fig. 29.1 shows the profile log-likelihood eqn. 29.4 for the above example. (The actual data is given in Appendix 29.7). From this, we see that

$$\hat{\beta} = 1{\cdot}26 \text{ (to two decimal places)}$$

for this set of data, with a corresponding value (in days) of

$$\hat{\theta} = 1140{\cdot}1 \text{ (to one decimal place)}$$

In practice, we need not draw the function eqn. 29.4, but may attempt to locate its maximising value of β numerically. For instance, Newton's iterative method, which attempts to locate a zero of the first derivative (with respect to β) of Lp (β), is easily implemented — even on a spreadsheet. The standard form of this method requires the first two derivatives of $Lp(\beta)$, and an initial estimate of the maximising value. Previous analyses may suggest a suitable initial estimate, but in the absence of any prior information, the starting value $\beta = 1$ has proved convenient in practice, and ensures some numerical stability. Calculation of the maximum likelihood estimates in this way is relatively quick, so that these parameter estimates can be recalculated each time the database of equipment lives and failures is updated.

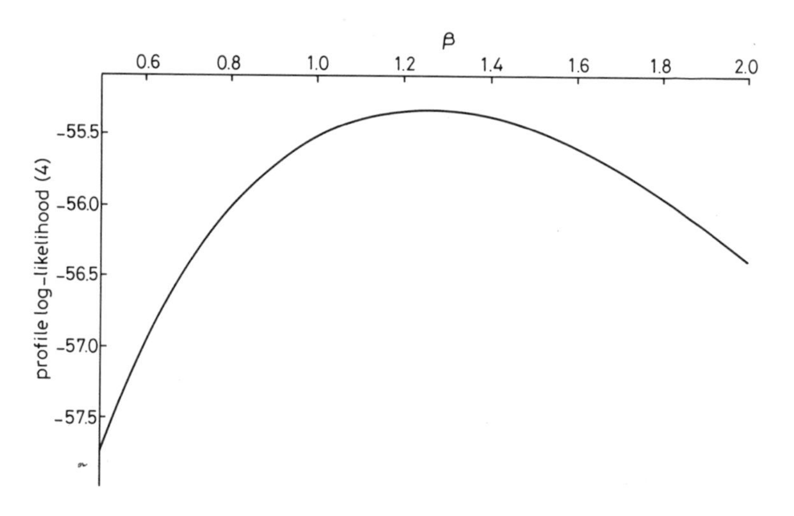

Fig. 29.1 Profile of log-likelihood [eqn.(29.4)] for the example data ($m = 6$, $n = 54$)

29.4 The role of relative likelihood

The maximum likelihood estimates are, by definition, those values of β and θ most in accordance with the observed data, and, as such, provide us with a statistically consistent guide to the true, unknown values of β and θ. This implicitly acknowledges that there will be some error in these estimates, and, as already noted, it is of considerable interest and importance to attempt to quantify such errors.

For instance, we can appeal to standard large sample theory and obtain approximations to the distributions of these estimators. This allows us to consider 'confidence' contours, which contain, with specified probability, the true value of (β,θ). A second approach is to use the idea of relative likelihood, as summarised by Kalbfleisch (1979). This then allows us to draw contours of equal likelihood, within which all points (β,θ) are, on the basis of our observed data, at least the specified percentage as likely as the maximum likelihood estimator. These approaches are not independent, but, in an intuitive sense, complementary. For instance, we interpret points within a 95% confidence contour in the same way as we interpret points inside a 5% relative likelihood contour. Somewhat more formally, we can employ the connection between relative likelihood and likelihood ratio test statistics to make the nature of this dependence clear for large samples.

Here, we adopt the approach of relative likelihood, which is valid for all admissible sample sizes, and requires no further assumptions. In the next Section we outline the principles for producing an accurate contour map of the relative likelihood function, based on eqn. 29.1 and our observed data.

29.5 Contouring the relative likelihood function

Given the observed data, the maximum likelihood estimates, and a set of values

$$p_1, p_2, \ldots \text{ with } 0 < p_1 < p_2 < \ldots < 1$$

for which contours on the relative likelihood surface are required, the algorithm has two stages:

(*a*) To search the – plane for a rectangle within which the contour corresponding to p_1 will lie.

(*b*) To draw this contour by first finding, and then joining, a number of points on it.

This second stage is repeated for each contour.

29.5.1 Finding the drawing area

We first consider the relative likelihood for a series of fractions, and multiples, of β. For given β, the value of θ which maximises the relative likelihood function is, from eqn. 29.3,

$$(s(\beta)/m)^{1/\beta}$$

thus, we can calculate the maximum relative likelihood for β, and hence find the minimum and maximum values of β that we need to consider.

The same idea is used for θ; that is, we attempt to find the maximum relative likelihood for a series of fractions, and then multiples, of the maximum likelihood estimate of θ. This requires finding that β which maximises the relative likelihood function for given θ; this value must be found numerically, as no analytical expression exists.

This search indirectly introduces the transformation

$$\beta = x\hat{\beta} \text{ and } \theta = y\hat{\theta}$$

and we will find it convenient to continue to work in terms of the new variables x and y thus defined. This transformation introduces a measure of numerical stability, as x and y are usually 0(1), and hence approximately equal at all points we wish to consider. The transformation also allows for θ — and $\hat{\theta}$ — varying considerably between data sets. We may note, firstly, that equivalent transformations are both possible and desirable for alternative parmeterisations of eqn. 29.1, and, secondly, that other, similar, transformations can also be considered; for instance, once the drawing area is defined, we can map it to a unit square.

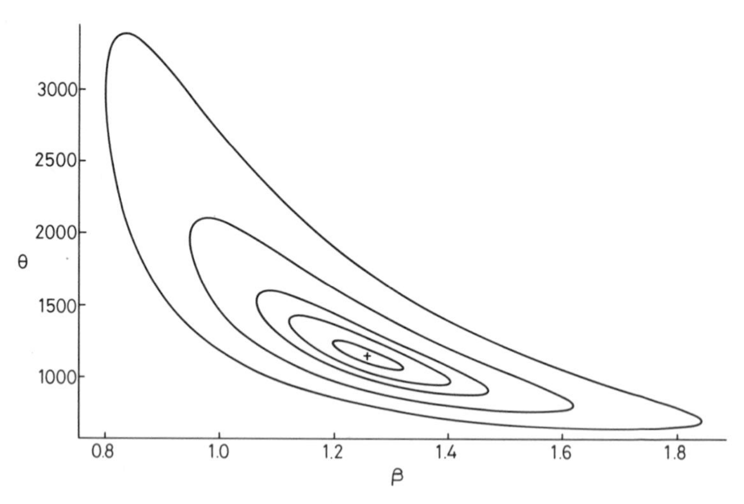

Fig. 29.2 Contour plot of the relative likelihood for the example data ($m = 6$, $n = 54$)

29.5.2 Drawing contours

Here, there are two points to cover: first, how to find an initial point on a contour, and then, secondly, how to move around a contour.

To find an initial point on a contour, we take $x = 1$, and search (again, numerically) for a value y such that the relative likelihood at

$$\beta = x\hat{\beta} \text{ and } \theta = y\hat{\theta}$$

has the required value.

To move along the contour, we first calculate the gradient of the tangent to the contour at this initial point on the contour; this requires the derivatives of the relative likelihood — or equivalently, the log-likelihood — with respect to both new variables x and y. Our first estimate of a new point is then found by moving along this tangent to

$$(x^*, y^*)$$

We now take x^* as fixed and find (once more, numerically) y such that the relative likelihood at

$$\beta^* = x^*\hat{\beta} \text{ and } \theta = y\hat{\theta}$$

has the required value. Having located this y, we now have our second point on the contour. We iterate the above, and so move around the contour. If, at certain points on the contour, we find we cannot locate a suitable value of y for fixed x, then we may adopt the computationally more expensive alternative of taking y as fixed and searching for a corresponding value of x; this may be necessary at the extremes (with respect to β) of the contour.

The above is a summary of the algorithm considered in detail in Watkins and Leech (1988), who give some relevant formulae and discuss additional theoretical considerations. We may, however, note here that a FORTRAN implementation of the algorithm has approximately 150 executable statements, including calls to SIMPLEPLOT routines to perform the actual drawing. This implementation produces Fig. 29.2 for the example data discussed above. The five contours correspond to relative likelihood levels of 0·5, 0·75, 0·9, 0·95 and 0·99, and the overall shape and scale of the contour plot now gives us some indication of the precision in our estimates.

We also obtain some idea of the accuracy in functions of the maximum likelihood estimates; for example, in the *mean* time to failure, which we estimate by

$$\hat{\theta}\,\Gamma(1 + \hat{\beta}^{-1}) \quad (29.5)$$

where $\Gamma(.)$ is the usual gamma function. For our example, this quantity eqn. 29.5 is, to one decimal place, 1060·4 (days), but, equally importantly, we now see from Fig. 29.2 that there is no real reason to suspect that the new sub-assembly has an average lifetime of less than three years. Hence, the operator of the fleet can be satisfied that the new sub-assembly is presently attaining the advertised performance level.

With its repeated use of numerical methods, it can be seen that the contouring algorithm is relatively demanding in terms of computational effort required, and, although our implementation has been successfully used with field data on vehicle component lives with $N = 500$, we would therefore envisage using this program comparatively rarely. The following factors directly influence the total computational cost incurred:

(*a*) the amount of data collected
(*b*) the number of contours to be drawn
(*c*) the accuracy in the numerical methods used — that is, the accuracy with which we find points on contours
(*d*) the number of points we find on contours.

Thus, the total cost in part reflects our desire for smooth and accurate contours. The first three factors above are fairly self-explanatory, but in connection with the fourth, we note that the difference between the first and last contour values may also need to be taken into account. This is because we will also want the final contour — which may now be based on a relatively small number of points — to be reasonably smooth.

We finally remark that, as the amount of data at our disposal increases, so the contours will contract about the maximum likelihood estimate, owing to the additional information available. However, the first contour may always occupy most of the drawing area, and we must then rely on markings along axes to gauge precision. To illustrate this point, we return to our earlier example, and consider the data available after three years of using the new sub-assembly. Now, a total of $N = 182$ items have been entered into a service, with $m = 55$ failures recorded and $n = 127$ items still in service at the time of this second census. Fig. 29.3 shows the profile log-likelihood eqn. 29.4 for this set of observed data, which is again given in the Appendix 29.7. We have

$$\hat{\beta} = 1{\cdot}30 \text{ (to two decimal places)}$$

with a corresponding value (in days) of

$$\hat{\theta} = 1190.1 \text{ (to one decimal place)}$$

and the contour plot in Figure 29.4 again suggests that the advertised lifetime is being attained. In Fig. 29.5 we have plotted the contours from Fig. 29.4 on the same scales as for Fig. 29.2, and on comparing Fig. 29.5

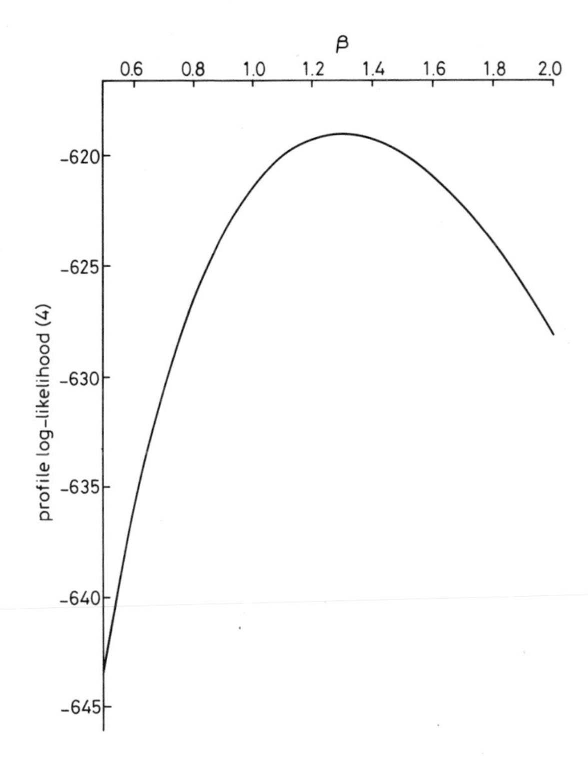

Fig. 29.3 Profile of log-likelihood [eqn.(29.4)] for the example data (m = 55, n = 127)

with Fig. 29.2 we see how these contours have shrunk about the (new) maximum likelihood estimator. Thus, with an extra two years' data, we can be correspondingly more confident about the precision in our parameter estimates, and hence in our estimate of average lifetime in service of items.

29.6 Discussion

Although we have emphasised the assessment of lifetime distribution of items — and hence their reliability — at a relatively early stage of the life of the bank or fleet of equipment, it is hopefully clear from our discussion above that the procedures and algorithms outlined are applicable in the analysis of more general forms of reliability data. In particular, such analyses are possible at all stages of the life of the fleet; that is, they permit

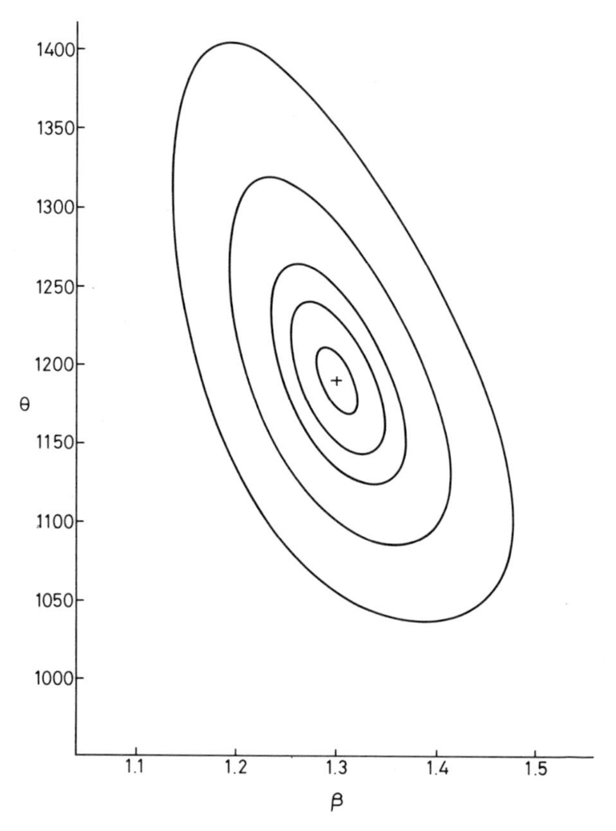

Fig. 29.4 Contour plot of relative likelihood for the example data ($m = 55$, $n = 127$)

the continuous monitoring of lifetime distribution. However, as previously noted, an early assessment of reliability is necessary if the owner or user of equipment wishes to challenge the specifications of manufacturer or supplier within a guarantee period.

We have assumed that observations on all items entering service are available at the end of the cutoff period. This need not be so; for example, it is possible that only data on failures will automatically come to our attention, and that it is then necessary to conduct a survey or census of those items still in service. This then raises the possibility of obtaining less than complete information; for instance, if the cost of a census is prohibitive, we may use a sample of items still in service. Alternatively, we may only know the number of censored items, or their *average* (censored) life, but not the individual times in service. Suzuki (1985) has considered the effect of sampling on the maximum likelihood estimator — which technically, is then transformed to a maximum *pseudo*-likelihood

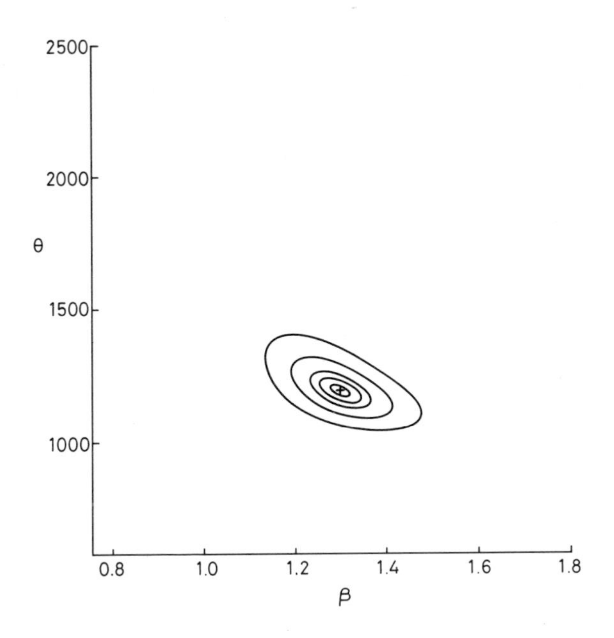

Fig. 29.5 The contours in Fig. 29.4 on the axis in Fig. 29.2

estimator, but it remains to fully assess the impact that different types of data imperfections can have on our analysis.

We can consider further refinements to the situation described above. Firstly, the lifetime distribution eqn. 29.1 may change over time, as use in service suggests modifications and enhancements to newer items. This then suggests that we should analyse data in stages rather than as a whole. Secondly, we have assumed that items which fail, are repaired and then re-entered into service, are unaffected by these failures and repairs; that is, repairs restore items to original condition. This is possible, but clearly need not be so, and it remains to assess the effect of different repair policies and effects on our data and analysis.

Finally, we note that we have assumed the Weibull distribution eqn. 29.1 as a basis of our discussion, and, although many of our points are more generally applicable, we should note that the outlined procedures for continuous monitoring of reliability data can be augmented by goodness-of-fit assessments.

Work on these, and other, refinements is continuing, and will be reported elsewhere.

29.7 Appendix

In this Appendix we list the data for the vehicle sub-assembly example in the above discussion. This then facilitates comparison with other algorithms and techniques. All times to failure and times in service are in days.

First year's data

The six times to failure are:

12 88 130 180 216 220,

and the 54 censored times in service are

10	12	16	24	32	36	42	44	54	60	70
72	78	80	90	96	100	108	114	115	126	141
144	147	156	162	166	176	180	185	199	201	209
215	220	227	233	236	263	275	284	287	294	299
305	311	316	323	330	335	345	347	351	359	

First three years' data

The 55 times to failure are:

12	47	88	98	107	107	108	114	121	130	136
139	140	153	180	208	215	216	220	237	268	269
280	288	289	306	317	324	330	331	381	384	404
408	410	413	427	439	451	468	494	515	535	599
635	638	652	668	689	701	742	774	777	834	892

and the 127 censored times in service are:

9	12	19	25	30	36	40	51	55	60	67
71	79	84	91	96	100	108	115	120	124	132
144	156	156	162	170	174	184	190	199	202	209
221	226	234	236	245	251	256	262	270	273	281
288	293	310	317	329	329	334	341	345	353	365
377	384	397	401	405	415	422	425	431	434	445
449	455	466	479	491	507	515	535	538	545	550
556	563	574	580	599	606	609	627	628	648	650
659	663	671	676	690	700	712	730	736	745	760
801	805	820	826	868	875	880	890	892	906	910
915	925	930	941	945	950	959	969	995	1002	
1010	1035	1049	1053	1060	1073	1083				

Chapter 30

Diagnostic systems for FMS

F. Gohari and U. S. Bititci

University of Strathclyde UK

30.1 Introduction

Modernisation of manufacturing plant often means installation of automatic equipment that machines and assembles parts. The benefits expected from automation are reduced labour costs, consistent quality and higher rates of production. The twentieth century has seen the introduction of many Flexible Manufacturing Systems (FMSs). The earlier systems consisted of a limited number of machine tools, typically two or three connected together via a simple material handling system. As technology pushed forward, data processing and control hardware costs decreased and larger scale and more complex systems were implemented. Today a small system, such as described above, may cost as much as £2.5 million. Larger systems requiring an investment of £10 million to £15 million are not uncommon. For these levels of investment to be justifiable it is necessary to maximise the availability of such systems. However disciplined and comprehensive planned maintenance procedures are, failure of such complex systems cannot be eliminated. To minimise the downtime associated with these failures, diagnostic systems, capable of identifying possible causes of failures and prompting on appropriate corrective action, are required.

Diagnostic and maintenance systems may exist in various organisation levels. These being Plant, Centre, Cell, and Sub-Station levels. Fig. 30.1 shows various diagnostic and maintenance systems available across these levels where the need for real-time diagnostic help becomes increasingly critical at the lower levels. This is because operations at plant and centre level are mostly planning types whereas operations at cell and station levels are carried out in real time.

There are a number of plant maintenance software packages available which would readily address the need at plant/centre levels. These systems are typically used for logging breakdowns, repair and maintenance procedures as well as aiding in the generation of planned maintenance policies. Examples of such systems are RAPIER, MAINSAVER, MAINPAC, MAINMAN, MAXIMO, COMAC etc.

At station level where a station is a CNC machine tool, robot or similarly sophisticated equipment comprehensive diagnostic facilities are available. Because in this category mainly standard equipment is used the diagnostic capabilities tend to be well proven and rarely require further modifications and enhancements. Some equipment suppliers, such as Siemens, provide the necessary information and training to allow full utilisation of the diagnostics facilities built in to their controllers. For example, the Siemens 850M controller is capable of diagnosing faults such as:

- part program fault
- tool offset parameter
- speed conditions at software limit
- PLC to CPU communications fault
- NC CPU stop
- PLC CPU stop

As various standard and customised workstations are integrated and made to work together under one control system manufacturing cells are formed. Owing to the diversity of applications, at cell level, provision of standard diagnostic systems have not been feasible. This fact is backed by a recent survey of cell management systems carried out by Bititci (1988)a. Owing both to the difficulty associated with providing generic/standard solutions for specific applications at cell level and the cost associated with purpose-built systems many users in effect steered round this problem.

As a result of experience gained with implementing a number of manufacturing cells, a recent publication by Bititci and Ross (1988b) identifies the features expected of cell level diagnostic systems as follows:

- integrate in real-time and at cell level
- provide information on all possible faults and stoppages
- analyse causes and prompt on corrective action
- allow Just In Time maintenance strategies
- assist in the generation of planned maintenance policies
- simple to understand by operators and maintenance personnel
- flexible for future expansion
- affordable.

Early FMS applications were associated with centralised computer systems responsible for the control and monitoring of all cell activities. Recent years have seen the development of distributed hierarchical control systems consisting of number of mini/micro computers and programmable logic controllers (PLCs) working together through local area networks (LANs). As PLCs with high processing, communication, colour graphics, maths computation and data storage capabilities became available a large number of flexible manufacturing cells are now managed by a combination of PLCs and microcomputers.

The remainder of the paper discusses various PLC based diagnostic strategies that may be available with reference to a case study, i.e. Connecting Rod (con-rod) Line FMS in Cummins Engines Co., Shotts, UK.

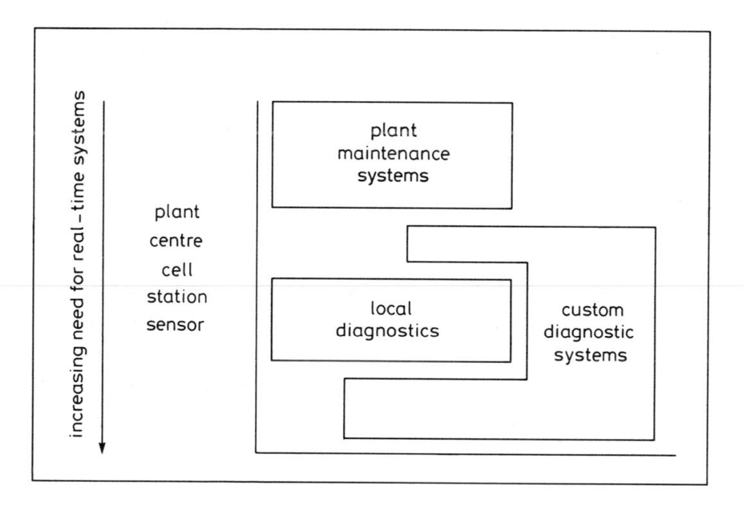

Fig. 30.1 State of the art in diagnostic and maintenance systems

30.2 Cummins Con-Rod Line FMS

Cummins Engine Co. Ltd. are currently operating a multi-cell FMS for machining and assembly of connecting rods used in diesel engines.

The forged and unmachined con-rods are introduced in to the system via a crack detect machine where rods are either rejected or accepted in to the system. Essentially, cells 1 and 2 are machining cells, cells 3 to 6 are robotic semifinishing, finishing, fitting and assembly cells, and cells 7 to 9 are currently manual cells for balancing, final inspection, demagnetising and packing.

A considerable amount of work has been carried out to provide diagnostic capability on a number of cells; however, for the purposes of this paper, the case study has been limited to the diagnostic system in cell 6 to enable detail discussion.

30.2.1 Cell 6

This cell carries out the lock-slot milling operations on the rod and the cap and the deburring operations on the joint and the pin-end faces. The detail configuration of cell 6 is shown in Fig. 30.2. It consists of an ASEA IRB60 robot for handling the con-rods between seven stations, one of which is a robotic deburring station which employs an ASEA IRB6 robot.

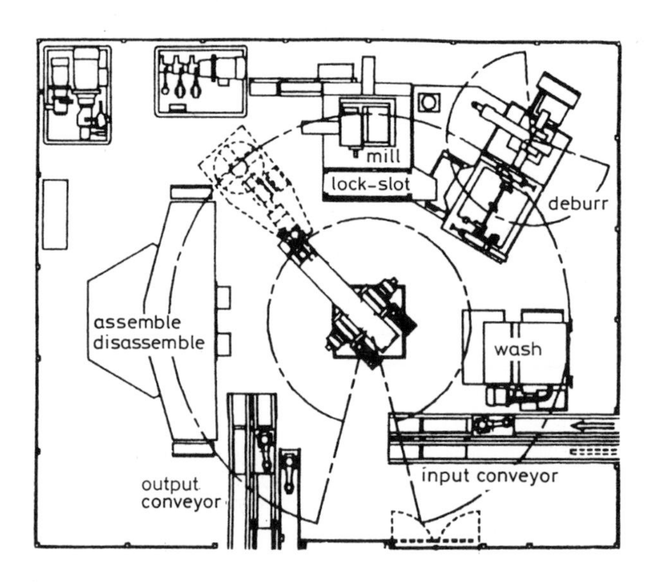

Fig. 30.2 Schematic layout of cell 6.

The rod and the cap arrive in to the cell on a pallet on the input conveyor. The IRB60 robot transports the rod and cap assembly to station 1 (disassembly unit) where the cap and the rod are separated. The robot then transports the separated parts to station 2 (lock-slot milling machine) where the lock slots are milled. The components are then placed on to the deburr unit (station 3 - IRB6 Robot). The deburred components are then transferred to the wash unit (station 4) and then to station 5 where the rod and the cap are assembled together. The finished workpiece is then placed on to a pallet on the output conveyor for transportation to cell 7.

30.3 Alternative PLC-based diagnostic strategies

In this Section, three alternative PLC-based approaches to cell diagnostics are discussed:

- one-shot method
- counter method
- parallel method.

Whatever the method used for any diagnostic system it will be essential to categorise potential faults under four levels:

Level 1: Faults which cause the cell to stop immediately (ie. stations are stopped at mid-cycle); e.g. air or hydraulic failure, guard open when not requested
Level 2: Faults which cause the cell to stop once all stations complete their current cycle; e.g. robot failed, coolant or tube overload
Level 3: Faults which result in failure of a station. The remainder of the cell carries on as normal. At the end of cycle the cell mode changes from automatic to manual. When the fault is cleared the cycle needs to be restarted; e.g. part not present, clamps not on
Level 4: Faults which result in cycle freeze. When the fault is cleared the cycle resumes from the point of interruption; e.g. external faults.

All forementioned methods require a system to be implemented which is capable of detecting the occurence of various faults. These errors may be monitored by employing various timers at different levels. These timers may be used to monitor cell, station or operation cycle times. When calculating the cycle time to be used, it may be necessary to extend the timer constants in order to prevent false alarms.

30.3.1 One-shot method

This method employs a function block (FB) program containing all the major errors, which is being scanned cyclically. The major errors take priority over any other error. The remainder of the errors may be monitored using the cell cycle timer technique. If, as a result of a non-major error an operation has not been carried out, then the actual cycle time for the cell would exceed the programmed time. The PLC would then be instructed to go through the function block softward for each station until an error is detected. The relevant function block will be scanned and the error would be identified and displayed. On a cycle without any faults, the timer would be reset with the receipt of the cycle complete flag. Fig. 30.3 shows the logical operation of the described system.

This method deals with one error at a time. The PLC would not look at any other errors until the one that is being displayed is attended to and cleared. To enable this method to function effectively the control program on the PLC should be written in an orderly manner to allow sequential checking of all the flags used.

When an error occurs, the PLC scans the diagnostic program and

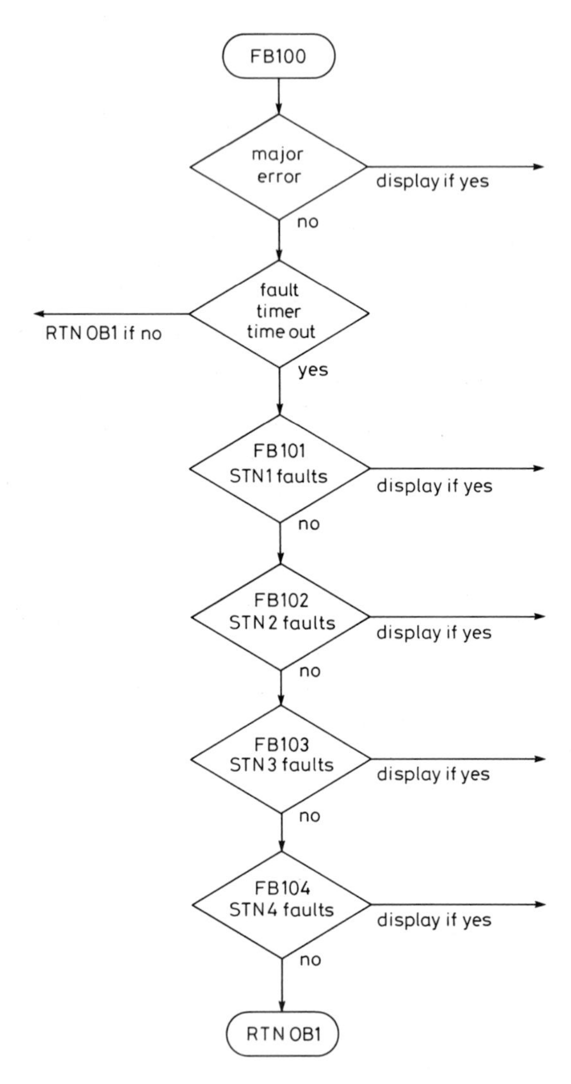

Fig. 30.3 Logistics of the counter method.

catches the signal that has not been made, thus identifying the error; on detection of a particular error the appropriate text message may be displayed on the screen prompting the operator.

In this method PLC drops out of the continuous cycle, but it allows the various stations (machines) to continue their cyclic operation. This is useful since, when the breakdown has been fixed, the operation of the particular station could be completed using a maintenance box, and the next cycle could be started without having to go onto manual cycle.

The advantages to be gained through the application of this method are:

- If the cell control routines are complex and there are memory constraints, this method uses up less memory than alternatives.
- Additional PLC scan time is minimised.
- Easy to understand by operators and maintenance personnel.

The main drawback of this method is encountered in the case of external faults, such as conveyor errors. In this case, the robot goes through the cycle and the timer which checks the cycle time would be reset without registering an error since the conveyors are controlled by separate PLCs. To avoid this, say there is a situation of a full pallet not arriving on the input conveyor. The robot will go to pick up the component; therefore the robot-in-position signal is present but the pallet-ready signal is not. Here the cycle timer would be reset but a second timer (which monitors the external faults) would be started using the signals mentioned above and the fault PALLET NOT PRESENT ON INPUT CONVEYOR would be displayed on the screen.

30.3.2 Counter method

This is an alternative method of developing diagnostics whereby counters are used to locate breakdowns. There would be different counters and data blocks allocated to each station. The different activities that take place within a station cycle, for all the stations, would be noted. For example, the wash station in cell 6 has the following moves:

(i) Load component.
(ii) Close lid.
(iii) Wash.
(iv) Stop wash.
(v) Air blast.
(vi) Stop air blast.
(vii) Open lid.
(viii) Remove component.

The counter which is allocated to this station is reset to 0 before the station starts its operation. Each of the different activities within a station cycle would be designed to generate a pulse in its particular counter. Hence, as the component is being loaded on the wash unit, a pulse is sent in to the counter which would increment the counter. During air blast, for example, the counter would have the value 5.

When an error occurs in the wash station, say the lid has not closed, the counter number 2 is transferred into the allocated Data Block (say DB100). In DB 100, Data Word 2 (DW 2), bit 2 would be set. This would then be transferred.

In this method as well as the complete cycle timer there is also a need for station cycle timers; the reason being if it is in manual mode, the cycle timer would obviously time out. The second timers used would then identify that the stations have reached end of cycle within the specified time and there are no errors (Fig. 30.4).

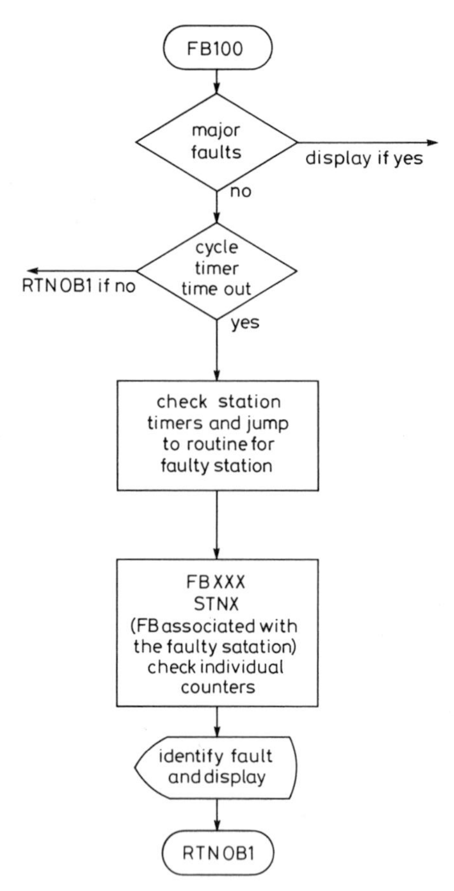

Fig. 30.4 Logistics of the counter method

When a fault occurs as in the one shot method the cycle timer times out. The PLC then looks at the counters; all would have different numbers, but assuming there is only one error, the rest of the counters would be at the number representing the last move. The diagnostic software will be written in such a way as to recognise that the fault is at the station where end of cycle has not been reached. This method allows the detection of more than one simultaneous error.

Here the diagnostics program associated with a loaded station should be scanned only. This is because, say there is a case where a component has been removed from station 2, this station immediately requests a cycle load. The robot then moves towards station 3 in order to pick up the component and load station 2. If the robot fails to load station 2, then the following conditions would be present:

- Robot tripped out.
- Station 2 empty.
- Station 3 empty.

In this case there are three different statements, but the only valid condition would be that the robot has tripped out. With the diagnostics program being scanned, when the stations have a component this problem would be avoided.

The advantages to be gained using this technique are similar to the one-shot method, with the additional advantage of being able to detect and display more than one error.

The major disadvantages of this method are:

- If cycle operation varies from one component to another then a separate dedicated diagnostic routine has to be written for the stations where the number of moves has changed.
- If there is a machine in which multiple slides move simultaneously, it will be difficult to generate a signal to identify slide movements.

30.3.3 Parallel method

This is probably the easiest method of diagnostics but not necessarily the best. In this method the diagnostics program is written in such a way as to by-pass the actual control software which is supplied by the vendors. Here, as with the other methods of diagnostics, the programmer must have a detailed understanding of the cell operation before attempting to write the software. A discussion with the operators can give valuable information about not just the cell operation but also the different faults that can occur.

A list of all the sensors and actuators used in all the various stations would have to be obtained. This could be obtained from the electrical diagrams of the cell, or an up-to-date set of symbolic addressing of the cell control program would give the programmer the same information. These signals would be used to monitor the state of different operations and form the basis of the diagnostics program. A list of all the possible errors that could occur would have to be prepared to enable the programmer to write the diagnostic routines. There are various signals from sensors and actuators that would be present for each of the different activities associated with the cycle operation of a station. These signals would be fed into individual timers and the time constant would be set according to the duration of each move. If an unwanted set of conditions exists for more than the specified period of time, then the output of the individual timer triggers a text message for transmission to the display unit.

In the majority of the cases an actuator is used to energise a particular move. When this operation is finished there would be a sensor which would identify that the move is complete. These two signals, which give the status of the mechanical move, would be fed into a timer with a preset time constant. If this move takes more than the specified amount of time a bit would be set in the appropriate data block which would in turn activate the error message display.

The diagnostics routine for each station would be written in a separate program block. This would result in an organised software as well as allowing the maintenance personnel to know exactly which program block to interrogate when an error occurs in a particular station. There are various progress blocks which contain the diagnostics program for all the stations, as shown in Fig. 30.5. The diagnostics software in this method is scanned cyclically, regardless of the type of errors. The main advantages to be gained through the implementation of this method are:

- Cyclic scanning of all faults.
- Easy to implement.
- Easily understood by the operators and maintenance personnel.
- Debugging possible in any mode.
- Multiple error detection possible.

The disadvantages are:

- Takes up too much memory space.
- Increase scan time considerably.
- Too many timers and counters would have to be used.

30.4 Operator interface

The various methods discussed above outline how a PLC may be logically configured to detect faults which can cause stoppages at various levels. However, this information needs to be communicated to the operators efficiently and effectively, and the operator should be able to interrogate the system to identify the exact cause and location of the fault.

A graphics package can be used in which drawings are designed and viewed. A separate graphics page can be set up for each station, which would include a drawing of the station with all the sensors and actuators

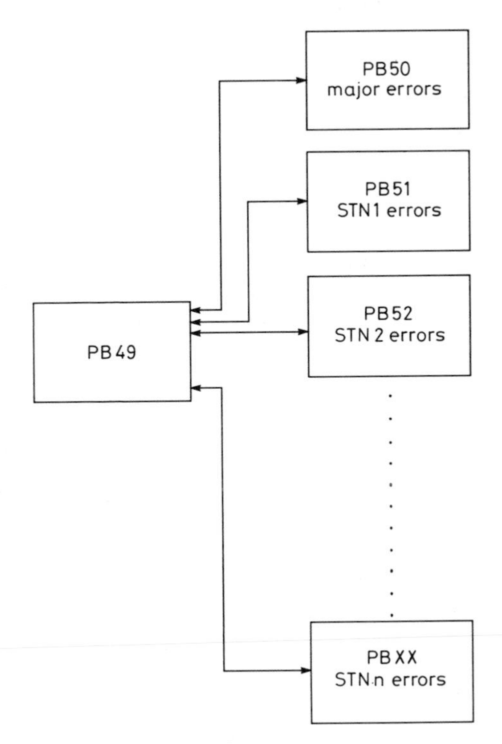

Fig. 30.5 Logistics of the parallel method.

indicated. Function keys may be provided to allow the operator to move from one page to another. If a sensor fails, then on the drawing of the relevant station it could be shown with a flashing colour. In this way, the operator could locate the fault immediately and use the function keys to call up the relevant explanatory text. Experience has shown that most of the faults are mechanical; faulty electrical devices and software take up a very low percentage of the total breakdowns. With a mechanical fault there would be a number of sensors and actuators associated which have failed to set. The station drawing would locate the area where the fault has occurred and then the operator sees the sensors and actuators that are not made and checks if:

- the sensor is faulty
- the actuator is faulty
- the mechanical part is faulty.

For the con-rod line FMS this type of approach has been adopted in order to provide an effective interface with the operator as well as the maintenance personnel.

30.5 Diagnostics in cell 6

As briefly discussed before, cell 6 primarily performs the machining of the lock slots and the deburring operation. This cell comprises of seven stations, i.e. disassembly, assembly, lock-slot mill, deburr, wash, handling robot, input and output conveyors. The diagnostics for this cell was one of the first to be implemented on the con-rod line at Cummins. The parallel method, being simply understood, maintained and developed, was employed as there were no problems regarding the scan time, memory size, timer and counter availability.

An analysis of the exact sequence of activities was made studying the action of each station in detail in order to identify possible sources of error. In total 80 possible fault conditions were identified. A sample of the 80 faults currently monitored is given below:

Component not present, stn. 1.
Rod clamp not on, stn. 1.
Nut runner not low/rtn., stn. 1.
Lube low level, stn. 2.
Fixture A to B not complete, stn. 3.
IRB 6 (robot) spintex failed, stn. 3.
Wash pump over-load, stn. 4.
SP1 (torque unit) unit failed, stn. 5.
Gripper not open/closed, stn. 5.
Robot awaiting pallet.
Air failed.
Input/output conveyor not running.

First of all the programmer was familiarised with the exact sequence of operation. The action of each station was then looked at in detail. A list of all the sensors and solenoids was obtained and the fault analysis software was then written.

For example, the PLC would have generated a signal to initiate the rod clamping operation in station 1. The same signal is also used to start a timer dedicated to the rod clamping operation in station 1. If the rod clamping operation fails the PLC does not receive the signal from the sensor which resets the timer. If the timer times out the PLC will then set a data bit corresponding to this particular fault. The relevant error message will then be displayed on the screen and an audible warning is also initiated. The operator will then look at the fault page to identify the fault. In this case the error message displayed will be:

105 ROD CLAMP NOT ON, STN. 1.

The operator may then interrogate the system further in order to get instructions to carry out the appropriate action. This is achieved by calling up the corrective action page corresponding to the error number.

The operator interface in the case of cell 6 is provided through the Siemens WF 470 colour graphics system which interfaces in real time with the PLC and is activated by the data bits set within the PLC. The actual graphics information and the text messages are stored within the memory module of the WF 470 board.

This system has now been implemented in cell 6 since August 1988. In general the system is running satisfactorily and it is found to be extremely useful, as without such detailed information the operator and maintenance personnel would have to check a large number of potential causes in order to identify the fault.

However, the implementation of the system was not 'painless'. During the development and introduction of the diagnostic routines the following difficulties were encountered:

- Existing vendor-supplied control software was over complex and ill structured. In order to maintain warranty, enhancements were carried out under the restriction of this program.
- The cell consists of complex assembly machines with many possible faults. All possible fault conditions had to be identified and decisions were made on the means of monitoring these conditions.
- The additional diagnostic routines had to be written in such a way that they were well structured and easy to understand as well as satisfying scan time and memory constraints.

The configuration of the software shown in Fig. 30.6 demonstrates how a logical, simple structure was achieved. The diagnostics section is clearly separate from the vendor program. The diagnostic part is divided into sections for each station.

30.6 Conclusion

Diagnostics are becoming an increasingly important feature of modern manufacturing systems. Although vendor supplied diagnostic systems are available with most off-the-shelf equipment, there is a distinct lack of such systems at cell control system level. In those installations where the control system is based on PLCs, a diagnostic system such as the application in Cummins Engine Co. may be implemented, making use of one of the three methods outlined in this chapter. Fig. 30.7 shows how various cell

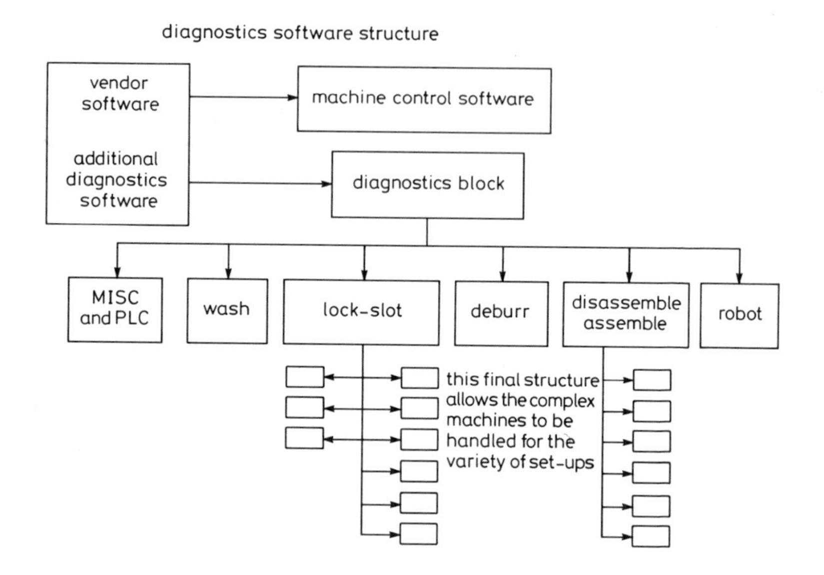

Fig. 30.6 Structure of the diagnostic program for cell 6

control PLCs will be integrated using a local area network facility which will eventually act as a gateway to the plant maintenance system (Rapier) employed in Cummins.

The diagnostic systems discussed in this chapter should enable an operator without specific knowledge concerning a particular manufacturing cell to perform effective trouble shooting and repair of the cell following the detection of a breakdown. After a malfunction, the diagnostics system should have the effect of

- locating the fault area immediately.
- decreasing the down time. The sooner a fault is detected and its cause identified, the sooner the operator or the maintenance personnel can attend to the fault.

However, it must be appreciated that the methodologies discussed in this chapter have been formulated round the facilities and limitations imposed by the PLCs used (Siemens Simatic S5 – U range and WF 470 colour graphics system). Applications using other equipment may have a different set of facilities and limitations which may require modifications to the methods discussed.

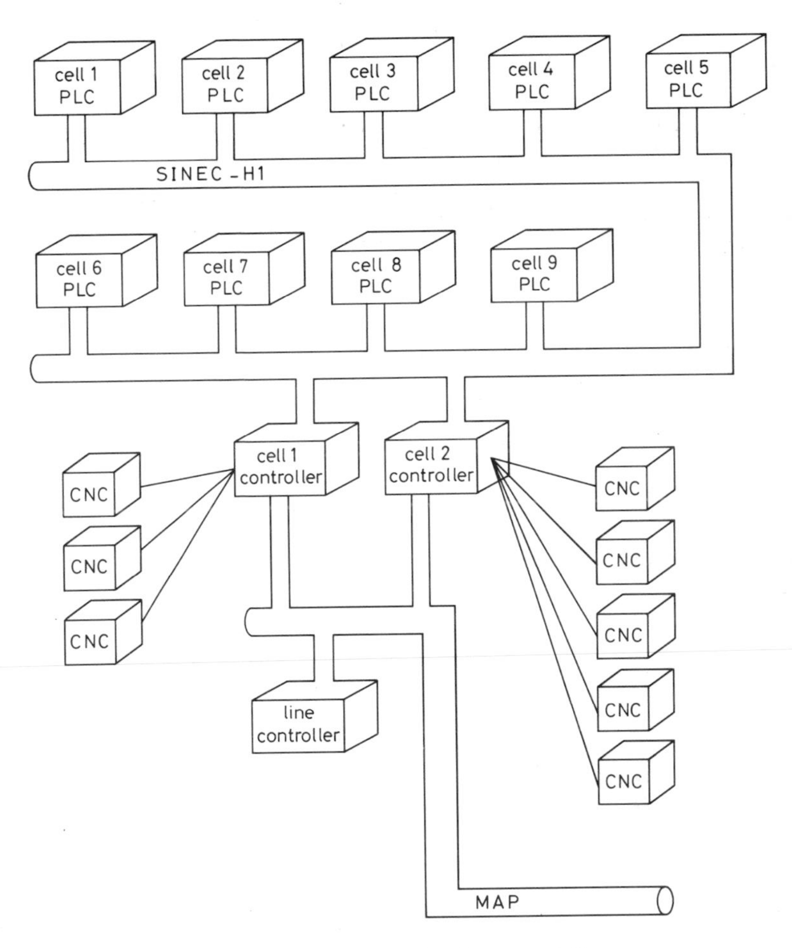

Fig. 30.7 Architecture of the control system for the con-rod line

Chapter 31

Automatic detection of surface defects in castings

M. P. Elliott, P. Hewitt, C. W. Rapley and D. C. Richardson
Sunderland Polytechnic UK

31.1 Introduction

The problem of surface defects in castings is widespread throughout the foundry industry and their detection is of paramount importance in maintaining product quality. As high product quality is seen as one of the keys to future success in the industry, the focus of attention is on non-destructive testing and the introduction of a high percentage of full component inspection, even 100% inspection!

Such a move will require a secure method of automatic inspection and detection of possible faults. This chapter concentrates on the planar surface defect aspect of this problem, outlining the range of defects likely to be encountered in castings, the common methods of detection available, previous work in the development of automatic computer-based methods of inspection and the main problems associated with such systems, and, finally, the work relevant to this problem now in progress at Sunderland.

31.2 The common surface defects in castings

Surface defects can occur at various stages in the pouring, subsequent heat treatment and machining of a casting. The most commonly found defects and the process or processes during which they are most likely to occur are summarised in this section. Also mentioned is the characteristic shape and appearance of each defect.

From a detection point of view, the size of the defect, in terms of length, width and depth, is obviously also important. However, whether the size of a particular defect will cause rejection of the casting or not will also depend on its location.

Hot tears occur in the primary casting stage, due mainly to non-uniform cooling and shrinkage in the solidification process together with the relief of thermally induced stresses, causing a rupture type void in the surface. The characteristic shape is irregular and inter-granular and they can be invisible in normal white light.

Lap marks or mis-runs are caused by interrupted metal flow at the casting stage, causing lack of metal fusion or re-joining. They are characterised by a wavy form which is delineated by surface topography. They are usually visible in normal white light and vary in depth from the surface.

Sand inclusions also occur at the primary casting stage, due to such causes as erosion or loose sand in the mould. Geometrically they are circular or semi-circular and tend to occur in clusters and are not always planar. They are usually visible in normal white light.

Fibrous cracks have a characteristic star shape after casting and are formed at the rough dressing stage after casting. They are usually irregular and develop in any size with a wide range of aspect ratio and are usually invisible in white light conditions.

Mechanical cracks can occur at any stage after casting, being primarily due to mechanical over-stressing. Geometrically they are linear and usually distinguished from other linear defects by their greater width and depth. They can be visible in normal white light conditions.

Quench cracks occur in the heat treatment stage. Geometrically they are predominantly linear and planar and generally larger than hot tears. They tend to be invisible in normal white light conditions.

Hydrogen cracks usually form at the heat treatment stage, although the underlying structural damage may have been initiated at the casting stage. They are specific to certain alloys and to certain positions on the casting surface. The geometry is generally linear and planar with sub-surface exfoliations and they are not acceptable in any form or size. They are not visible in normal white light.

Mechanical grinding marks occur at the dressing stage and consist of a series of parallel grooves or gouges of varying length. They are normally visible in white light, although, when isolated, can appear similar to quench cracks.

Edge burrs can form at any process stage, being due to such causes as pattern-core joint lines, mechanical dressing and shot peening. They have the geometric characteristic of being parallel to the component edge, although they can appear to be similar to grinding marks or quench cracks.

The photograph in Fig. 31.1 illustrates an obvious mechanical crack and inherent features on part of a complex casting, whereas the photograph in Fig. 31.2 illustrates a typical surface found in castings, with a hot tear, grinding marks and edge burrs all in evidence.

Fig. 31.1 Mechanical crack

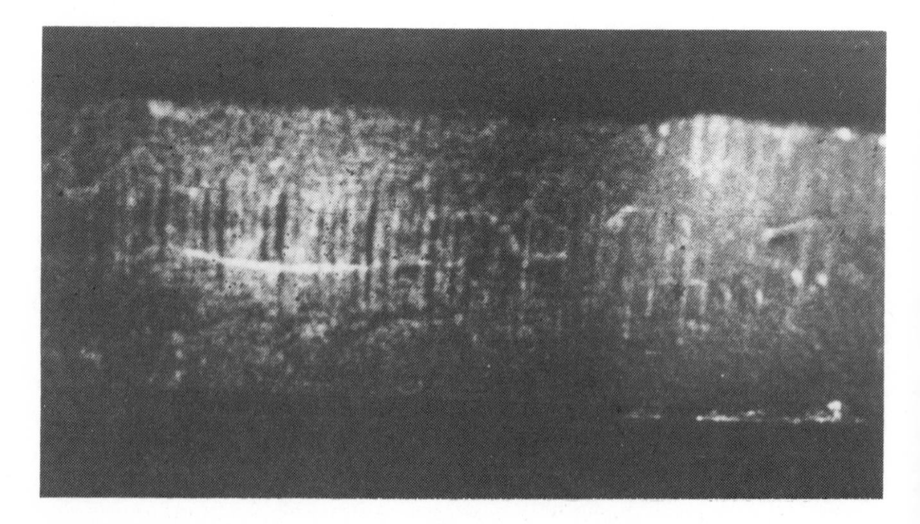

Fig. 31.2 Hot tear, grinding marks and edge burrs

31.3 Methods of detection enhancement

Many of the above defects often occur in a form making them invisible to the naked eye, so that a method of enhancement is required in order to detect them all clearly.

Over the years many methods of enhancement and detection have been tried, some of which are confined to surface effects, whereas others have been developed to detect interior as well as surface faults. Typical of the latter in current use are ultrasonic methods and film radiography (X-ray) methods, whereas typical of the former are dye penetrant methods and magnetic particle methods. A brief outline of each of these methods will be given here to provide a useful base for consideration of possible automatic systems.

Ultrasonic transducer of 'search unit' consists of a piezo-electric crystal, electrically charged to vibrate at high frequency so that it emits ultrasonic waves, plus a detector which receives any reflected waves returning to the transducer. With the transducer in contact with a component, the size of any defects can be predicted from a study of reflected signals. An intimate contact, usually from a liquid film interface or from total immersion, is needed between the transducer and component surface, and the detector and transmitter must, of course, be on incident planes.

Film radiography has been in widespread use now for nearly a century and many industrial applications have been developed. X-ray images can be used to reveal faults in components, with two or more used in parallax or triangulation arrangements to determine locations. Careful setting up is needed to produce sharp images, and image processing can be used with each film or more directly in real time exposure.

Dye penetrant enhancement of surface defects has been in use since the 1920s and is still widely used, particularly for non-ferrous components. As the name implies, a liquid borne dye, often fluorescent, is applied to the surface and then removed after a dwell period, to leave only residual amounts that have penetrated surface defects. A 'developer' is then applied, the action of which is to draw out the residual dye to mark and enhance any defect. Careful preparation of the surface is required to ensure good cover and wetting of the dye whilst avoiding damage to the surface or the peening over of any defect openings to inhibit dye penetration.

Magnetic particle detection is naturally confined to ferrous components since the component is magnetised, usually by passing through it a very high current. Any surface defect will cause a small local leakage magnetic flux field at its location and these are revealed by coating the surface with magnetic particles which are drawn to the leakage flux regions. The strength of the local flux leakage depends on the amount of magnetic reluctance present, due to such cause as air in a crack, and to the orientation of the defect to the main magnetic field, giving a maximum at 90° and zero when parallel. Fluorescent substances are often added to the magnetic particles to further enhance the effect.

31.4 Previous work on automatic casting inspection

Although there have been a number of publications on various aspects of identifying defects in castings, very few are on the development of an automated system. The possibility of the use of the magnetic particle or dye penetrant defect enhancement techniques in an automatic inspection system for metal components was outlined in principle by Strauts and Flaherty (1981) and also by Porter and Mundy (1982). The system proposed consisted of a method of surface defect enhancement, a video camera which provided an image of the enhanced surface of the component and a computer with software which processed the image in such a way that the surface defects could be detected. Video image capture boards were becoming commercially available for the commonly used microcomputers, together with software (and firmware) for processing the image so that a system could be put together without having to make all the development from scratch.

One of the first reports on the development and application of an actual automatic system for defect detection in castings appears to be that of Yoshikuni Okawa (1984) at Gifu University in Japan. An industrial television camera was employed to scan the entire casting as a single image. A normal white light, in the same direction and at a small angle to the camera, was used to illuminate the casting surface which appeared as a light image against a darker background. A reverse lighting set-up was also tried, giving a dark silhouette of the edges of the casting against a light background. To minimise image distortion, a window covering the casting was chosen in the middle of the video picture which was stored in a microcomputer as an 8 bit (256 grey levels) image 192 × 192 pixels.

The objective was set at detecting the large, visual defects of casting fins or notches and textural effects in the surface so that defect enhancement was not necessary. Simple circular cast iron pulley wheels of about 12 cm diameter were used as the main test components. These were taken from the same mould and shot-blasted before entering the automatic inspection unit.

To detect fins (an increase in outer radius) and notches (a decrease in outer radius) a method of detecting and storing the outline contours of the casting was developed. The image was binary thresholded (grey levels set to white below, and black above, the chosen threshold) and an edge-detection/contour-tracing algorithm used together with a method of establishing a centre and radius for the main circle of the contour. Significant deviations from this radius represented a fin or notch.

Extensive testing and comparisons with manual inspections established that the system worked well with large (typically 1—2 cm) fins and notches but gave poor results with small (less than about 5 mm) defects. This is not surprising since the 192 × 192 pixel resolution image of the whole casting represents a coarse image from which only gross defects can be expected to be detected with any certainty.

The flat surface of the casting was illuminated from the front and side, so that the coarse textured surface gave relatively larger and more widely spaced shadows than the normal more finely textured surface. Five simple, statistically based, parameters were developed to analyse the grey levels found in a pixel scan of the surfaces of both normal and rough casting surfaces. These parameters included the mean and standard deviation of grey level frequency plots and ratio and average of smoothed and un-smoothed grey level differences.

The surface was divided into 64 sub-regions and tests were conducted with 30 normal surfaces and 30 abnormal surfaces. From the computer image the five parameters were calculated for each sub-region of each pulley surface and the results for the normal surface were compared with those for the abnormal surfaces so that regions of abnormal surface texture could be identified. These results were in turn compared with results from manual inspection of the same castings. From the example given, the automatic inspection method tended to detect slightly more regions of abnormal texture than were apparent from manual inspection, although, as the author concluded, their capabilities were almost comparable. The overall conclusions were that large scale defects could be detected by the system, and by magnification of the image, small scale defects should also be able to be detected.

The development and application of an automatic system for the detection of surfaced flaws in machined components was reported by Sid-Ahmed, Soltis and Rajendran (1986). An 128 × 128 pixel, 8 bit video image of the component surface was obtained and processed in a minicomputer. The image was binary thresholded and then analysed with a border following algorithm to determine the size of all the enclosed areas in the image.

The controlling computer program for the system was menu driven from a touch video screen and successful applications of the system to a number of different problems were reported. One of the applications was to detect surface defects on a machined piston head. After thresholding and border following, template matching was used to eliminate inherent features to leave the remaining detected enclosed areas as defects. A final report was given on the number and total area of the faults detected.

Also reported by Sid Ahmed *et al.* (1986) was the development of a liquid crystal surface defect, depth measuring, device for use on smooth flat surfaces (such as the machined piston head).

The basis of an automatic inspection system has been described by Sanderson, Weiss and Nayar (1988) using video image processing on a super-microcomputer together with commercially available firmware and software. Much of the paper is devoted to the problem of dealing with smooth reflective surfaces and on the lighting of small shiny surfaces, in particular with simulated applications of the system to soldered joints. However, application of the system to machined component surfaces is also mentioned although no specific cases are studied. The general conclusion was that accurate mapping of smooth surface contours could be achieved, although two cameras are needed to obtain surface orientation.

31.5 Features of an automated inspection system for surface defects in castings

Current trends in manufacturing quality control demand greater features from inspection systems. No longer is quality control a matter of simple accept – reject according to certain measurement criteria. With the present escalation in the use of statistical process control methods, inspection systems need to monitor trends in the measured product parameters in order to provide feedback which can be used to effect process control and thus minimise product rejection and subsequent scrap levels.

In this particular application, the system requirement is a method of detecting planar surface effects, determining their location and characteristic size and shape, classifying the defect as to type and probable cause, recording the details and reporting whether size and location of the defect are outside the bounds of acceptability.

From the above review of proposed systems, the essential components of such a system using available technology would be *surface enhancement* equipment, a *video camera* and a digital *computer* which is programmed for frame grabbing, image processing, image analysis and structured filing of quality data.

The *surface enhancement* equipment must be able to make clearly 'visible' (by enhancing contrast between a defect and background) surface defects with a wide range of size and shape and which may not be visible in normal white light conditions. This must be achieved in a way that can be machine controlled and is consistent over the range of surface topology and texture found in a particular casting.

The *video camera* will need to be able to scan the areas of interest on the casting surface and produce an image of good contrast in what may be low non-white light conditions. The camera must be able to operate for long periods in shop-floor conditions.

The digital *computer* must have provision to receive, capture and digitise video images with adequate resolution and to run the necessary image processing and analysis routines that enable classification, recognition and recording of all prescribed defects. All this must be done within the pre-set inspection time allowed for the component.

31.6 Progress at Sunderland

The work on automatic inspection at Sunderland, which is at a relatively early stage, is centred on the detection of surface defects in ferrous castings and is being carried out jointly with George Blair Plc, who are manufacturers of high integrity castings which are mainly used in defence and transport vehicle equipment.

The aim of the work at Sunderland is to develop an automatic system that will surface inspect pre-set regions on a particular casting in order to detect, recognise, locate and record all planar surface defects that may lead to failure of the component. This means including all the defects listed above, some of which are very small and are invisible in normal white light.

Although there has been much development over the years of the individual parts that could be employed in an automatic surface inspection system, very little has been done in a systematic way to determine how they can be best put together. As with most efforts in creating an effective and reliable system, the main aim must be simplicity and to make the optimum use of individual components.

The first requirement is that of a reliable and consistent method of defect enhancement. The methods of dye penetrant and magnetic particle are both widely used in manual inspection systems, and would be suitable as the first stage in an automatic system. However, much work is needed to identify the optimum parameters such as the dye and developer type(s) and mix and the dwell and development times for the dye penetrant method, or the magnetic field strength and orientation, dye/particle types and mix and application method for the magnetic particle method.

With the present work being based on ferrous castings, attention is focused on the magnetic particle method. A test rig is being developed around a commercially available magnetic field generator. Associated with this test rig is the lighting system and video camera. Since many of the defects are not visible in normal white light, ultra violet lighting, together with a filtered video camera and fluorescent dye and magnetic particles, is being used. A solid state camera is preferred for compactness and reliability in an industrial environment. Initial tests have emphasised the problem of specular reflection with metallic surfaces and the care needed with the distribution and intensity of lighting sources, camera positions, fluorescent dye and camera filters. All this will need systematic study to determine optimum operating conditions.

Initial image capture and processing software is being developed utilising one of the current 16 bit microcomputers, equipped with a high resolution analogue RGB monitor. Image capture and basic processing, such as binary thresholding and edge detection, have been effected with one of the widely used commercially available systems. A range of picture resolutions can be selected; the one presently being used is 512 × 512 pixels, 8 bits deep. The further processing software for pattern recognition and classification will need to be developed. The full range of defects are to be detected and classified with a high level of consistency and within an acceptable process time limit. The classification will be based on the planar image, with respect to length, width and defect orientation.

Accurate location of the video camera will be required, with either a number of fixed cameras or a single mobile camera required to cover the areas of interest on surfaces of complex castings. In some cases, it may be possible to leave location of the defect to the image analysis stage, by referencing to selected inherent features in the casting image.

A line diagram of the proposed set-up for the automatic inspection system is shown in Fig. 31.3. The requirement of accurate location of each defect detected is an important one, not only for routine identification of a particular rejected casting, but also as input for the formation of a database of information on defect occurrence and location. The location and frequency of each type of defect will be important feed-back information to an SPC system in the identification of problem areas, which in turn should lead to significant improvements in product quality.

31.7 Conclusion

Significant improvements in the quality of castings will require an effective automatic inspection and surface defect detection system. Such a system will most likely be computer based, employing a video camera with computer image processing and analysis. One of the current generation of 32 bit mini- or super-microcomputers will be needed to provide the level of computational speed and storage required for a viable industrial system.

The individual components of such a system exist at present and some of them have been put together by previous workers to form a simple automatic inspection system capable of detecting large defects. A programme of work has been initiated at Sunderland to develop an effective and consistent automatic surface inspection system for castings, capable of detecting and recognising a range of planar defects, including

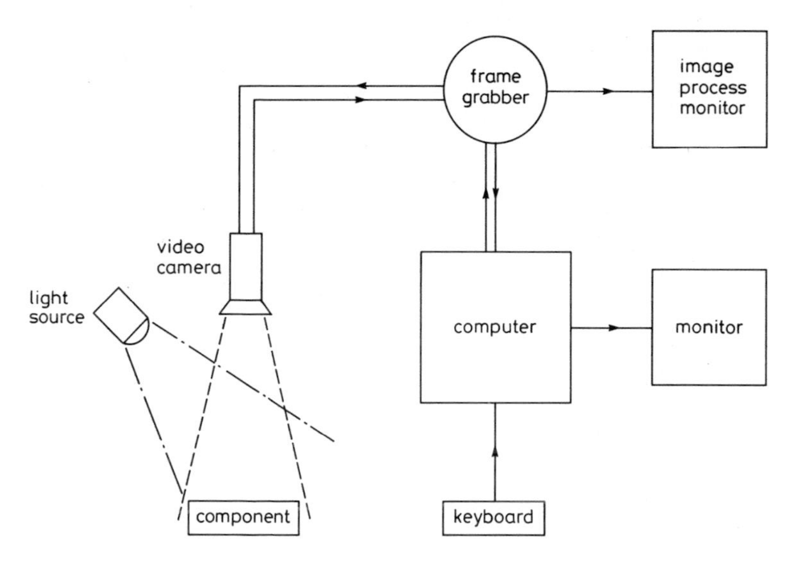

Fig 31.3 Test set-up for the automatic inspection system.

the small but critical defects that are invisible in normal white light conditions.

The system will make contributions to quality control at both ends of the casting process. As well as acting as a filter to reject castings with critical surface faults in the final stages, the database of information on defect occurrence recorded by the system will be used to pin-point areas of improvement at the primary casting and subsequent processing stages to eventually reduce or even eliminate the defects.

Chapter 32

Using modern quality control techniques: some case studies

J. Hensser and R. Murray-Shelley

Mycalex (Motors) Ltd. and The Polytechnic of Wales, UK

32.1 Introduction

'Quality' is often a somewhat abstract concept to which is ascribed different meanings depending on circumstances. High 'product quality' can imply the use of expensive materials — as in some jewellery — or the application of extraordinary skills — as in sculpture or painting. In the more down-to-earth world of manufacturing engineering quality is often synonymous with reliability and a 'quality product' is often regarded as being one which will almost certainly operate within a defined specification once installed, and which will continue to work effectively throughout its useful life — whose length is itself defined.

Far from being simply 'desirable', the ability to make reliable products is essential for many manufacturers if they are to survive and prosper in what is now a very competitive and quality conscious environment. This is particularly the case for those companies which venture into the high volume automotive component and computer system component fields. In these areas the penalties for producing components which do not conform to agreed quality standards can be severe. Clearly customers will reject whole batches of items in cases where their goods inwards checking procedures indicate that the agreed specification has not been met. Of more significance is the quite severe warranty implications which are increasingly passed on to component suppliers by major manufacturers. Thus a maker of, say, a simple low cost engine management computer system could be involved in a warranty claim many times the value of his component should its failure result in serious engine damage — or indeed an accident. Without doubt 'quality' cannot be instilled into a product simply by testing it, no matter how exhaustively this testing is done. All that testing can do is to demonstrate that at the time the test was carried out the device operated as expected. True quality must be designed into any product and this design process must reach out to encompass not just the physical dimensions and structure of the item but the materials and production processes used in the whole manufacturing operation.

This chapter examines how a judicious application of modern quality control techniques and computer-based testing equipment can help to ensure that this apparently impossible goal of 'zero defects' can at least be approached very closely — even if it can never be absolutely attained. Two product types will be considered. The first is a group of electromechanical solenoids which are typically used within conventional vehicle fuel systems to prevent 'run on' when the engine is stopped. Secondly, we will consider the quality assurance procedures needed for a more complex product — a motor assembly for a computer disk drive.

32.2 Solenoid manufacture and test

In many ways the solenoid is the simplest of all electromechanical components, but this apparent simplicity belies the considerable skills which are required to design units suitable for mass production which are at the same time cheap to make and reliable. Within a vehicle they are used for a bewildering variety of purposes and worldwide consumption figures are staggering.

Though we tend to regard the electrical systems of most vehicles as being 'twelve volt', the actual voltage available varies from perhaps three quarters of that figure when the car is being started on a cold winter morning to 14 vs or more under other circumstances. In addition to operating over this wide voltage range the solenoid must not generate unduly high voltages when it 'drops out', which could affect the electronic control systems which now grace almost all modern vehicles. Such solenoids must open reliably when their energising voltage is removed even if this happens quite slowly — they must not stick closed — and their electrical parameters, particularly resistance, must be controlled within quite close limits. The operating environment for components of this kind is one of the worst to be found anywhere. It is quite possible for the component temperature to change by over 100°C between the time that the engine is first started and when it has warmed up, and the micro-climate within the mechanism of the unit may be extremely hostile — due, for example, to the presence of 'sour fuel'.

Tests on items of this kind fall into two basic categories. In the first are the long-term 'type tests' which include, in the case of solenoids, operating single units many hundreds of thousands of times and then examining them for wear and conformity to the agreed specification. Other tests include subjecting components to extremes of temperature, humidity, vibration and corrosive atmospheres. Clearly such tests are often destructive, take many hours to perform and can really only be carried out at the product design and development stage and at intervals during the product's life cycle to assure continued conformance.

As far as actual production units are concerned, the degree of testing which can be performed clearly depends on the time available to carry out the tests and their nature. Typical volumes for the sorts of devices which we are discussing might be around 4000 per day or 500 per hour assuming single shift working. Such volumes are high enough to tempt the use of statistical techniques alone with only a proportion of the product volume actually being tested — at least to any significant degree. In fact both logic and experience dictate that even with these volumes 100% testing is really essential bearing in mind the demanding nature of the customer. This does not mean that the lessons embodied in what we now know as 'statistical process control' (SPC) must be thrown away. A combination of 100% testing and the judicious application of even quite simple control charts and concepts such as process capability indices produce a far better quality assurance capability than either simple testing or sampled testing and SPC can provide alone. SPC is also applicable to these products in areas not associated with 'measurement' as such; for example failure mode and effect analysis and in the control of both measured and attribute data on sub-assemblies and components.

Typically solenoids such as these have to be subjected to a battery of electrical and mechanical tests after assembly. Before being put together, the component parts and bought-in items generally would have received a thorough goods inwards inspection of the type required in any organisation aspiring to operate a quality system such as BS 5750/ISO 9001. The electrical tests would involve checks for open and short circuits and a resistance measurement to ensure that the coil resistance falls between upper and lower limits. Electromechanical tests include operating the solenoid at normal rated voltage and measuring the pull-in and drop-out times, ensuring correct operation at reduced voltage and measuring the 'drop-out voltage'. This is variously described by different customers but is basically the voltage at which the solenoid will drop out, having been energised typically at 14 Vs when the supply voltage is slowly (or at least relatively slowly) reduced. Success in this test — which is particularly onerous since even the smallest amount of excess friction can cause failure — means that the component is unlikely to stick open in service and will actually drop out when the ignition is turned off even though rotating alternators and motors mean that the electrical system remains slightly energised. At the same time as this battery of tests is carried out, an opportunity is taken to exercise the component several times.

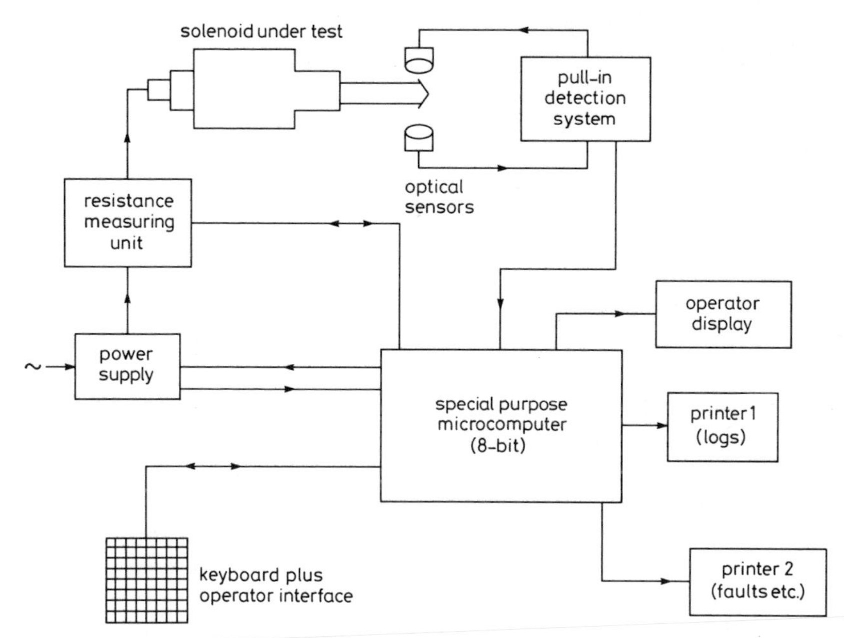

Fig. 32.1 Functional representation of production test equipment

32.3 Production test equipment

Fig. 32.1 shows a functional representation of the type of test equipment used for this task. It is very much a 'one off' unit and is able to check components against ten different test parameters as well as exercise the solenoids two or three times all in the space of four seconds. The equipment is fitted with two 'nests' to hold components under test and the operator is able to unload and re-load one station while the other contains a device being tested. At the same time the operator can carry out a visual inspection looking for obvious mechanical damage or imperfections. We have found that great care has to be taken in designing the operator interface in situations such as this. There is little value in providing operators with extensive instruction books or lists of 'dos and don'ts'. Such books or lists get lost, are ignored or forgotten. Equipment of this kind has to be simple and obvious to operate with great care being taken about all aspects of operator safety — even a 'low voltage' solenoid of this kind can generate many hundreds of volts across its terminals if it is removed from a test jig whilst energised. Clearly a balance has to be drawn between ease of operation, safety and the assurance of proper testing. There is always a temptation, for example, for an operator to open a jig too soon before the equipment has indicated pass or fail. One way to avoid this is to provide mechanical locks; the other way, which we adopt, is to make the machine halt in such a way that trying to 'beat the system' causes the operator more difficulty than 'doing it right'. Such a 'Pavlovian' approach to the interface between the operator and the test equipment, with a real attempt to make it 'user friendly', does seem to pay dividends and, evidence shows reduces significantly any unintentional — or intentional — operator abuse.

Behind the basic test routines and accessible only via a keyswitch reside various routines which can provide a variety of management information. Over the years as a number of these systems have been developed, we have found that providing too much test data is counter productive. It is necessary to consider very carefully the realities of the manufacturing environment when designing equipment of this sort, and in particular the software which it contains. The process control techniques which are of most real practical value are those which enable problems to be anticipated before they become disasters. 'Two level' systems consisting of a background logging arrangement with a foreground re-settable log which the patrol inspector can use over a short period to check small batches or to investigate short-term trends seem to be specially useful. The ability to print test results on single components is also a valuable feature in any test equipment of this kind.

Great care has to be taken in the way in which reject components are handled after failure. Clearly they have to be effectively 'quarantined' in some way, and many customers insist that they are deposited in a locked box to ensure that failures cannot be mixed in with good production. By equipping such a box with a funnel containing a suitable sensor — typically an infra-red light source and photo-detector — the test equipment can be made to 'lock up' until the offending component is properly disposed of. Obviously such a system can be defeated in various ways, but again the philosophy is to make the easiest course the one which assures compliance. A further development will be an interface between the test equipment and the management information system which actually identifies the offending component with its perceived failure mode for later analysis.

Increasingly companies claim to be using quality assurance systems which meet, typically, BS 5750/ISO 9001 — even though they may not be actually accredited to such a standard. Clearly this is a desirable situation so long as the procedures which are introduced actually do improve the perceived quality of the product in question and are not simply 'paperwork exercises' which do little more than fill filing cabinets or — what is worse — fill computer disks. Standards such as BS 5750/ISO 9001 do not in themselves prescribe, in detail, the procedures and systems which should be adopted in individual cases. Our experience shows that the greatest benefit from such disciplines is gained when they are kept as simple as possible. Any information provided by test equipment, for example, should be directly useful in improving the product or process to which it is related and too much data merely masks problems. It should always be remembered, however, that some major automobile manufacturers, for example, do lay down their own SPC/test routines which naturally have to be followed. Thus some flexibility in the available data logging techniques must be available to allow for the preferences of individual customers.

Few companies can really honestly claim to be operating a true computer integrated manufacturing (CIM) plant with all that such a project involves in terms of shop floor data collection and analysis. However, the judicious use of data collection from various points within a single manufacturing unit has been shown to be of value provided that such a system is sufficiently flexible to allow feedback 'in real time' to warn of potential problems as soon as they can be detected.

32.4 Maximising yield by intelligent testing

The purpose of testing is to 'assure' quality — not to generate it as we discussed earlier. Clearly if components pass on final test then all is well, but if they fail then the work of the test equipment is only just beginning.

The solenoids referred to above, because of their construction, can rarely be re-worked and thus reclaimed if a fault is detected. It is nevertheless clearly essential to know just what the fault was, so that suitable remedial action can be taken should it recur. In circumstances where re-work is possible, however, suitably designed testers can greatly assist in minimising product loss. A good example of this approach is the wound stator of a small computer hard disc drive motor which was manufactured in quite large volumes of many hundreds each day. This particular stator carried four windings and, after it had been wound, a plastic injection moulding was formed completely obliterating all the copper wire.

Obviously the eight ends of these four windings were labelled prior to the moulding operation and they were then soldered into a small printed circuit board which formed an integral part of the assembly — and in the process necessarily lost their identification. Microprocessor-based test equipment at the end of the production line checked a variety of characteristics and was able to suggest possible causes of any failures which were detected — often due to reversed coil connections, bad solder joints and so on. In the event that simple re-work could not produce an acceptable component, the reason was probably due to a major confusion between the identities of the eight wires. Here it should be explained that our experience shows that, even with the most carefully contrived inspection and checking procedure, mistakes will always happen such is the nature of the human being as an assembly tool.

In this example, it was found possible to re-identify the exact nature of each of the wires using equipment which is described in schematic form in Fig. 32.2 This was able to first determine which groups of two wires constituted a pair and then, by pulsing the newly discovered coil and analysing the magnetic signature produced by the component as a result, could identify the wires exactly. In this way failed components which, at this point in production had considerable value, could be rescued and restored to full specification.

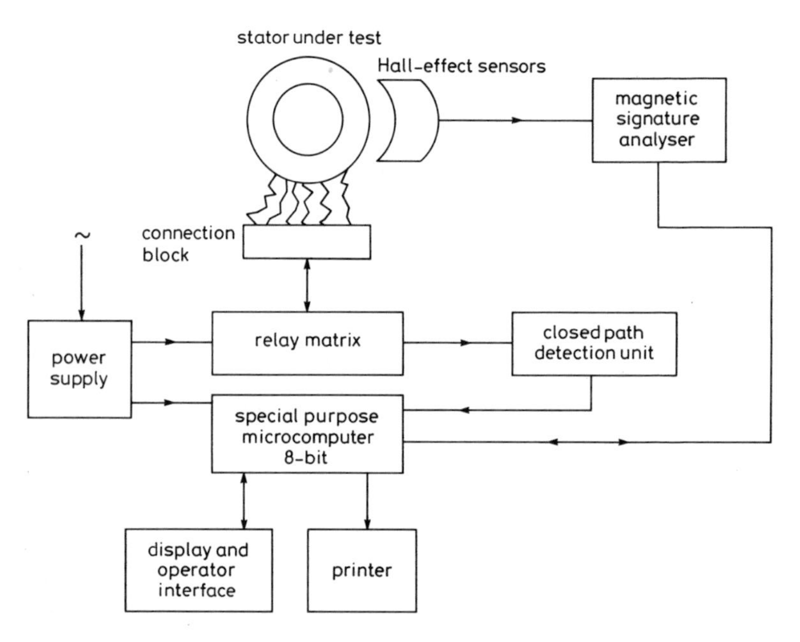

Fig. 32.2 Equipment for intelligent testing

32.5 Data presentation

One of the main advantages of using a computer-based 100% test system is that all the data can be logged and analysed in many ways. The system used currently displays the 'as tested' individual result as each test takes place and retains this information on a screen until the next test is instigated. The data is then stored and presented as batch information in a printed report at pre-set batch quantities, e.g. 20, 100 etc. From this data a sample of readings, typically the last five, are used to plot an average and range chart against control limits which are constantly updated each ten batches in line with the 'performance based' SPC systems in use today. Additionally, a calculation of 'capability index' is made from the data to allow a comparison of the calculated total spread to the specification limits to be determined. This figure gives a confidence level in the process and provides a quick capability 'snapshot' for the Quality Engineer or customer as required. Both the plots can have alarm indicators to alert the operator that the process is out of control. A further advantage lies in the potential for the transfer of data by 'electronic mail' or other communication systems utilising direct links and for storing it using discs or magnetic tape. More and more customers are asking for data to back up the original Certificate of Conformity and it is every Quality Manager's nightmare that he will produce more paper than product with little chance of retrieving vital data should the need arise.

We have attempted to show that the benefits of a careful application of 'intelligent' test equipment can be significant in assuring the quality of a well designed product and in maximising product yield. Slavish adoption of 'textbook' SPC procedures will probably not be as useful, as a critical application of some of these techniques together with 100% quality assurance systems should be kept as simple and thus useful as possible.

Chapter 33

Stitch quality monitoring using digital signal processing

J. L. Catchpole and M. Sarhardi

University of Durham, UK

33.1 Introduction

Industrial sewing machines used for the mass-production of garments are individually operated and run at speeds of up to 6000 to 8000 stiches per minute in a stop – start manner as garments are fed in. Under these conditions, a machine, even when properly maintained and correctly set up, misses stitches at random intervals. One faulty stitch in 10 000 is considered poor.

The accepted practice for detecting these faults is for experienced staff to examine the seams by eye. Quality control practices generally involve either checking every finished garment or spot-checking the in-process work on a statistical basis to guarantee a minimum number of faults. Mis-stitches are a correctable fault — the bad seam is unpicked and stitched again. This takes extra time and material to complete a batch.

Since the 1960s a large body of work has been carried out on the monitoring of sewing machines in operation. Much work in particular has been done on the subject of stitch formation and the tension variations during it (Price & Rae, 1982*b*; Deery & Chamberlain, 1964; Wira, 1981; Price, 1982*a*). The published work concentrated on average tensions over a series of stitches or on the variations during very low speed or simulated operation. The tension in general has been measured using strain gauges, but the data was observed visually and recorded by hand or recorded on film (Wira, 1976). This work was mostly carried out before the advent of cheap flexible digital electronic systems.

The work described here is an investigation into the monitoring of the tension variations during individual stitches using electronic data gathering and subsequent analysis. The purpose of this research is to facilitate the in-process detection of such faults, allowing immediate action to be taken to correct them and giving a minimum of wasted time. It would also allow faster recognition of poorly performing machines and should therefore reduce the number of faults overall.

The tension on the two threads and the looper of an industrial lockstitch sewing machine have been monitored using strain gauges and the data recorded using an analogue-to-digital convertor and 68020 based microprocessor. The data collected has been analysed by three different techniques to identify the variations relating to missed stitches. This analysis indicates thus far that detection of mis-stitches is possible and a viable real-time algorithm is being developed.

33.2 Theory

The three main approaches tried are based on Fourier analysis, autocorrelation and adaptive linear prediction (Gold & Rader, 1969; Proakis & Manolakis, 1988).

Fourier analysis is a technique for breaking down a periodic waveform into a set of sine-waves which would recreate the waveform if summed linearly at all points in time. The Fast Fourier Transform (FFT) is an algorithm for taking a set of samples of a waveform (which is assumed periodic) and calculating the set of coefficients which, when multiplied by the set of sine-waves of frequencies ranging from the sampling frequency down to the sampling frequency divided by the number of samples in the set, will (approximately) recreate the waveform.

The recreated waveform is only approximate since there are a number of errors inherent in the use of digital processing. The most important in the algorithm itself are those resulting from the representation of the spectral content of the waveform by a set of discrete frequencies only. Any components at other frequencies must be represented by the components allowed, which tends to spread the spectral peaks. The range of components allowed is also limited to half the sampling frequency, a result of representing the waveform by a set of samples. There are also those resulting from representing the range of values of waveform amplitude by a set of fixed levels (quantisation), a process which results in a small error in each sample value.

The form of the FFT used here is that defined by

$$\hat{z}_k = \frac{1}{\sqrt{N}} \sum_{j=0}^{N-1} x_j \exp\left[-i \frac{2\pi jk}{N}\right]$$

This transformation is useful since variations in a waveform are often more easily observed in the changes in its frequency content than in the waveform itself. For more detail consult Brigham (1974).

The techniques of linear prediction are widely used in data compression and modelling. The basic approach is to attempt to predict the future values of a series of data values by summing the past values as a weighted series, a process very similar to an FIR digital filter (Widrow, 1971; Makhoul, 1975). The basic equations for this are given below.

The estimate $y(n)$ produced by the predictor is

$$\hat{y}(n) = \sum_{k=0}^{M-1} h(k)\, x(n-k)$$

If the correct value for the next member of the sequence is $y(n)$ then the error $e(n)$ in the predicted value is

$$e(n) = y(n) - \hat{y}(n)$$

The adaptive linear prediction technique used here is one in which the weights on the past samples, equivalent to the tap weights in the filter, are automatically adjusted after each estimate to give a better estimate of the next value. The criterion taken is the minimisation of the mean squared error. The mean squared error can be expressed as a quadratic function of the predictor coefficients which can be solved to yield the set of linear equations known as the discrete time Wiener – Hopf equation

$$\sum_{k=0}^{M-1} h(k)\; \gamma_{xx}(l-k) = \gamma_{yx}(l),\; l = 0,1,\ldots, M-1,$$

where γ_{yx} and γ_{xx} are the statistical cross-correlation with the reference and the auto-correlation of the data, respectively.

The problem then is to solve this for $h(k)$. This is commonly carried out iteratively in practice, a process known as 'training' the filter. Since the origin of the equations is a quadratic function there is a unique minimum, the 'ideal' solution. This can be estimated by obtaining an estimate of the gradient of the function and moving the coefficient values towards the minimum. The gradient can be estimated as

$$g_k(n) = -2e(n)\, x_k(n)$$

By adding a fraction of this product to each of the weights the predictor can be made to converge on the ideal solution, i.e.

$$h_k(n+1) = h_k(n) + 2ux_k(n)\, e(n)$$

This, known as the Widrow LMS algorithm, is one of a class of algorithms known as the stochastic-gradient-descent adaptive algorithms. While it is not the most accurate it is the simplest from a computational point of view.

The size of this fraction (the convergence coefficient) used when adjusting the weights gives the speed of convergence to the ideal output; however, there is a minimum mean-square error on the output which is approximately proportional to the coefficient. Thus a larger coefficient will give a predictor which converges to the ideal model of the signal faster, but is less accurate (Widrow, 1976). Some methods exist to get around this, such as reducing the coefficient as the predictor errors decrease, but these are outside the range of this discussion. They are also irrelevant since the predictor is used in an unusual way by considering the errors as the useful output.

The sample autocorrelation function $r(k)$ of a set of data values $x(i)$ is defined as

$$r(k) = \frac{\sum_{i=1}^{n-k} ((x(i) - x)(x(i+k) - x))}{\sum_{i=1}^{n} (x(i) - x)^2}$$

where k has a value between unity and the number of samples n in the set. It can be seen that the function cannot have any values above 1 or below -1 and that for small values of k the function will be large. At $k = 0$, the value must be 1. K is the delay or offset of the data set against itself; thus the value of 1 indicates a perfect match between the data set and the delayed set of values.

This function describes the characteristics of the data set in question, particularly the rate of change and the periodicity of the values. Any data set will give a peak at $k = 0$, since it must match itself. The width of this peak and the slope of its sides give an indication of the rate at which the data values change. The occurence of further peaks gives an indication of the periodicity of the data values. A periodic waveform has well defined peaks, whilst, say white noise would give an indeterminate set of function values round zero. An inversely periodic waveform gives inverted peaks (down towards -1) (Gold & Rader, 1969; Proakis & Manolakis, 1988).

The algorithm was applied here since it can indicate variations in the nature of a waveform and thus could show changes in the tension profile from stitch to stitch. It can also be used on the data to give indications of the speed of convergence of predictor required to track the data. It is also simpler to compute compared to the others.

33.3 Data collection

The largest problem here is the comparative rarity of the mis-stitches among the good stitches. Since the mis-stitches occur so rarely, a large number of stitches have to be monitored to capture just one and the quantities of data involved make this a slow process.

The 68020 based system used can sample at rates up to about 40 kHz and has a memory of 1 MByte. This allows data to be collected for just over 5 s at maximum sampling speed and for longer periods at slower rates. With the machine at 500 rpm, the data of interest is all located below 2 kHz, and mostly below 1 kHz. Operating at this sampling frequency will therefore allow the collection of data for machine speeds up to 2500 rpm if the frequencies containing the data increase linearly with machine speed. The collected data has been processed on a mainframe computer.

The sewing machine used is a two-needle lockstitch machine which was fed a standard seam binding tape for stitching. The machine was observed when stitching correctly and also when the thread tensions had been deliberately unbalanced to increase the possibility of mis-stitching. The capture of a mis-stitch was then a matter of waiting, no other action being undertaken to disturb the machine's operation. The comparison of the tension profiles for perfect and disturbed profiles allowed confirmation that the profiles were not badly perturbed by the unbalancing of the tensions. The observed occurence of a mis-stich in a sequence of stitches monitored was accurately indexed against the data collected by marking the start point of the machine on the tape and triggering the start of sampling using the signal from an opto-electronic switch on the drive shaft of the sewing machine. The sampling process then started immediately the machine began stitching. This was felt to be valid since most commercial machines in factories are operated in a stop – start fashion as garments are fed through.

33.4 Data analysis

The data collected can be divided into several groups: firstly, that for good stitch sequences, taken as a reference; secondly, that containing full mis-stitches, where the looper fails to engage one of the threads; lastly, some occurences were found of faults which were damaging to the seam in lesser ways but where no major upset occurred in the tension profile. An example of such would be a stitch in which the looper hooked around a loop of thread from one of the needles, but did not pass through it. As a result, the stitch is weak and could unravel if tugged sharply or repeatedly. This chapter will concentrate on the good stitches and the full mis-stitches, although the minor faults are also under investigation.

The sequences of events for a good stitch as indicated on a tension profile need to be discussed first (Fig. 33.1). The profile for the thread on one needle only (the outer needle — on the left from the operator's point of view) is illustrated but the two cases are very similar and thus the discussion applies to both. First, the tension is high at the start of the cycle for both needles (point A), since both are fully raised. The tension then drops rapidly as the needles come down, the threads going slack just before the needles hit the tape (point B). The tensions then rise as the threads are dragged through the tape and they pull taut the loops held by the looper arm from the last stitch. The threads show a sharp drop then a second climb at this point (region C), when the looper arm disengages the previous stitch loops and the thread goes momentarily slack before the needle's progress tightens it up again, closing the previous stitch up. The inner thread tension shows the same drop in tension, but does not recover as strongly since the outer thread is pulled more strongly by the bobbin thread. Both tensions rise to a peak as the needles begin to retract. It is just before this that the looper arm moves back in and slips through the loops of thread beside the needles. The threads then become taut again as the needles rise and the looper holds the new loops steady at the bottom (region E).

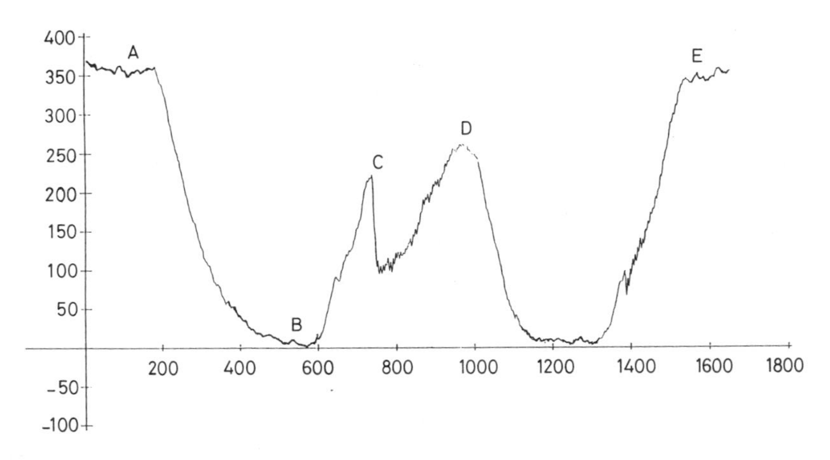

Fig. 33.1 Tension profile of good stitch

Most of the data concerning the stitch formation is then contained in the first half and middle of the sequence, where the looper is engaged and a stitch is tightened up. A failure by the looper arm to engage the thread correctly during the previous cycle causes a mis-stitch and the profile collapses at this point (Fig. 33.2, point A). The peak following this (point

B) corresponds to the peak between C and D on the profile of the good stitch (the 'recovery peak'). This is the only region for which the looper interactions are not important. The looper engages the thread correctly again following this (peak D) and stitching continues normally. The rough edge to the following peak (region C) is a result of the slack in the thread being taken up as the needle rises again. These are the profiles for the outer thread. The inner thread is largely unchanged and does not appear to be affected by the presence of the mis-stitch on the other thread. Comparing the correctly balanced seam tensions to those for the unbalanced seam, it can be seen that the main peaks are much smaller, as are the recovery peaks after the looper has disengaged. Since the unbalancing was accomplished by reducing the tension on the outer thread, it can be deduced that these two peaks are largely tension dependent. The initial peak as the thread tightens against the previous loops does not fall in the same way and is therefore independent to some degree of the set tension. This peak collapses completely during a mis-stitch. Since the previous loops will be wedged through the tape, and must be pulled loose as the needles rise, the preceding main peak does not collapse completely. From this it becomes apparent that a mis-stitch on the tape shows up mainly in the tension profile on the following stitch (as a stitch is defined below).

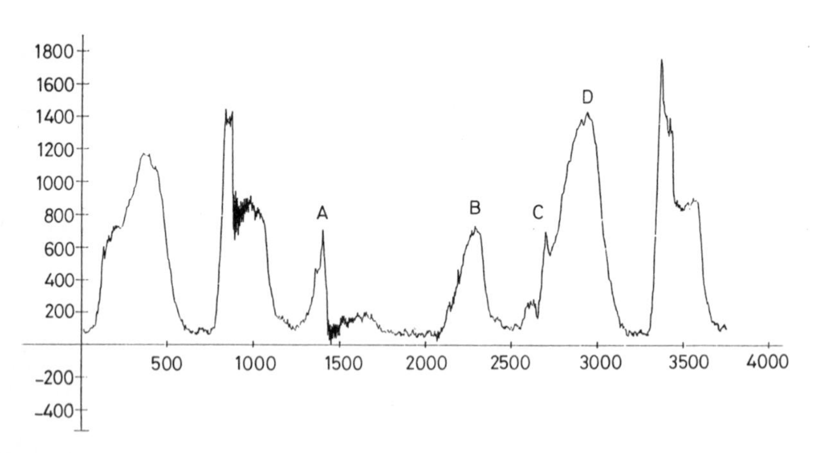

Fig. 33.2 Sequence showing mis-stitch.

A full mis-stitch, then, is easy to identify and the problem is to find a simple robust method of doing so.

The minor faults present more of a problem in that they do not appear to be so consistent in either cause or result and they lead to much smaller changes. They are also more of a problem in that similar events can occur in the tension profile without a fault resulting, such as momentary snags in the thread. This makes them harder to identify and the discussion of their detection correspondingly more complex. These will not be considered in such a short account.

To allow the analysis of a single stitch, when such an entity is only part of a continuous sequence, meant arbitrarily defining breaks in the data collected where one stitch could be said to finish and the next begin. For the most part, this partition was carried out at the point in the sequence when the needles were at their highest. However, for some of the later work it was found to be convenient to split the data at the preceding low point, when the needles were about to leave the tape. As a consequence, all the data relating to a mis-stitch fell into the middle of the segment.

33.5 Fourier analysis

The Fourier analysis has been performed on the data in two ways. The first is the calculation of the frequency spectrum of individual stitches in the seam; the second of a block of data covering a number of stitches.

Owing to the small variations in tension occuring from stitch to stitch even during good stitching, which would otherwise lead to discontinuities where the data steam was divided, windowing was performed (Harris, 1978). The first window used was a simple triangular system, which was good enough for the original study and was kept for some of the earlier work. This was replaced by a cosine bell on the first and last 10% of the data in the window. This was found to give better results generally and particularly for the larger faults.

In both cases, good stitches were found to give a spectrum with a pronounced peak at very low frequencies, which appears to carry most of the information, and a single secondary peak which is about 15—20 dB lower than the main peak (see Fig. 33.3.) The large variations in the tension profile due to a full mis-stitch were found to lead to a drop in the main peak of around 5 dB, and are therefore detectable. The spectrum is otherwise basically unchanged.

The lesser faults do not appear to lead to easily identifiable variations on the spectrum since they appear to be overwhelmed by the variations due to the changes in the tension which occur from stitch to stitch in normal operation. These are capable of introducing large variations into the spectrum and are the cause of the range of heights for the secondary peak mentioned above. The Fourier analysis technique therefore appears to be able to detect major faults but not minor ones.

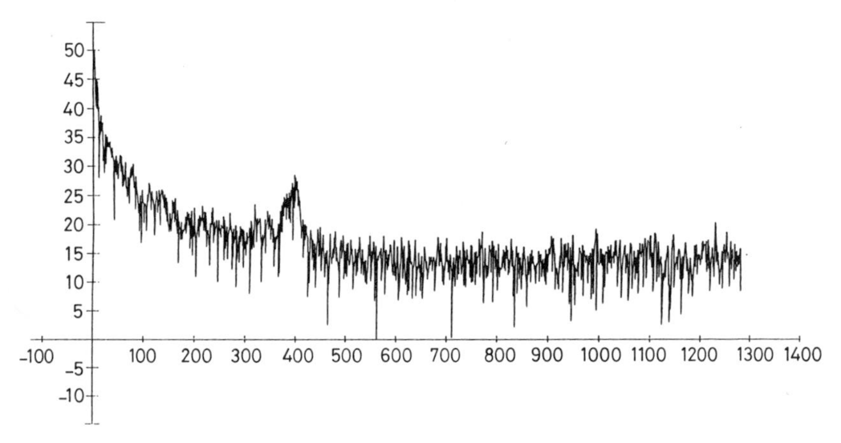

Fig. 33.3 FFT of good stitch.

33.6 Adaptive linear prediction

The normal approach when using this technique is to consider the estimates $y(n)$ made by the predictor as the useful output. In this case, the errors $e(n)$ were initially considered instead. Later the modifications made to the predictor coefficients $h(k)$ were also considered, as explained later.

By using the collected data as both reference and input for the predictor, but delaying the reference input by one stitch cycle, any fluctuations in the data from stitch to stitch cause the predictor to be innaccurate and give large errors compared to the average, and are thus, in theory, easy to spot.

The two different approaches to observing the output give differing, but in a way complementary, results. The predictor errors indicate directly the scale of any variation between two stitches, whereas the scale of the modifications made in updating the filter indicate how badly the predictor was out on its last estimate. Thus, with a small convergence coefficient, the predictor errors will remain high throughout a stitch if there is substantial difference between it and the reference stitch. The modifications to the filter will also persist for some time, but will tend to be smaller. With a large coefficient, the errors will be minimal since the predictor will adjust to compensate for the change, whereas the modifications will be large. Thus the two outputs will differ for the same variation.

The size of the convergence coefficient, in determining the rate of adaption, determines the frequency of the variations detected. In other words, as the coefficient used becomes larger, more transitory errors are detected, but, longer term, often larger errors are missed. Trying to use a coefficient which will give large errors directly for a mis-stitch is not possible since it would require little or no adaptation during the mis-stitch, so the predictor retained its memory of the previous stitch. This would then mean a very long convergence time (several dozen stitches). The range of available coefficients which will converge in a reasonable time tend to adapt to the mis-stitch as it occurs, giving smaller errors since the signal variations are smaller during the mis-stitch and the predictor is better able to track it. Spikes occur at regular intervals which are not due to the minimum MSE, but result from a period in the stitch profile where the tension varies in an irregular fashion, when the looper arm disengages the threads. The predictor errors are therefore much larger than for the rest of the cycle since the signal in this region changes every cycle. These spikes pose a problem in that they are often larger than the outputs due to variations of interest, particularly minor faults. They are basically noise and define the upper noise level which must be surpassed by the useful output signal. For full mis-stitches, this criterion is achievable, but, for the minor faults, other approaches may be necessary, such as removing the spikes completely.

The problems in detecting a mis-stitch using the predictor errors directly were the reason for considering the modifications made in adaption as well. This offered an easy solution to the problem. A mis-stitch can be clearly identified by the predictor using a 50-tap sample set using this approach with the smaller convergence coefficient (see Fig. 33.4.) Reducing the number of predictor taps to 25 reduces the clarity of the output but the mis-stitch is still picked out. Reducing the number of taps still further to 12 produces no useful output. This sets a bound on the simplicity of the filter which can be used. The ratio of the mis-stitch peak height to the noise peak heights can be improved by setting a threshold and summing the areas of the peaks which pass it. Doing so allows a reduction of the number of taps to as little as four or five. This is still about an order of magnitude more complex computationally than the autocorrelation described below, however. The predictor will also respond to faster variations, such as those from thread snags, or from minor faults, especially if the convergence coefficient is increased appropriately.

Simulations carried using perfectly periodic sequences built from the collected data, with and without faults included, and modification made to the original data to isolate the causes of particular output peaks have confirmed the above results.

In summary, this approach is much more sensitive to small variations than Fourier analysis. The algorithm appears to be successful at detecting the fluctuations associated with minor problems when the convergence coefficient is large, and can detect large changes such as those caused by full mis-stitches when the coefficient is smaller.

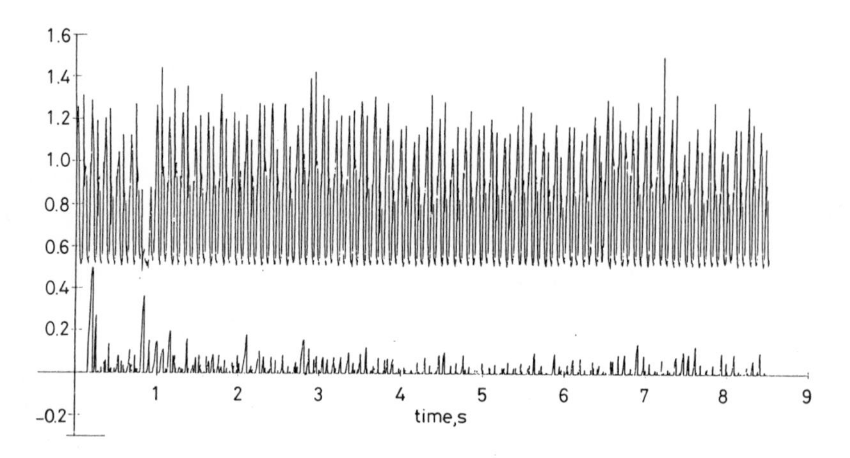

Fig. 33.4 Adaptive predictor analysis of stitching sequence.

33.7 Autocorrelation

The autocorrelation function taken for the single stitches shows that good stitches are highly correlated, slowly varying and periodic (see Fig. 33.5). The comparatively minor variations in the tension over a stitch cycle due to a minor fault do not show up in the autocorrelation function taken over the cycle, but the result of a full mis-stitch is sufficiently major to cause a large change in the function (see Fig. 33.5). The function indicates that it is much less correlated and has less periodicity.

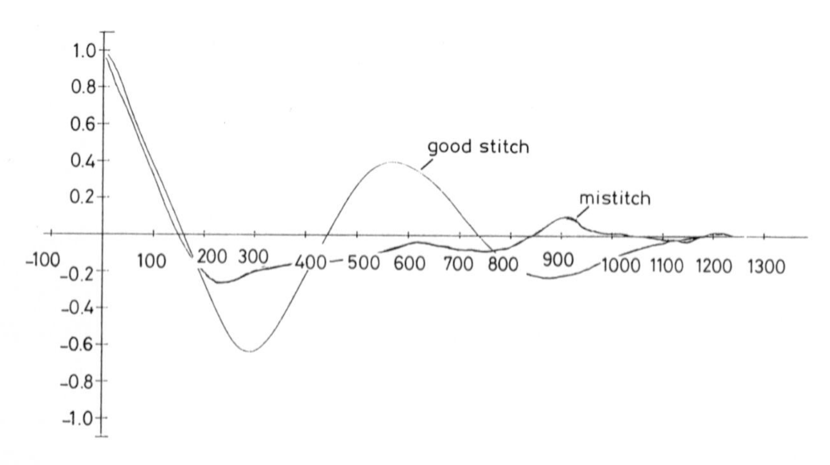

Fig. 33.5 Autocorrelation function of good and bad stitches.

This is also important for the adaptive predictor since it immediately indicates that a filter capable of tracking a good stitch will be able to handle a mis-stitch which has lower values for the function and changes slower. In this case, the autocorrelation provides a fairly robust method of detecting such a fault with much less processing. The approach investigated has been to consider one point on the function where a substantial difference exists betwen the values for a good stitch and a mis-stitch, at about the halfway point in the set of values for one stitch. The value of the function for each stitch is calculated at this one value of the delay k. By comparing the values from a sequence of stitches to a threshold, any mis-stitches will immediately show up (see Fig. 33.6). Other moderately large but more transitory changes, such as snags in the threads, also affect the output to a smaller degree. Minor faults do not appear to measurably affect the output.

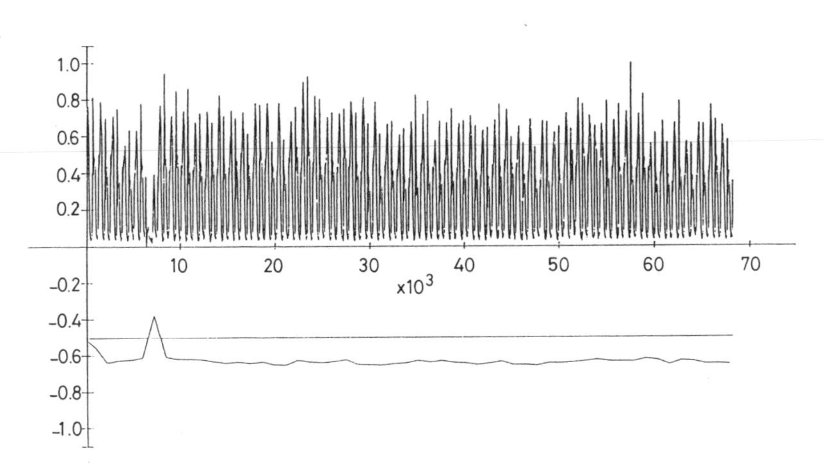

Fig. 33.6 Autocorrelation analysis of stitching sequence.

33.8 Conclusions

The conclusions drawn from the work so far are as follows:

The autocorrelation approach so far offers the best method for identifying major faults, being computationally simply and apparently robust. It also can give some indication of abnormal spikes as result from snags on the thread. These, however, do not appear to generally result in a fault, although a knowledge of their occurrence might be useful. The FFT has been largely a failure, being only able to detect major faults, and is considerably more complex. The adaptive linear prediction is able to detect major faults at a considerably greater cost in computation time.

Since it is also able to detect more minor variations, it is of greater interest and will require further work to realise its full potential.

In closing it should be mentioned that large variations of the order resulting from a full mis-stitch can also be detected by simpler algorithms such as a peak periodicity check. These, however, are likely to be less robust than the autocorrelation and the latter can be implemented in hardware (analogue or digital) very simple. This, then, is the best choice at present for the mis-stitch detector out of the above, and is likely to be as good or better than any customised simply algorithm. The longer term interests for research into further diagnosis based on this tension-oriented approach must centre on the adaptive prediction.

33.9 Acknowledgment

The authors gratefully acknowledge the financial support of the ACME directorate of the SERC for this work.

Part 4

Planning and scheduling

Chapter 34

Introduction

H. M. Ryan

Sunderland Polytechnic, UK

The final part considers planning and scheduling and the papers presented will deal with the following: requirements for MMS companion standards, planning in FMS, including production planning in high tool variety environment of FMS, a contribution to time scheduling for real-time control systems, the use of a generic model to capture process planning functional information, evaluation of a knowledge-based planning system for assembly, a feature-based approach towards intelligent reasoning for process planning, hybrid-simulation/knowledge-based systems for manufacturing, implementation of MRP II production planning and inventory control, and experience with spreadsheets in production planning.

Chapter 35, by Pandya *et al.* deals with requirements for manufacturing messaging and specification (MMS) companion standards for Production Planning and Control (PP & C) systems, and looks at their applicability to future systems such as a distributed PP & C processing facility. The conditions necessary to develop such standards are outlined and are classified primarily as the operations which the PP & C undertakes, and their names of implementation in a distribution environment.

In Chapter 36, Perera reviews several UK-based FMS and highlights the varying nature of tool management problems encountered in production planning for high tool variety environments. Operational problems are discussed together with the development and implementation of a tool flow simulator and tool availability strategies for FMS.

Jan Kuvik, CSSR (Chapter 37), presents a useful contribution to the analysis of the 'correct function' for the time scheduling of real-time working controls systems. Both ideal and practical synchronous models are considered and the present position regarding their practical application to time scheduling is discussed.

Chapter 38, by Baines and Colquhoun, describes the derivation of a Generic IDEFo Model (GIM) for process planning (with activity diagrams and 53 activities) and demonstrates how such a model can be used effectively to provide a specific 'as is' UDEFo 'model' of process planning (14 activity diagrams and 44 activities) for a company designing and manufacturing a range of complex products. The chapter describes how these techniques provide effective strategies to analyse the process planning activities of companies.

Thaler, in Chapter 39, describes the evolution and evaluation of a knowledge-based planning system for assembly which was developed in an on-going research project of the IAO, which started in 1986 in co-operation with four German companies and a software house. An overview of the project strategy is presented. To date, two prototypes have been developed and built:

(i) a research prototype, with focus on further possibilities of development

(ii) an industrial prototype with focus on the relevant company interfaces.

The paper presents examples and outlines how the functions of the prototype were assessed, including feedback suggestions from users.

Chapter 40 by Juri *et al.* discusses the use of a feature-based approach to the product data requirements of an intelligent planning system. It attempts to demonstrate that complex reasoning involved in process planning can be partially automated, by modelling the product, at a level appropriate to the task, and by providing the system with the necessary knowledge.

Jones, in Chapter 41, reports on experience with spreadsheets in production planning. Practical examples are presented, using spreadsheet-based software, covering price lists, stock calculations, material requirements planning and scheduling. The author claims that customer reaction to the work has been very good and that studies to date have demonstrated the potential for the software already available.

Chapter 42, by Farimani-Toroghi and Peck outlines the preliminary work carried out on hybrid simulation/knowledge-based systems for manufacturing. It presents a simulation perspective and goes on to incorporate knowledge-based rules at the operation and flow control level of decision making. The approach uses PROLOG language and incorporates an event-orientated approach to the description of the simulation dynamics. The work is still in a preliminary phase and further study is required to develop a consistent methodology with appropriate tools to support it.

Chapter 43 by Shafransky and Akimov considers the questions of planning in FMS, and developing intelligent program systems aimed at modelling, designing and planning of automated manufacturing strategies. The authors describe the general aims of the system development, the main tasks of the modelling procedures and the general characteristics and possibilities of the AIMGO (Aütomated, Modelling Interactive Graphics Optimisation) system.

The final chapter in this Part (Chapter 44 by Walker and Maughan) discusses some of the problems associated with the implementation of Manufacturing Resources Planning (MRP II) production management systems within manufacturing organisations, which use small — medium batch sizes and non-flow line production technologies. An architecture is proposed in which expert systems and computer simulation techniques are used in order to generate MRP and production schedules — which are both mutually consistent and feasible within the constraints imposed by the production system.

Chapter 35

Requirements for MMS companion standards for production planning and control

P. R. Bunn, K. V. Pandya and R. Sotudeh
Paisley College of Technology, and Teesside Polytechnic, UK

35.1 Introduction

The services and operations offered by the Manufacturing Messaging Specification (MMS) protocol of MAP (or TOP) are described in the DIS 9506 parts 1 & 2. These operations and services are applicable to all the programmable devices on the factory floor, e.g. robots, numerical controllers, etc. and other application processes. Development of the Companion Standards for the robots, NCs and PLCs are already under investigations in the USA.

The Companion Standards is a specific set of standard commands which provides detailed interpretations of data and variables for specific classes of devices and application processes. In this chapter we look at some of the requirements that need to be considered when developing Companion Standards for Production Planning and Control (PP&C) facilities, and how these requirements have to be applicable to the future PP&C systems, e.g. a distributed PP&C processing facility.

35.2 Production Planning and Control systems (PP&C)

It cannot be understated that the reasons for efficient PP&C implementations, to reduce lead times, work in progress, achieve the least inventory levels etc., in any manufacturing system are complex. The efficiency of the PP&C is even more paramount in automated systems for the following reasons:

(i) In automated systems the lead time are considerably shorter than those of manual systems. Thus for efficient productivity the activities of the resources have to be closely controlled and scheduled.

(ii) Highly automated manufacturing systems are expensive. To ensure a rapid payback for the high capital investment a high level of system utilisation must be maintained.

In automated systems, PP&C decisions are aided by computer based software packages, sometimes called MRP and MRP II. They may be classified as an application process which may utilise MAP (or TOP) networks. These software package are supplied by many different suppliers, Industrial Computing (1987) lists 87 packages. Each package has a wide range of separate modules. Each module will perform a task which will contribute towards the total planning and control of the manufacturing system. The different modules may be purchased in stages depending on the systems requirements and implementation. Examples of these modules are: Bill of Materials, Work in Progress, Stock Control etc. The nature of the data input to the PP & C will include details of orders, parts lists, work centres data, stores and purchase requirements, costing etc. This data will be input from a wide range of work stations (this includes the terminals which allow input/output facility at the different departments, e.g. Sales, Stores, Purchase etc.). These packages vary widely in range and in complexity. This research is looking at the modules that are involved in the direct planning and control of manufacturing environment (i.e. the research does not look at other aspects such as Finance, Payroll, etc.), and uses a particular package MICROSS from Kewill Systems.

35.3 The interface structure

A program was written to undertake simulation of a remote MAP (or TOP) user input data to an MRP package, on a MAP node, and retrieve (or even send data to a third user) output in a required format. This structure is shown in Fig. 35.1. This simulation is an example of an interface application.

The user will run the BAT program. The BAT program will execute the C program which prompts the user to input data. This data is stored in the IN file. The data stored in the file is of the format that is required to be input (as commands) into the MRP program module. In this case the MRP package selected is MICROSS. The BAT program will then input these commands, in the IN file, to MICROSS. MICROSS is thereby activated and will act according to the commands received from the IN file. The output from MICROSS can be displayed on the screen or can be

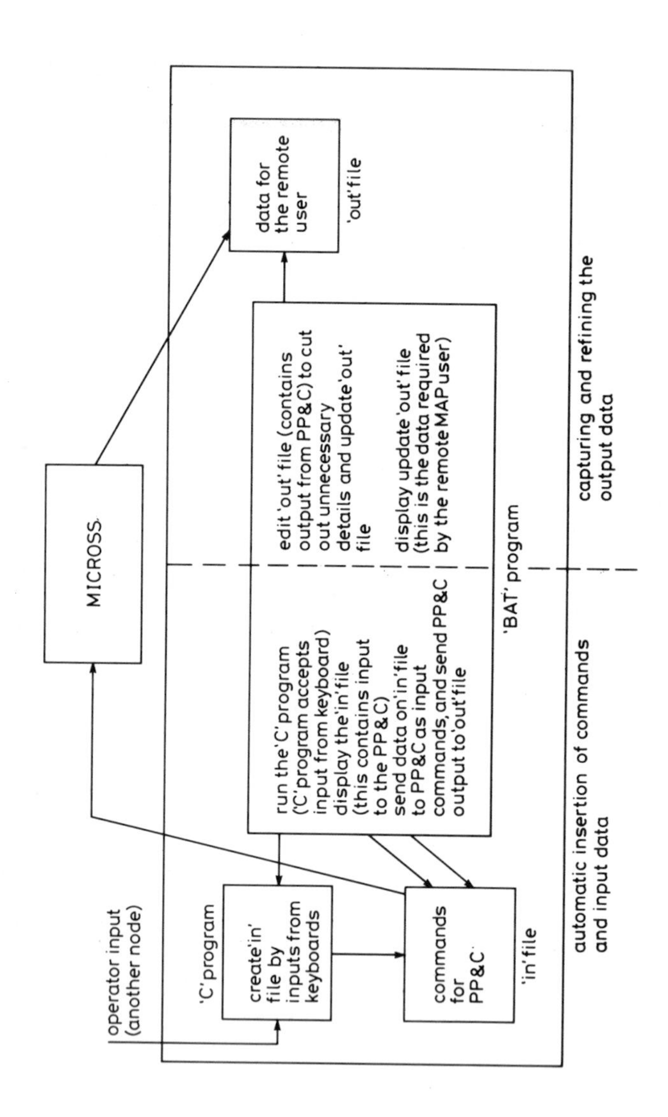

Fig. 35.1 An example of the interface scenario

sent to the OUT file. The OUT file is edited and used as necessary. The OUT file can also be sent to a remote node (which may be a third node on the MAP/TOP network).

BAT program: BAT Program is a sequence of steps carried out that allow for interactive input to the MICROSS, and capture of the output from the MICROSS. The steps are; run the C program, input the IN file, refine and trim the OUT file.

C: C program is a program written in the standard C language. It prompts the user to input data from the keyboard, according to the command format of the MICROSS. The reason for using C language was to achieve compatibility with the MMS protocols which are written in C language.

35.4 Example of the interface

A production manager (who is linked to the MAP network through a terminal) may wish to send a work-to-list file (which is in MICROSS, on a MAP network) to a Cell Control node. The manager will run the BAT program, which will prompt him to input the data according to MICROSS's requirement. The input from the keyboard is stored in the IN file. The BAT program will transfer the data (IN file) to MICROSS which will act upon it. The output will be sent to the Cell Controller.

35.5 Implications of the interface structure

The interface structure outlined here is a simple application concept as shown in Fig. 35.2 (Bunn & Pandya, 1988). All the programs, i.e. BAT and C, are resident at the same node.

The following extensions can be made:

(i) The input (IN file) is created at the manager's terminal. The information transmitted to the MICROSS node over the MAP network are only the commands in the exact format as required to execute MICROSS for the required output, i.e. the work-to-list file. In this case the BAT program, and the C program and the IN file are all resident at the manager's node.

Alternatively, the C program and a part of the BAT program may be resident at the manager's node. The IN file created may be resident at the MICROSS node. In this case a set of IN files might be stored in a queue at the MICROSS node. The files are then input to the MICROSS by another BAT program.

(ii) Once MICROSS has executed the output it is sent to the OUT file. This file may be displayed at the terminal of the MICROSS node, or sent

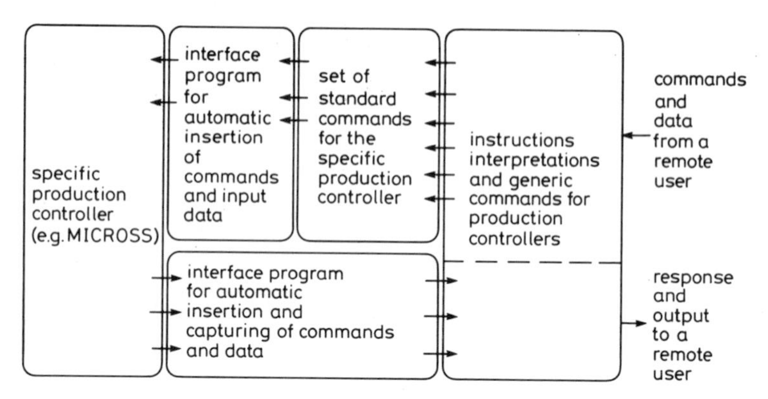

Fig. 35.2 Concept of the interface structure

either to the manager's node or to the Cell Controller node. The decision as to which node the output is sent to may be made by the manager at the time of initiation of the communication with MICROSS, or it could be a restriction placed when the system is initially implemented.
(iii) The editing and display ('trimming and refining' in the Figure) of the OUT file can take place at any of the nodes.

The interface illustrated here shows how the MRP module in MICROSS can be remotely activated and the results sent to another site. This is the concept of the interface required to 'plug-in' MRP to the MAP network. The BAT program has shown how the MRP can have its input treated and have its output refined. It also showed how in the future a distributed PP&C facility can be achieved. A certain way forward is to allow some form of Controlling and Scheduling facility at each of the nodes on the network that deal directly with the manufacturing activities. This is explained in the following sections.

The developments for Companion Standards (CS) for PP&C must take into consideration the following:

(i) The BAT program is dependent on the operating system of the computer on which it is based. In this case the operating system is MSDOS.
(ii) There is a good possibility that the MRP program module will be based on a different operating system to the BAT program. This problem is amplified when considering the distributed PP&C facility.

35.6 Hierarchy levels of PP&C

The interface concept developed in this research is related only to *Conventional* PP&C. In this type of PP&C all the decisions of the planning and scheduling are performed as one stage of the manufacturing process cycle; e.g. in almost all companies the decision of the production planning and control (i.e. what is to be made at which machines at what time) is made at the office level. The control and scheduling at the work centre levels and the cell level is very minimal. In some companies even some planning decisions necessary after a breakdown are taken by the senior manager's office.

The trend in the near future is to look at the Production Planning & Control in five Hierarchical Levels (Barkermeyer *et al.*, 1986; O'Grady, 1986), namely Corporate, Factory, Shop, Cell and Equipment levels. This research is concerned with the bottom four levels of the hierarchy and considers that the controlling and the scheduling in these four levels can be arranged in the two extremes shown in the Figs. 35.3 and 35.4. In practice, the factory environment may be a combination of the two extremes shown. For example, a company may not have a shop level, or its factory level will consist only of Sales and Production facilities, or in the shop level a company might only have one function shop (for instance Milling shop), or a company many manufacture only one product. There is also a reasonable possibility that an inspection cell may be shared by two or more shops.

The time scale of the levels range from a week or month for the factory level to the seconds or milliseconds for the equipment level. The decisions made at each level will be dependent on the planning and control system implemented. In some implementations, a cell controller will make decisions on every aspect of planning and control within the cell. In some implementations the cell controller will only make decisions about maintenance; it will not be permitted to re-schedule an order in case of a beakdown.

35.7 Generic commands

The set of specific commands that will be developed for the conventional PP&C will have to be adapted to the distributed PP&C described above. All the generic commands developed will not be available at all levels. This is dependent on the decisions allowed at that level.

35.8 Conclusion

In this chapter we have outlined the considerations necessary to develop the Companion Standards for PP&C. They can be classified as primarily the operations which the PP&C undertakes and how these operations are implemented in a distributed environment.

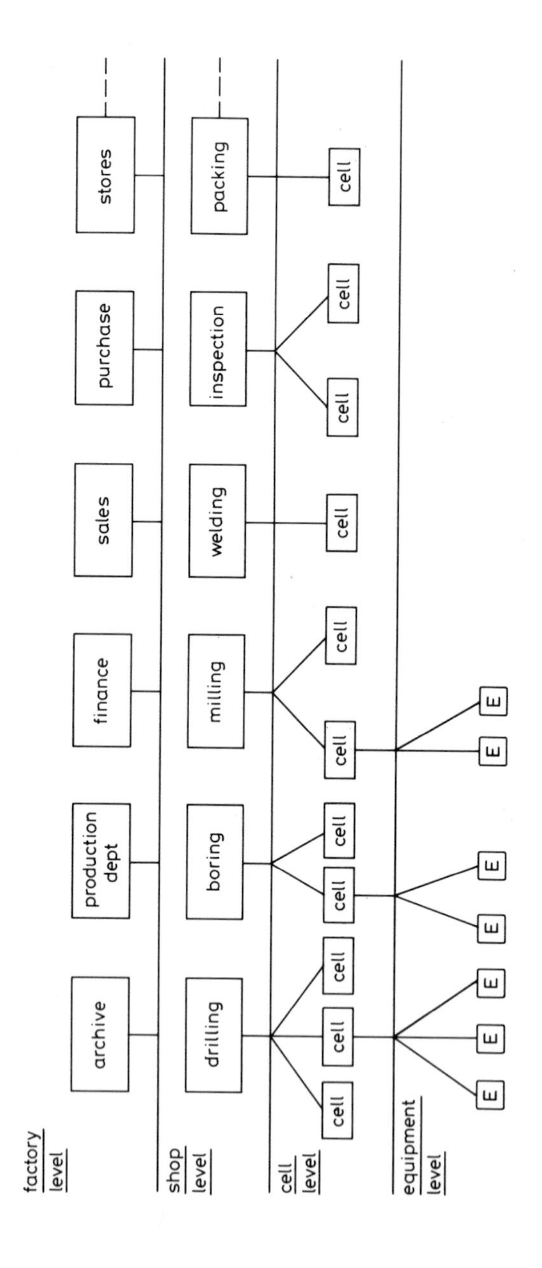

Fig. 35.3 Control and planning arrangement based on 'functional grouping'

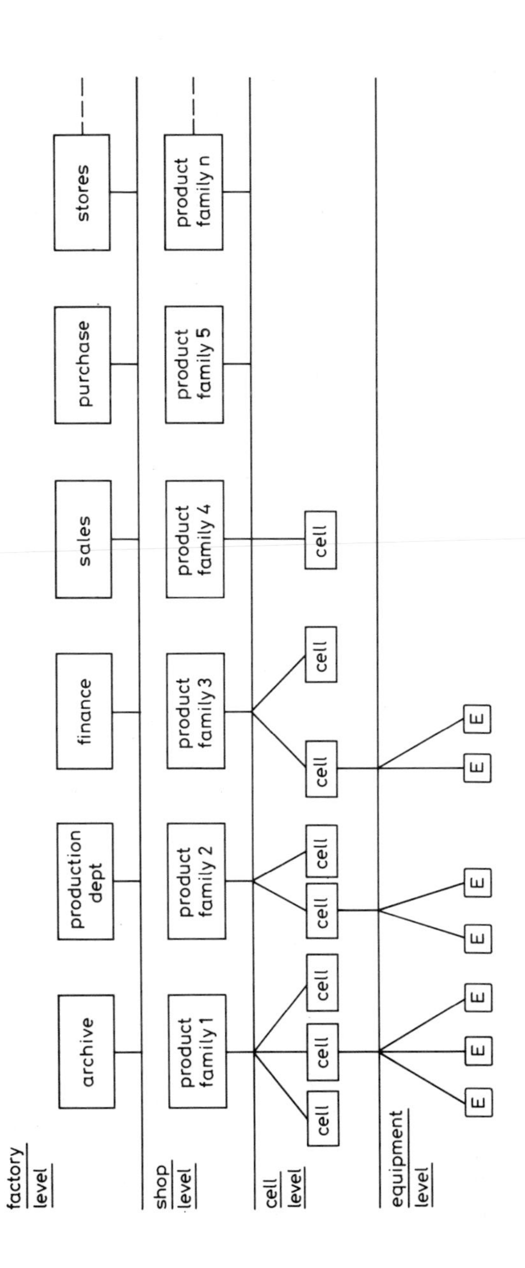

Fig. 35.4 Control and planning arrangement based on 'product grouping'

Chapter 36

Production planning in a high tool variety environment of FMS

D. T. S. Perera

Sheffield City Polytechnic, UK

36.1 Introduction

As flexible manufacturing systems (FMS) have evolved through several generations, technologies associated with these systems have been constantly enhanced to solve the various operational problems generated by these systems. Sadly, not all of these technologies have received adequate attention from the system designers, users and researchers. One of the lagging areas is that of tool management. Failure to perceive the importance of tooling aspects of the system has led to poor performance in many systems.

In this chapter, several UK based FMS are reviewed to highlight the varying nature of the tool management problems. In particular, operational problems. In particular, operational problems in a high tool variety environment are discussed. The development of a tool flow simulator is presented and the way it is used to develop tool availability strategies for a real FMS is explained.

36.2 Varying complexity of the tool management problem

There are about 40—50 flexible manufacturing cells (FMC) and FMSs in UK. Although all FMSs encompass similar hardware sub-systems such as workstations, material handling systems etc., they are operated in different ways. In particular, in the management and control of tool flow, a wide variety of operational strategies are adopted. In the following four examples, tooling aspects of systems are discussed to highlight the varying complexity of the tool management problem.

36.2.1. System A

This is a relatively small FMS for the production of valve components. The system includes four identical machining centres each with a free standing tool magazine with space for 100 tools. One special feature of this system is that FMS is used in assembly related batch manufacture, so that the sets of parts for a valve assembly are produced in line with the assembly programme. The tool types at a workstation are left in the magazine until the product mix changes, at which point some of the tools in the magazine are replaced by new tool types. This approach of bulk tool exchanges keeps the tool management problem relatively simple.

36.2.2 System B

This company manufactures a range of diesel engines with powers up to 8700 kW. At present the system consists of two workstations and the company expects to add two more machines, in the next stage of implementation. Batch sizes vary between 1 and 150. The high tool requirements forced the company to install an automatic tool delivery system between the tool store (180 stations) and tool magazine. A rail guided truck with an integral tool transfer device takes worn tools out the magazine (72 ports) and replaces them with new tools from the store. Although the system consists of two workstations, it has a quite complex tool management problem owing to high tool variety. A set of 20—25 different types require 300 different tool types.

36.2.3 System C

This is one of the most advanced FMSs in UK. It consists of seven machining centres each with an 80 station tool magazine and auxiliary 24 tool carousel, a co-ordinate measuring machine and two washing stations. The system includes several AGVs which can deliver and remove 24-tool turrets at each machining centre. The system has a capacity to machine 28 part types and their variants at a rate of 1400 parts per week. Worn out tools are removed from the turrets at a centralised tool maintenance and presetting facility. New tools are replaced in the turret ready for an AGV to deliver them back to the workstation concerned. A second tool changer at each workstation transfers tools between the turret and magazine. There is a computer link between the company and the tool supplier to ensure replenishment of all tools and tooling components within 24 hours. The integral tool handling system and the nature of production create a complex tool management problem.

36.2.4 System D

This is the largest, most sophisticated and intricate flexible manufacturing facility in UK. This is capable of machining 1500 part types in steel, titanium and aluminium alloys in bathes of 5—10. It consists of two FMSs with different hardware configurations and operationsl features. Notably, different methods of tool handling systems are adopted. In one system, crates of tools are delivered to workstations by AGVs. At a workstation, the AGV exchanges the new crate for the old crate which is then delivered to the central tool storage area. A tool crate can accommodate 63 tools, which may be sufficient for several different parts. The second system is dedicated parts manufactured from hard materials which consume cutting tools quickly, necessitating more frequent replacements. An AGV deliver crates with new tools to the system and then they are transferred to a robot trolley which replaces the new tools with used tools in the magazine. Tool management is a major task. There are several thousands tool assemblies which have to be managed by the system.

It is clear from the above discussion, that different tool management policies need to be implemented in different systems. Specifically, in high tool variety environments, more attention must be focused on tool management as tools can impose additional constraints on production planning and control strategies.

36.3 High tool variety environment

There are several system parameters (such as the number of tools, the number of tools required per operation and the tool exchange rates etc.) which influence the management and control of tool flow within the system (Perera, 1988*a*). The nature of tool management problems are highly dependent on the ratio between aggregate tool requirements (at workstation) and the tool magazine capacity.

$$\text{Tool variety index (TVI)} = \frac{\text{aggregate tool requirement at workstation}}{\text{tool magazine capacity}}$$

As aggregate tool requirements depend on the product mix, this is a dynamic parameter. If TVI is less then unity during a particular planning period, then tools are exchanged owing to wear and breakages only. However when TVI exceeds unity (high tool variety environment), additional tool types exchanges occur owing to product variety (i.e. certain tool types may not be readily available at workstations). Worn or broken tools are replaced by their duplicates and new tools can take up the same location in the tool magazine. On the other hand, tool exchanges due to product variety force some other tool types out of the magazine. The selection of tools to be unloaded is a crucial factor. For example, if a tool used by many print types is taken out, the number of tool exchanges is unnecessarily increased. This could be further complicated if different sizes of tool exist. When the tool exchange rate is high owing to product variety, two different strategies can be considered to improve the situation.

36.3.1 Increasing the tool magazine capacity

Tool magazines which have modular architecture may be extended to accommodate more tools. Although this is technically possible, it is quite difficult to implement in practice. This option must be considered when there is an upward trend in the number of tool exchanges. If excessive numbers of tool exchanges are observed only in certain planning periods, perhaps the operational policies can be enhanced to reduce the number of tool exchanges.

36.3.2 Changing operational policies

In this approach, the number of tool exchanges is reduced by controlling the product mix and/or adopting more effective tool availability strategies at workstations. Although it is desirable, to reduce the number of tool exchanges due to product variety to zero, it is not essential. What is important is that the number of exchanges is reduced to a manageable level. In the following Sections, this approach is discussed in detail with reference to a real FMS.

36.4 A flexible manufacturing system

This system produces components (gear box housing, motor housing etc.) of heavy mining equipment (Carrie *et al.*, 1984). The system consists of five indentical machining centres (with a 100 station tool magazine) and a special workstation. These machining centres, together with two load/unload stations, are served by a single automated guided vehicle. Identically tooled machine groups provide alternative routes for the parts which have very complex processing characteristics. Every part types requires two or more fixtures. Processing times can vary from a few minutes to several hours.

In this system tools impose severe constraints on management decisions. The following system data highlights the complexity of the tooling problem.

36.4.1 System data

At present 12 different part types and their variants are processed and require 372 different tool types. The growth of tool requirements at workstations is shown in Fig. 36.1. As the Tool Variety Index exceeds unity, tool exchanges due to product variety are unavoidable for certain product mixes. The situation is further complicated by the fact that tools vary in size. The following classes of tools are found in the system:

Class I tools which take only tool position.
Class II tools which take up only one position, but, because of their size,

prevent another class II tool being positioned in the pocket next to it.
Class III tools which take up three positions: the pocket in which the tool is placed and the positions on either side.

The distribution of tool classes in the system is shown in Table 36.1.

Table 36.1 Distribution of tool classes

Class I	Class II	Class III
293	34	45

The average number of tools needed per operation is about 18 tools.

36.4.2 Operational problems

As expected, for certain product mixes, the number of tools exchanges due product variety was very high. The traditional approach to this problem is to split production requirements into batches of parts with similar tool requirements. These batches are released to the system in a sequential manner with bulk tool exchanges between batches. However, in this system, production requirements cannot be grouped owing to several reasons. The major factors are small batch sizes, long processing times and random processing of parts. Owing to random processing and small batch sizes, the product mix changes rapidly, and as processing times are long the system may not reach a steady state. Thus, instead of bulk tool exchanges, tools are changed constantly to support the varying product mix.

In the case of bulk tool exchanges, the number of tools involved can be estimated by aggregating the tools required for each batch. In this case a simulation model is required to compute the number of tool exchanges.

36.5 Tool flow simulation

As a part of failure to perceive the importance of tool management aspects, most simulation studies are confined to part flow simulation. Although there are numerous simulation systems available for part flow simulation, there are no proven tools to study tool flow within manufacturing systems. Furthermore, many simulation systems do not provide essential features required for tool flow simulation (Perera, 1988*b*). For example, extensive data handling capabilities are more important in tool flow simulation, as even a simple FMS can claim a massive tool database.

A tool flow simulator (Perera & Carrie, 1987) was developed for this system to study the nature of tool exchages in detail. It was done in two stages:

36.5.1 Tool post-processor

The post-processor reads a file of work flow data, and, by referring to the part routing and tool requirement files, maintains a list of tools which would be present in each tool magazine. Hence the occurrence of tool changes due to product variety can be deduced. By aggregating the cutting time of each tool on each operation performed, the program computes the occurrence of tool changes due to usage.

36.5.2. Tool flow simulator

Although the tool post-processor can output essential statistics about tool exchange rates, it has a major drawback, i.e. it cannot be used to study interactions between part flow and tool flow. Thus program modules written for the tool post-processor were enhanced and linked to a part flow simulator. A separate user-interactive module was also constructed to set up the system tool configuration before simulation. These tools were used to evaluate various tool availability strategies which are explained below.

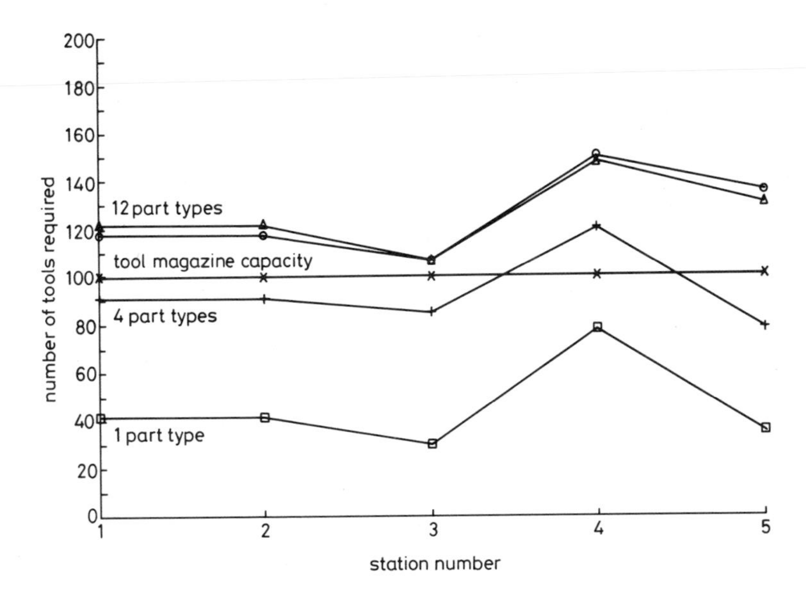

Fig. 36.1 Growth of tool requirements

36.6 Tool availability strategies

As different sizes of tool are involved and aggregate tool requirements exceed the magazine capacity, tool loading must be done logically to minimise the number of intermediate tool exchanges due to product variety. Different strategies are described and evaluated below, using the data drawn from the FMS.

36.6.1 Strategy I: Load all class III tools first

Class III tools are the most difficult ones to handle, as they need three consecutive empty pockets on the magazine. When tools are exchanged, it is easy to change a class III tool with another class III tool rather than a set of class I and class II tools. Therefore all class III tools are loaded first and remaining pockets are filled with class I and class II tools alternatively. (Note: Class II tool cannot be placed in neighbouring pockets.)

The system was simulated for a period of six weeks (the output data corresponding to first two weeks are omitted to eliminate transient effect). The results are shown in Table 36.2.

Table 36.2 The rate of tool exchanges

Rate of tool exchanges (exchanges/week)	Week No.			
	3	4	5	6
Due to wear	128	114	131	140
Due to product variety	409	360	415	387

The number of tool exchanges due to product variety is very high and some changes in tool availability strategies are required to reduce it to a manageable level.

36.6.2 Strategy II: Strategy I with a set of permanent tools

It was thought the number of tool exchanges due to product variety can be reduced if a set of tools is held permanently in the magazine. The tools which are used by many part types can be given a higher priority. The distribution of the number of parts assigned to each tool type is shown in Fig. 36.2. It is evident that most of the tools are used only by one part type. There are also many tools consumed by many part types. The tools are ranked in the order of decreasing usage by part types and a set of tools taken from the top of the list is classified as permanent tools. The results are shown in Table 36.3. Some improvement in the number of tool exchanges was note, but it was not very significant.

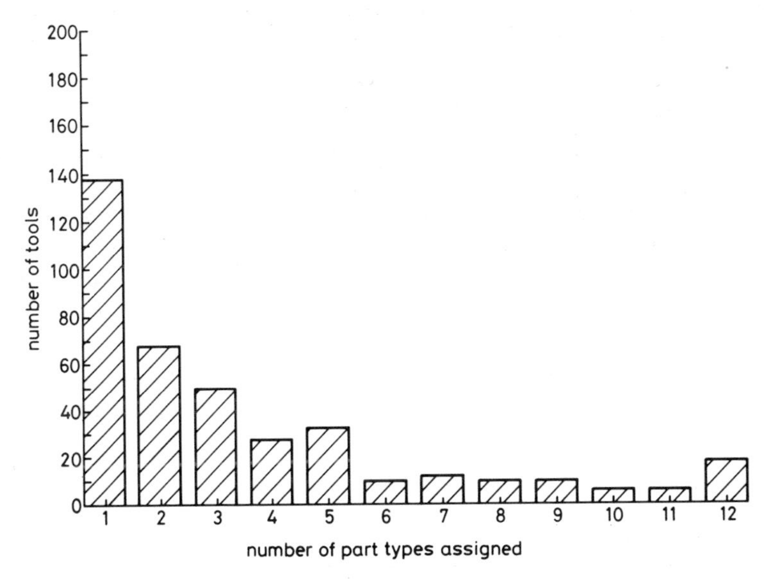

Fig. 36.2 Number of part types assigned to tools

Table 36.3 The number of tool exchanges due to product variety

	Week No.			
	1	2	3	4
Strategy I	409	360	415	387
Strategy II				
20 permanent tools	394(3.7%)	346(3.8%)	394(5.1%)	366(5.4%)
30 permanent tools	389(4.9%)	327(9.2%)	378(8.9%)	347(10.3%)
Strategy III				
(no permanent tools)				
10 Class III tools	340(16.8%)	319(11.3%)	335(19.2%)	298(22.9%)
Strategy IV				
(10 Class III tools)				
20 permanent tools	324(20.7%)	298(17.2%)	313(24.5%)	279(27.9%)
30 permanent tools	323(21.0%)	301(16.4%)	311(62.6%)	275(28.9%)
Strategy V	196(52.0%)	181(49.7%)	155(62.6%)	194(49.8%)

(#) Percentage reduction

36.6.3 Strategy III: A limited set of class III tools

As class III tools take up three pockets in the magazine, there may be quite a number of pockets which cannot be used. Thus, if the number of class III tools held at the magazine is reduced, more tools from other classes can be loaded.

In order to avoid a class III tool being changed with a set of class I and class II tools, a certain number of class III tools have to be maintained at the magazine. The tool requirements for operations are grouped under each class and the maximum number of tools used in each class at each workstation are shown in Table 36.4.

Table 36.4 The maximum number of tools used in a single operation

Station	Class I	Class II	Class III
1	36	7	8
2	36	7	8
3	33	5	2
4	39	4	5
5	35	3	8

For example, although about 20 class III tools are assigned to station 5, there is no need to hold all of them at the magazine, as the maximum number of class III tools used in a single operation is only eight. A significant reduction in the number of tool exchanges was noted (Table 36.3).

36.6.4 Strategy IV: A limited number of class III tools with a set of permanent tools

The last two strategies are blended here and results are shown in Table 36.3.

36.6.5. Strategy V: Strategy IV and the part release control

In the above strategies, the tool configurations at tool magazines were changed but it was realised that further improvements could be achieved if part input sequence is controlled. An algorithm based on a mathematical model is used in conjunction with the simulator (Perera & Carrie, 1987; Perera, 1988*a*). This alogrithm communicates with the simulator to obtain possible future tool changes. Significant reduction in tool exchanges was achieved (Table 36.3).

36.7 Conclusion

The management and control of tool flow within the system can be a complex problem. More attention must be given to tooling aspects of the system at the design stage.

Owing to variety of reasons, the part types assigned to an FMS can be changed and can decisively change the way tools are managed within the system. Tool availability strategies and other control policies (Perera, 1988*b*) must be refined to overcome these new problems.

Chapter 37

A contribution to time scheduling for real-time working control systems

J. Kuvik

System Engineering Institute of Slovak Industry, CSSR

37.1 Introduction

Real time control represents in general an asynchronous stochastic system of activities being initiated by various requests from process, operator, algorithm real time module etc. The activities are done by computer programs. From the time scheduling point-of-view there exist two groups of program. In the first one it is necessary to guarantee that each a program is completely processed within a special interval following the occurrence of the initiating signal. Such programs are referred in Serlin (1972) as time-critical processes (TCPs). This objective can be met if, for example, each TCP is executed in a separate, 'sufficiently' fast CPU, or, if a single CPU of some high, but finite, execution speed is provided. We consider a single CPU.

The second group of programs ought to be done in a 'reasonable' but not a critical time. They are referred to in this paper as Time Non-Critical Processes (TNCP's).

Our first aim is the design of individual program time scheduling, to ensure all TCP's being carried out in the requested time. Besides, we need to gain time for carrying out the programs concerning the TNCPs. Serlin (1972) defined the idle time, and it has been minimised. In our case, the time during which the CPU does not serve any TCP is not idle; we say it is 'the rest' time from the TCP point of view. In our design we use only the necessary time for the TCP, in order to get maximum rest time.

We now introduce an approach to time scheduling by a synchronous model.

37.2 Ideal synchronous model

The values of critical times for the activities are given by the dynamics of the controlled object and its parts. They are often very different; e.g. for temperature measurement a longer time period is demanded than for preasure measurement, etc. Corresponding programs can be initiated periodically or asynchronously. The model generally represents an asynchronous stochastic system. Because of its stochastic character such a system does not enable us to analyse properly the proposed control system from the time scheduling point of view.

Therefore we allocate to each activity a critical time in which the corresponding program ought to be executed. From Serlin (1972), a basic assumption underlying the study of TCP scheduling is that the external world is insensitive to the actual time at which the process is executed, so long as the process receives the time of CPU somewhere between the occurrence of the associated interrupt and the deadline. This means that the results are actually transferred out of the computer at the requested time.

For asynchronous processes being initiated by interrupts, it is the minimal time which can occur between two interrupts. For synchronous processes it is a time less than the repetition period.

Let the critical times of individual activities be ${}^it_c$ ($i = 1, 2 \ldots$ n) and the corresponding maximal CPU times for programs executions be ${}^it_p$ ($i = 1, 2, \ldots, n$). We substitute the initiation of activities (asynchronous or synchronous) by periodic ones, their periods being ${}^iT_c < {}^it_c$ ($i = 1, 2, \ldots, n$). Assuming that individual programs are, during the times ${}^it_p$ ($i = 1, 2, \ldots, n$), interruptible and form an ordered set of activities with priorities of execution according to growing critical times, we get a synchronous model for time scheduling:

$$M = \left\{{}^1T_c, {}^2T_c, {}^3T_c, \ldots\right\}, \text{ where } {}^1T_c \leqslant {}^2T_c \leqslant {}^3T_c \leqslant \ldots,$$

Fig 37.1 is an example of a synchronous model for $i = 3$ activities ordered in a set of growing critical times: $\left\{{}^1t_c; {}^2T_c; {}^3T_c\right\}$.
The initiation of individual programs and the transitions due to the priorities are represented by dotted lines. The transition times from one priority level to another one are short in comparison with the critical times of industrial processes, and they are neglected here. However, it is possible, if necessary, to add them to the corresponding CPU times.

This model was constructed for the worst case, because:

- the shortest times of occurrence of requests were taken into account
- the maximal values of program durations were assumed.

To find the right conditions for the time scheduler we define in Fig. 37.1 the rest times ${}^it_r$ ($i = 1, 2, \ldots, n$). For the highest priority level the rest time in interval

$${}^1T_c \text{ is} \qquad {}^1t_r = {}^1T_c - {}^1t_p. \tag{37.1}$$

The rest times on lower priority levels ($i > 1$) are influenced by computation times of higher priority levels: i.e. for $i = 2$, the rest time on interval ${}^2T_r$ is

$$ {}^2t_c > {}^2T_c - {}^2t_p - 1 . {}^1t_p \tag{37.2} $$

for the i the priority level it is

$$ {}^it_r > {}^iT_c - \sum_{j=1}^{i} \sum_{k=1}^{l(j)} {}^jt_{p,\,k} \tag{37.3} $$

where
j = index for the same and higher priorities
$l(j) = {}^iT_c/{}^jT_c$ = number of computations on the jth priority level and for the time interval ${}^iT_c$ (the result is rounded up, if the ratio is not integer).
k = index for the summation of time computations on the individual priority level

The rest time of all TCP processes is then

$$ {}^nt_r > {}^nT_c - \sum_{j=1}^{n} \sum_{k=1}^{l(j)} {}^jt_{p,\,k} \tag{37.4} $$

and can be used for serving the TNCPs. All the rest times can consist of more individual parts (Figure 37.1).

The system will guarantee the serving of all TCPs if, for each priority level, the following are fulfilled:

- The computation must be finished earlier, then the next request on the same priority level will occurs, i.e.

$$ \sum_{j=1}^{i} \sum_{k=1}^{l(j)} {}^jt_{p,\,k} < {}^iT_c \text{ for } \forall i \tag{37.5} $$

where $l(j) = {}^iT_c/{}^jT_c$ (rounded up if the ratio is not integer)

- The rest time of the ith level corresponding to the i + lth level period is greater than the computation time of i + lth level, i.e.

$$ \sum_{m=1}^{q} {}^it_{r,\,m} > {}^{i+1}t_p \text{ for } \forall i \tag{37.6} $$

where $q = {}^{i+1}T_c/{}^iT_c$ (rounded down, if the ratio is not integer).

- Rest time of the ith priority level must be the greater than the sum of the computation times on the all lower priorities corresponding to the interval of the lowest priority of TCPs, i.e.

$$\sum_{m=1}^{q} {}^{i}t_{r,\,m} > \sum_{j=i+1}^{n} \sum_{k=1}^{l(j)} {}^{j}t_{p,\,k} \tag{37.7}$$

where $q = {}^{n}T_c/{}^{i}T_c$ and $l(j) = {}^{n}Tc/{}^{j}Tc$ are integers as mentioned above.

The above model divides the TCPs into priority levels so that only one TCP belongs to each level. Furthermore, each level can be initiated in practice by an asynchronous interrupt signal. The values of ${}^{i}T_c$ in the model need not be general integer ones. These circumstances lead to the maximum rest time of TCPs, which can be used for the computation of TNCPs. Here maximum rest time does not mean the maximum on account of the TCP program (minimisation of ${}^{i}t_p$) quality. It means only the correct organisation of program initiations. Therefore we refer here to the synchronous model as ideal one.

The grade of CPU time utilisation in a given period can be due to (37.1) represented by a load factor L which for ith level is

$${}^{i}L = \frac{{}^{i}t_p}{{}^{i}T_c} \tag{37.8}$$

In our model the load factor of the whole TCP system must be

$$L = \sum_{i=1}^{n} {}^{i}L < 1 \tag{37.9}$$

in a period ${}^{n}T_c$, because, in the opposite case the failure of all scheduling policies would occur. If this were the case for this type model, it would be necessary to use faster or parallel processing.

37.3 Practical synchronous model

However, it is practically impossible in real control systems of some hundreds of activities, to initiate each by a separate interrupt signal. Experience has taught us to minimise number of interrupts from a controlled process to the computer; there is a danger of electrical noise caused by switching on/off the powerful engines etc.

The grouping of individual activities is made in such a way that critical times are relative to the same values in one group. Let

$$\{t_o, t\} \quad (t = 1, 2, \ldots\ldots, u) \tag{37.10}$$

$$\{t_p, t\} \quad (t = 1, 2, \ldots\ldots, u) \tag{37.11}$$

be a unordered set of maximal computation times corresponding the set (37.10). The elements of (37.10) can represent the time necessary for input signal sensing, the time in which a computation must be done, the time in which an output signal must be generated etc. Ordering (37.10) for increasing values of time we get

$$\{t_o, t\} \quad (t = 1, 2, \ldots\ldots, u) \tag{37.12}$$

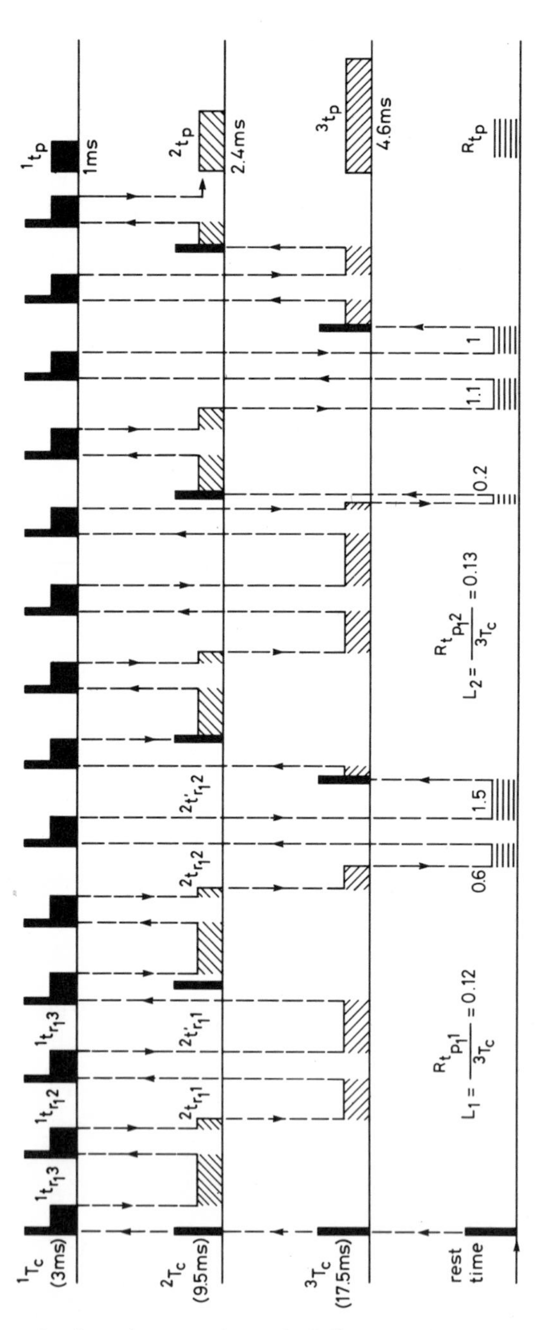

Fig. 37.1 Example of synchronous time scheduling model

where the first element has least time value and the last element has the greatest. To an individual elements order of (37.10) we form the corresponding order of elements in a set:

$$\{t_{p}, t\} \quad (t = 1, 2, \ldots\ldots, u) \tag{37.13}$$

Dividing (37.12) into n subsets in which the times are relatively comparable, we get

$$\begin{aligned} &\{^{1}t_{o,\,t}\} \quad (t = 1, 2, \ldots\ldots W) \\ &\{^{2}t_{o,\,t}\} \quad (t = w + 1, \ldots\ldots x) \\ &\{^{n}t_{o,\,t}\} \quad (t = y + 1, \ldots\ldots u) \end{aligned} \tag{37.14}$$

Similarly we get the subsets of (37.13):

$$\{^{1}t_{p,\,t}\} \quad \{^{2}t_{p,\,t}\} \quad \ldots \quad \{^{n}t_{p,\,t}\} \tag{37.15}$$

The first elements of each subset (37.14) are minimal and represent the critical time of the group. Let us transform the subsets in the n activities, being represented by their critical times and by their computation times. Let

$$^{1}t_{o,\,1} = {}^{1}t_{c},\ {}^{2}t_{o,\,w+1} = {}^{2}t_{c},\ \ldots {}^{n}t_{o,\,y+1} = {}^{n}t_{c} \tag{37.16}$$

We construct the ordered set

$$\{^{1}t_{c},\ {}^{2}t_{c},\ \ldots\ {}^{n}t_{c}\},$$

where $^{1}t_{c} < {}^{2}t_{c} < \ldots < {}^{n}t_{c}$ and get the equivalent model as described previously. Corresponding computation times are then

$$\begin{aligned} ^{1}t_{p} &= \sum_{t=1}^{w} {}^{1}t_{p,\,t} \\ ^{2}t_{p} &= \sum_{t=w+1}^{x} {}^{2}t_{p,\,t} \\ &\ \ \vdots \\ ^{n}t_{p} &= \sum_{t=y+1}^{\upsilon} {}^{n}t_{p,\,t} \end{aligned} \tag{37.17}$$

However, the difference with previous model is in the fact that only the first activities in groups are done in an effective way from time scheduling point of view. The rest activities in each group are served more often than necessary and on account of rest times. By increasing n we approach the previous model and get a growth in times. However, this is on account of the growing number the priority levels. It is appropriate to find the minimal number of levels where the system guarantees the correct serving of TCPs, and analyse whether the rest time is sufficient for serving the TNCPs.

We will give an example of practical model construction.

Example: The set of 12 activities (with their computation times in brackets) is given by their minimal times of occurrence (in milliseconds) as follows:

t_o, 1 = 2.3 (0.1) t_o, 5 = 28.3 (0.5) t_o, 9 = 25.7 (0.6)
t_o, 2 = 11.2 (0.3) t_o, 6 = 34.4 (1.4) t_o, 10 = 5.3 (0.3)
t_o, 3 = 12.5 (0.4) t_o, 7 = 3.9 (0.2) t_o, 11 = 35.2 (1.8)
t_o, 4 = 17.2 (0.3) t_o, 8 = 14.3 (0.2) t_o, 12 = 19.1 (0.5)

Let us construct a practical synchronous model consisting of three priority levels.

In accordance with (37.12) and (37.14) we get:

$$\{{}^1t_{o,\,i}\} = \{2.3;\ 3.9;\ 5.3\} \qquad (i = 1, 2, 3)$$

$$\{{}^2t_{o,\,i}\} = \{11.2;\ 12.5;\ 14.3;\ 17.2;\ 19.1\} \qquad (i = 4, 5, \ldots 8)$$

$$\{{}^3t_{o,\,i}\} = \{25.7;\ 28.3;\ 34.3;\ 35.2\} \qquad (i = 9, \ldots, 12)$$

Owing to (37.17) we get the computation times for individual levels ${}^1t_p = 0.6$ ms, ${}^2t_p = 1.7$ ms, ${}^3t_p = 4.3$ ms. Taking the first elements as the critical times for the priority levels, we get synchronous model

$$M = \{{}^1t_c,\ {}^2t_c,\ {}^3t_c\} = \{2.3;\ 11.2;\ 25.7\}$$

From a practical point of view, it is advantageous to choose the critical times on lower priority levels as integer multiples of the critical times on the highest priority level. After rounding the times down and using the above-mentioned advantage we get model

$$M = \{2;\ 10;\ 24\},$$

which fulfils the conditions (37.5) to (37.7) and guarantees the right serving of all TCPs. The graphical interpretation is shown in Fig. 37.2.

37.4 Practical realisation

One way of time scheduling realisation has already been mentioned. Each priority level is initiated by its own asynchronous interrupt signal. It leads

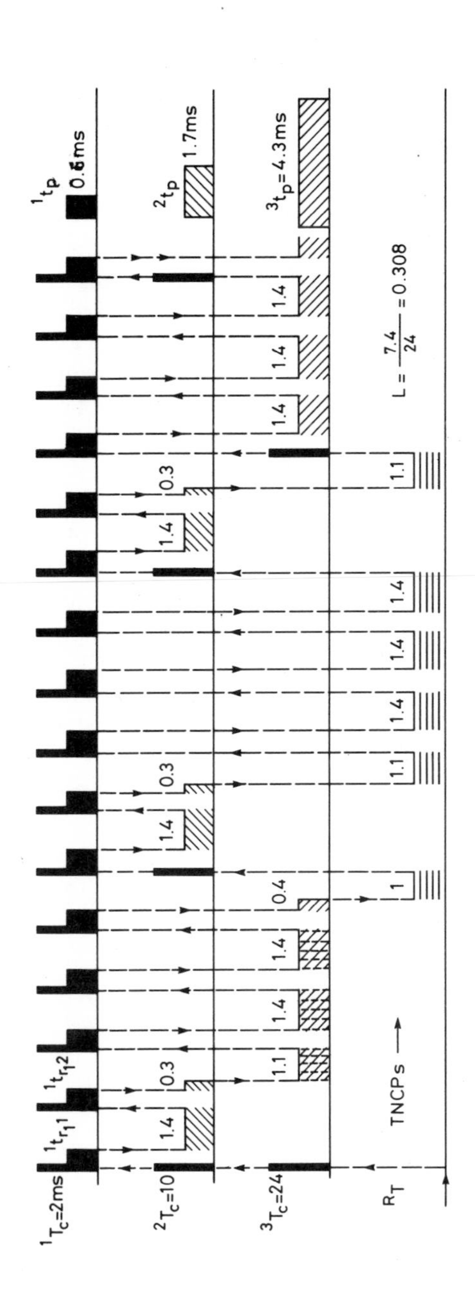

Fig 37.2 Example of practical model

to maximum load factor but needs a lot of interrupt signal.

The opposite method was applied in a control system for handling wood logs (Kuvic *et al.*, 1983) and his system is referred to by Hruz (1984). It used the synchronous model of three priority levels {4, 24, 48}, where only the highest level was initiated by a synchronous interrupt signal of an internal multivibrator with a period of 4 ms. On the first level were the fastest activities: the program counters for lower priority levels and the programs realising the time scheduling. The second priority level served mainly for electronic signals of photocells. The third priority level was reserved mainly for DDC of engines. The TNCPs optimisation, balances, protocols, on-line testing of main control programs etc. were done in the rest time of TCPs.

Generally the initiation of any priority level can be done either by synchronous or asynchronous signals. The asynchronous interrupt signal, connected only with its individual activity, need not occur in critical time. This leads to the growth of rest times. However, when the interrupt represents a group of activities, it must occur as often as necessary for serving the other activities of the group. The use of an interrupt signal for priority level initiation is advantageous also for processes where a fast reaction is necessary, but the period of occurrence requirement is great.

37.5 Conclusion

This chapter has presented primarily heuristic extensions of one of the possible time scheduling strategies. Much work remains to be done in proving some of the assertions in solving the optimal activity grouping. The chapter represents a contribution to the analysis of the correct function for time scheduling of this type.

Chapter 38

The use of generic IDEFo model to capture process planning functional information

R. W. Baines and G. J. Colquhoun

North Staffordshire Polytechnic, and Liverpool Ploytechnic, UK

38.1 Introduction

IDEFo models particularly for the general case of production planning and Control (CAM-I, 1980) have been well documented. The purpose of this chapter is to take the technique further into an area not previously modelled in depth, that of process planning. The chapter reports on the derivation of a Generic IDEFo Model (GIM) for process planning (16 activity diagrams and 53 activities) and demonstrates how such a model can effectively be used to provide a specific AS IS IDEFo model of process planning (14 activity diagrams and 44 activities) for a company which designs and manufactures a range of complex products. The paper concludes by reporting how the AS IS model in conjunction with the Generic model provides an effective way to analyse an organisation's process planning activities. Owing to the large volume of activity diagrams for each model it has only been possible to include a representative sample in Figs. 38.1 — 38.6.

38.2 IDEFo and the generic IDEFo model (GIM)

The IDEFo technique provides a structured analysis methodology capable of representing complex functional relationships graphically and identifying the information and objects that interrelate functions (CAM – I, 1980; CAM – I, 1984). The technique is now gaining acceptance for CIM modelling applications (Banerji, 1986; Hughes & Baines, 1985; Maji, 1988; ISO, 1986).

An IDEFo model consists of a series of related diagrams organised in a hierarchical manner stemming from the subject of the Top Level activity diagram. Subsequent diagrams decompose activities within the Top Level progressively. Complex systems can be decomposed to whatever level of detail is considered necessary. Boxes represent activities, are arranged diagonally, numbered sequentially and are connected by arrows see Figs. 38.2 — 38.6. The IDEFo system defines four types of arrow; Input, arrows which enter from the lest, represent information or objects required to carry out an activity and are converted to outputs as a result of that activity. Control arrows, which enter from the top, represent information or objects that govern an activity. Output arrows, which leave from the right, represent information or objects produced by an activity. Finally mechanism arrows, which enter from below, represent the means by which the activity is carried out; for example a person, hardware, software etc.

The completed diagrams do not represent a sequence of events or a time relationship; activities within a diagram can be simultaneous provided the input/output, control and mechanism requirements are met.

A GIM of process planning embodies all the functions of process planning as separate activities related by the information or objects necessary to support those functions. Any specific or AS IS process planning situation encountered in manufacturing can thus be identified in the GIM. The full extent of the derived GIM for process planning can be appreciated by viewing the GIM node index in Fig. 38.1 where all the activities are listed. By its very nature the GIM identifies no mechanisms, and therefore is equally valid for manual, constructive, retrieval and generative Chang and Wysk (1985) and process planning based applications.

In addition, the inputs, controls and outputs throughout the model are defined in generic form providing a common understanding and enabling the model to be used as the basis of interviews in any specific process planning situation. The GIM, as this chapter shows, can considerably enhance the basic IDEFo concept of combining structured analysis with human judgment to capture the functional information defining a complex situation and to indentify those responsible for it. In turn this approach quickly leads to the establishment of an AS IS model portrayed in IDEFo. Such AS IS models provide a structured way to link design and manufacture in a CIM environment (Colquhoun *et al*, 1989).

38.3 Capturing process planning functional information in an IDEFo AS IS model

To provide an AS IS model, reference is first made to the 'context' A – 0 activity diagram Plan Manufacturing of the GIM and the person

Node	Title
A – 1	Produce new product
A – 11	Market and manage product
A – 12	Plan manufacturing
A – 13	Execute manufacturing program
A – 0	Plan manufacturing
A0	Plan manufacturing
A2	Plan how to manufacture
A21	Plan product assembly methods
A211	Analyse product
A212	Establish assembly technique
A213	Establish assembly requirement
A22	Plan component manufacturing
A221	Analyse component
A2211	Retrieve component analysis
A2212	Separate features
A2213	Derive feature dependant geometry
A2214	Determine geometric component elements
A222	Derive manufacturing method
A2221	Determine raw material form
A2222	Select process options
A2223	Select process
A2224	Select process sequences
A22241	Select machines
A22242	Derive process sequence options
A22243	Derive capacity requirement
A22244	Assign sequence priority
A223	Establish auxiliary requirements
A2231	Select tools
A2232	Retrieve tool information
A2233	Select auxiliary tooling
A2234	Establish workholding requirement
A2235	Retrieve workholding information
A224	Establish operation information
A2241	Establish operation details
A22411	Retrieve operation data
A22412	Analyse operation
A224121	Establish operation component criteria
A224122	Derive machining parameters
A224123	Establish operation characteristics
A22413	Update operation data
A2242	Generate programs
A22421	Establish cutter path requirements
A22422	Produce machine movement data
A22423	Produce part programs
A224231	Establish system requirements
A224232	Post process
A224233	Update part program data
A224234	Prove part programs
A2243	Derive operation times
A22431	Derive machining times
A22432	Retrieve synthetic times
A22433	Produce operation times
A2244	Format information

Fig. 38.1 Generic IDEFo model (GIM) node index.

responsible for the mechanism which the single activity Plan Manufacturing is identified. Hereafter in this chapter such people are referred to as 'experts'. In the case where the mechanism is software then the person responsible for its operation is identified. The identified 'expert' in turn identifies those responsible for the mechanisms which achieve A0 activities A1, A2, A3 and, in particular, activity A2 Plan How To Manufacture (see Fig. 38.2). Activities Al Design Product and A3 Plan When To Manufacture, although an essential part of the context, will normally be outside the scope of process planning and thus do not form part of the AS IS model. The 'expert' responsible for A2 activities identifies those responsible for A21 Plan Product Assembly Methods and A22 Plan Component Manufacturing activities (see Figure 38.3). 'Experts associated with A221, A222 (see Fig. 38.5), A223 and A224 are next identified. The GIM concentrates on process planning for manufactured components and thus activity A21 Plan Product Assembly Methods is not decomposed. The titles of these activities are listed in the GIM Node Index (Fig. 38.1). Similarly all 'experts' who can contribute to defining process planning are identified and subsequently interviewed.

The interviewing process can either take place immediately an 'expert' has been identified or after all 'experts' likely to contribute have been found. Naturally, the latter has the advantage that conflicts and overlaps can be dealt with before the interview process starts. In practice the 'expert' is questioned using the appropriate activity diagram. The objective of the questioning is primarily to make the controls, inputs and outputs in the GIM company specific. This means finding precisely the document or terms used by the company. At the same time additional or missing inputs, outputs and controls may be identified. During the interview process the 'expert' is introduced to, and then views, the appropriate activity diagram. The 'expert' is prompted by the interviewer for the company specific information activity by activity. This process is completed all the activities within the GIM. In simpler process planning situations the number of activities may be less than the GIM but should not be more. As activities and activity diagrams change through the different levels, so may the 'experts' change. Indeed it is critical that 'experts' specifically appertaining to each activity box are those interviewed so that the true AS IS model can be identified, which in many cases may not be how the process planning system was originally conceived.

By definition the GIM of process planning should be able to cope with any type of manufacturing situation. In using the model in a company, which produces white goods, employs 800 people and has a turnover in excess of £0.30 millions, to derive an AS IS model this certainly proved to be the case. The company had comprehensive process planning functions, concentrating mainly on presswork for sheet-metal components. In all,

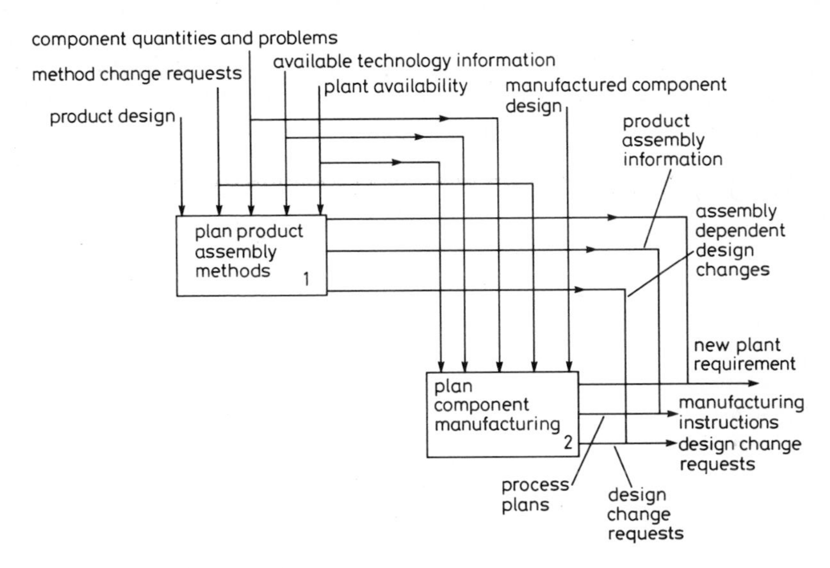

Fig. 38.2 A2 Plan How to Manufacture

five 'experts' were identified and subsequently interviewed. The study took a total time of 5 hours. It is also significant that the process planning functions were completely recorded by researchers with no previous experience of the company and in collaboration with 'experts' who had no previous knowledge of IDEFo techniques. During the interviewing exercise 14 activity diagrams from the GIM were used comprising a total of 47 separate activities. The same 14 company specific activity diagrams, but this time with only 44 activities, were required to fully define the process planning of manufactured components in the company and provide the AS IS model; these were C – O, CO, C2, C22, C221, C222, C2224, C223, C224, C2241, C22412, C2242, C22423 and C2243 all of which correspond to their counterparts in the GIM (A – 0, A0, A2 ... etc), which of course also have the same title (see Fig. 38.1).

38.4 Conclusion

Using the GIM proved invaluable in identifying those 'experts', at various levels in the organisation, that were able to assist the interviewer in the data gathering process. Three 'experts' by which the GIM activity A22 (see Fig. 38.2), were identified in the AS IS model were the Work Study Engineer, the Production Engineering Manager and a Project Engineer. The subsequent decomposition of activity C22 to activities C221 to C224, (see Fig. 38.4) showed the same 'experts' were directly responsible for lower level activities such as C2221 Determine Raw Material Form (see Fig. 38.6). In the lowest levels of the AS IS model some devolution of responsibility took place. For example, the company relied on a machine setter as the 'experts' responsible for activity C224122 Derive Machine Parameters.

Individual activity diagrams in the GIM were found to be an excellent way of stimulating the 'expert' into providing reliable and concrete information. The approach also aided the recording, critique and correction process directly using the box and arrow notations of structured analysis. Throughout the GIM specific company terms were identified as replacements for the generic terms. For example in C2 Plan How To Manufacture activity C21 is subjext to controls Design Change Note and General Arrangement Drawing, whereas in the equivalent GIM activity A21 (see Fig. 38.2) these controls were described generally as Method Change Requests and Product Design. Similarly in the AS IS model activity diagram C22 (Fig. 38.4) activity C224 has an ouptut Master Op. Card which has replaced the Operation Information used in GIM activity A224 (see Fig. 38.3).

The technique proved to be surprisingly good at drawing attention to weak practices in the process planning functions across the organisation. For example, it quickly became obvious that process planning at the company studied was rather unstructured with only weak links between the functions carried out by the Production Control Department and Work Study Departments. Also that the most recent process planning activities relating to CNC programming were developing entirely independently of the manual system. People with the title manager were also found to be working very much at an operational level in the process planning activities.

The AS IS model exposed extensive use of heuristics; for example, activity C224 Establish Operation Information (Fig. 38.4) has a control Process and Material Information (Heuristic) in contrast to the equivalent GIM activity A224 (Fig. 38.3) Process & Material Information, thus establishing that personnel carrying out the activity made no reference to established standards or company data, but relied solely on experience to optimise decisions within the activity. Too many heuristics flags a potential vulnerability of the company on key staff and also a potential implementation problem if the company wished to use computer aided process planning in the future.

If there are missing controls or inputs it suggests that the system is operating in an open loop mode which may lead to uneconomic modes of operation. For example, in establishing the AS IS activity diagram C222 (Fig. 38.6) the information Aggregate Annual Demand (Verbal) was identified as a control on activity C2221 based on the equivalent GIM

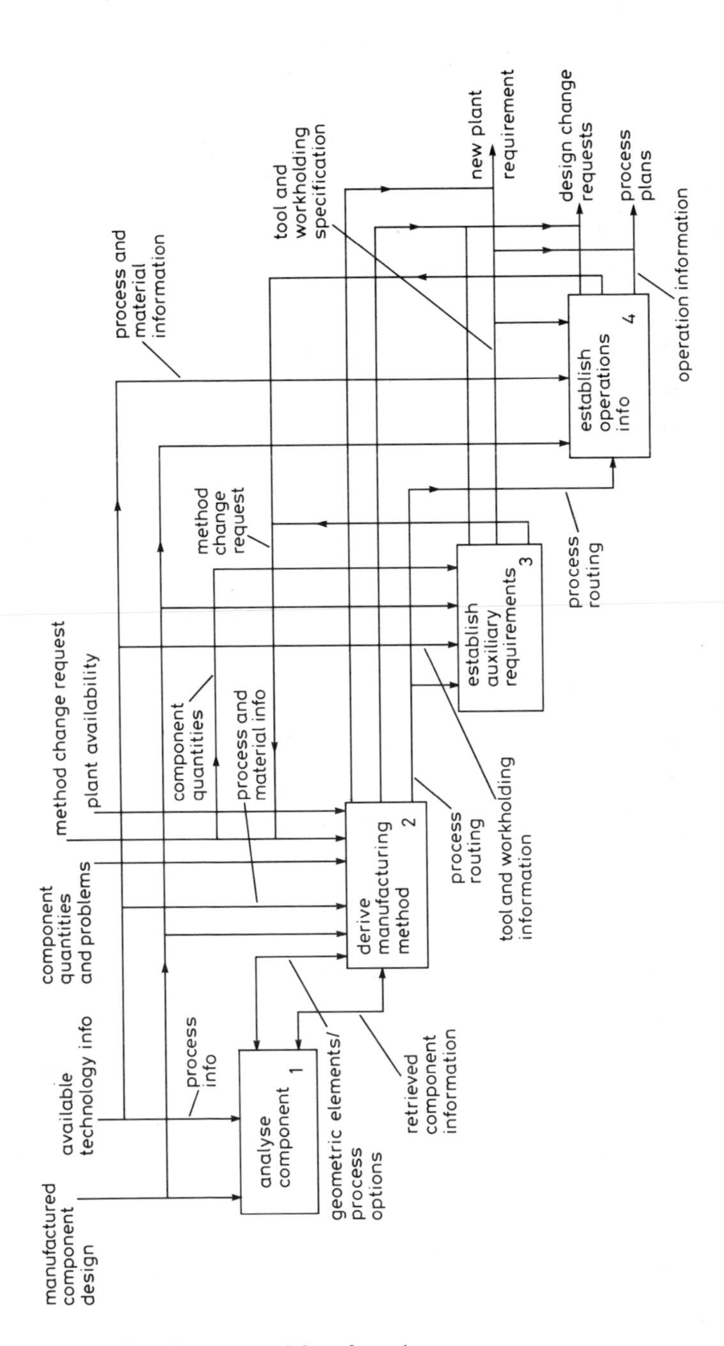

Fig. 38.3 A22 Plan Component Manufacturing

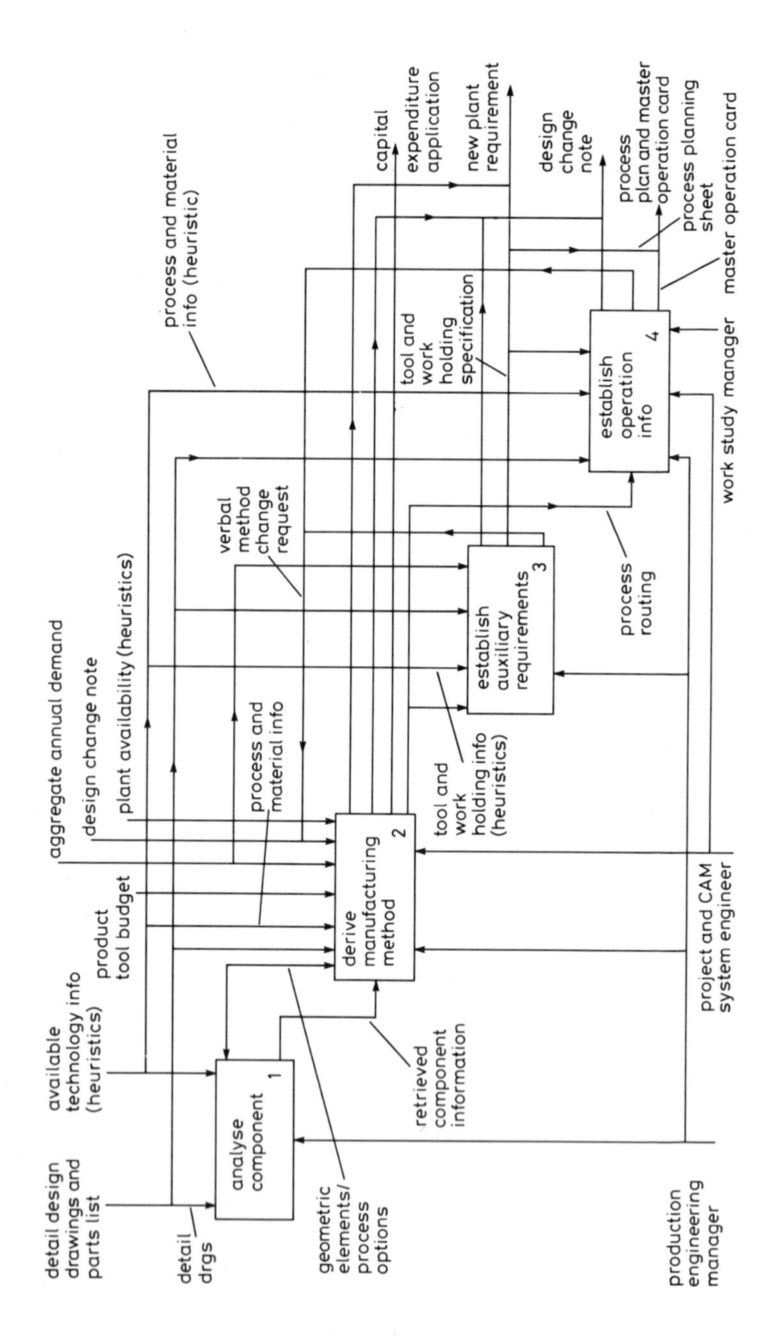

Fig. 38.4 C22 Plan Component Manufacturing

information Component Quantities and Problems (see Fig. 38.5). This comparison exposed that the source of the GIM information was as a result of activity A3 Plan How to Manufacture based on actual batch sizes whereas the company researched relied solely on aggregate quantities based on marketing estimates of annual product demand.

The GIM also highlighted the company's use of process planning activities to produce outputs not normally associated with process planning. For example, activity C2224 (Fig. 38.6) has an output Capital Expenditure Application and a control Product Tool Budget; no direct equivalent control or output exists in the GIM (see Fig. 38.5). The company specific activity involved financial evaluation of capital plant and resulted in a formal documented application for capital expenditure, a process not considered generally part of the routine process planning activity.

The GIM also highlighted the company's use of process planning activities to produce outputs not normally associated with process planning. For example, activity C2224 (Fig. 38.6) has an output Capital Expenditure Application and a control Product Tool Budget; no direct equivalent control or output exists in the GIM (see Fig. 38.5). The company specific activity involved financial evaluation of capital plant and resulted in a formal documented application for capital expenditure, a process not considered generally part of the routine process planning activity.

A comparison of the number of activities in the Generic and AS IS model revealed that the AS IS model contains three fewer activities. For example, in the GIM A22411 Retrieve Operation Data had no equivalent in the AS IS model. The reason for this is that the company re-defined its operation data each time a new component was planned. For the same reason, activity A22413 Update Operation Data also had no equivalent. Similarly A22423 Produce Part Programs in the GIM contains activity A224231 Establish System Requirements, but since the researched company had only one programmable machine and thus a single system, the activity or its equivalent does not appear in the AS IS model.

Previous knowledge of IDEFo was not found to be necessary for those interviewed nor was there a need for the researchers to know the company methods before undertaking the appraisal. Also the use of the GIM will limit the variability of different researchers if they were to address the same problem.

The final AS IS model can be contrasted and compared with the GIM to assist in the rationalisation of process planning activities as well as an information base to assist in software selection and implementation of computer aided process planning and in the development of GIM strategies to link design and manufacture.

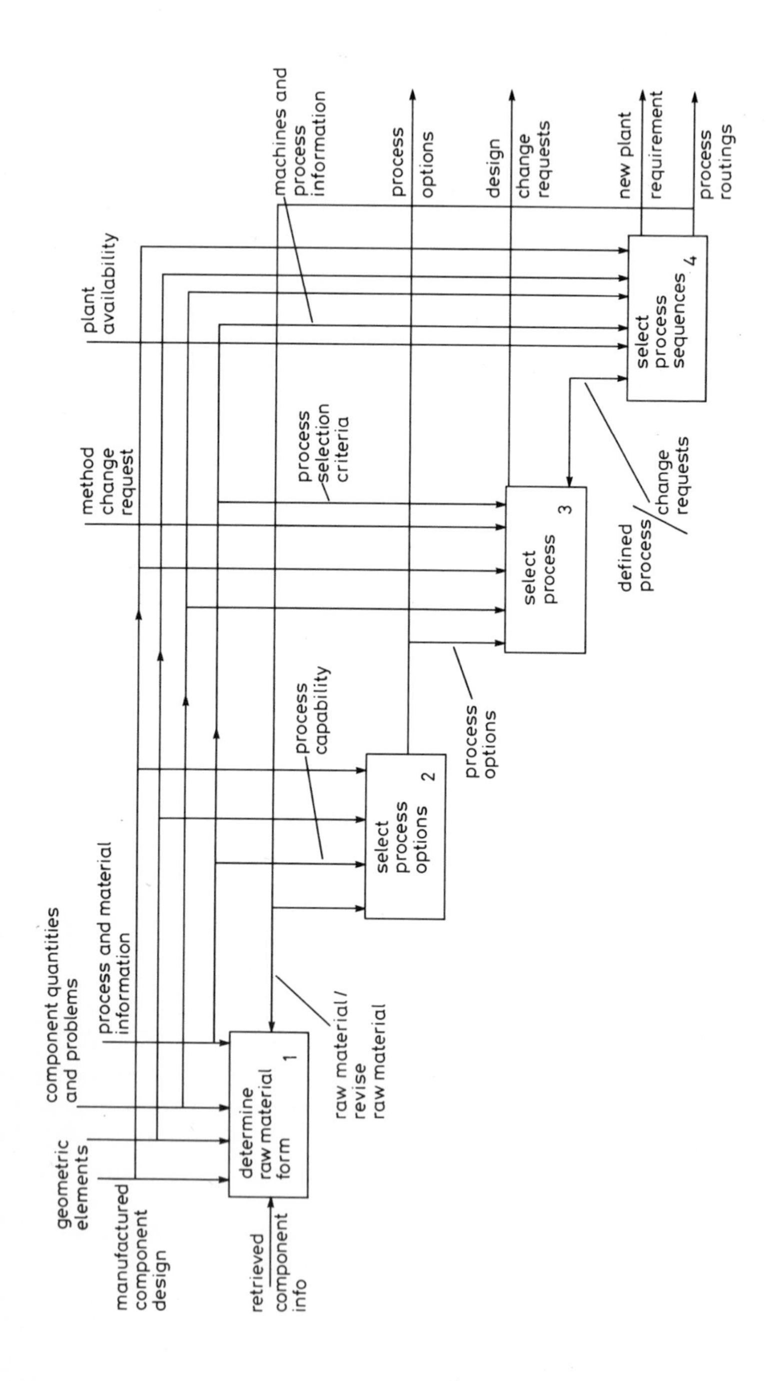

Fig. 38.5 A222 Derive Manufacturing Method

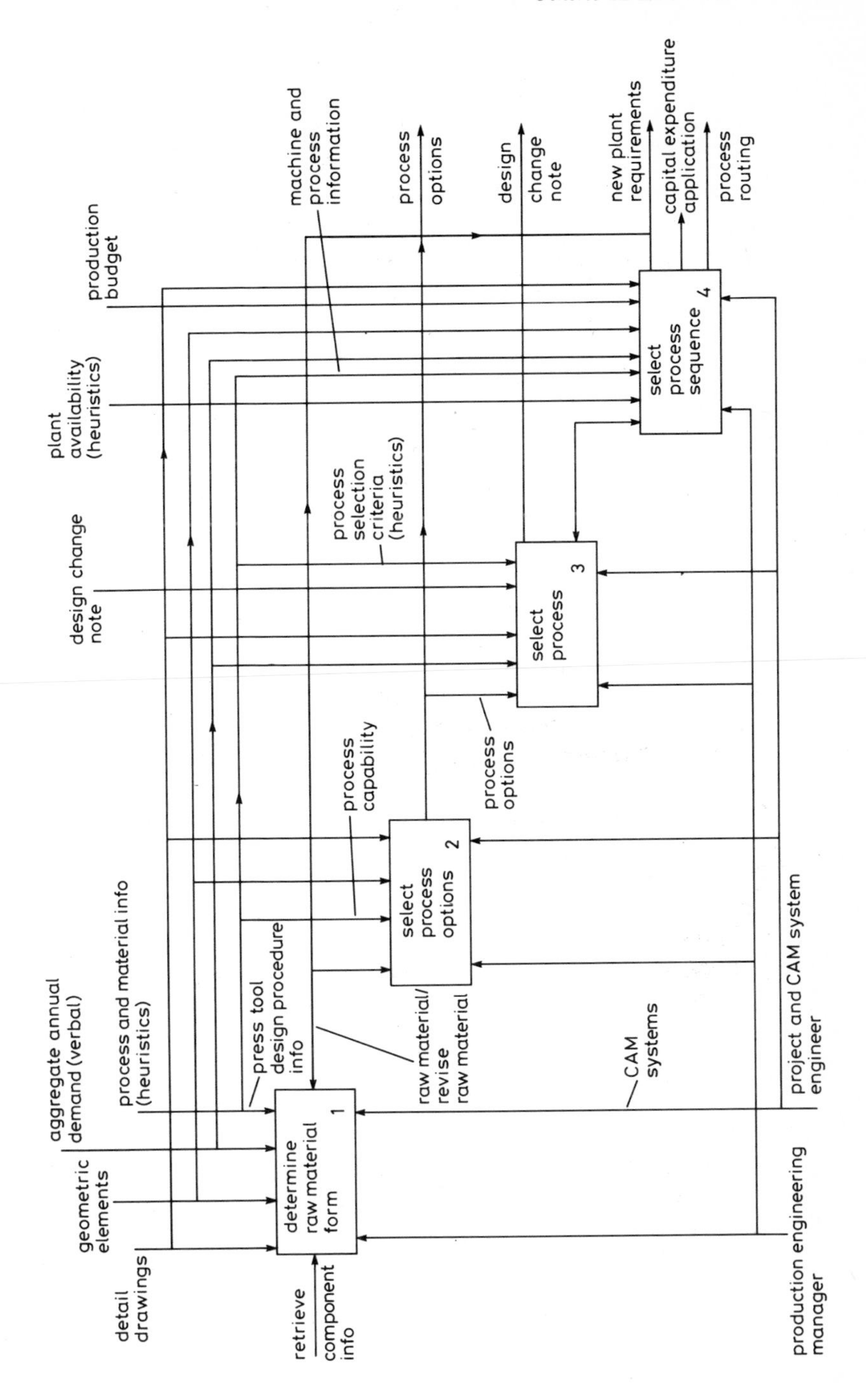

Fig. 38.6 C222 Derive Manufacturing Method

Chapter 39

Evaluation of a knowledge-based planning system for assembly

K. Thaler

Fraunhofer Institute for Industrial Engineering (IAO), FRG

39.1 Introduction

Future Assembly systems are expected to be characterised both by new planning philosophies and new manufacturing principles (Bullinger, 1988). Prospectively it can be anticipated that the planning tasks themselves will change with the evolution of these new strategies. Important requirements can be seen for a future process planning methodology (Fig. 39.1):

- demand for more flexibility, both in quantity and quality of planning methods
- demand for more harmonised integration, e.g. considering CAD or MRP
- use of dynamical planning and real-time scheduling approaches
- adaption to new scheduling approaches, e.g. JIT or Kanban
- adaption to principles of group technology and flexible automated systems.

Most of these requirements heavily depend on concepts and developments in th field of advanced computer technology, which launches a synergical effect on their technical and economical realisation. A few developments may demonstrate this:

- integration of knowledge-based systems into real world applications and connection to databases
- knowledge acquisition tools
- MAP and TOP Standards for open systems
- local Area Networks and workstation concepts.

On a technical basis these components allow the construction and implementation of CIM systems with a huge variety of distinct integration levels (Scheer, 1988). Material and information flow can be coupled closely. Reduction of non-productive time, the increase of reaction capabilities, and hence a signifcantly better competitiveness, are assumed. Ironically enough, the study of vendor concepts of hardware and software sometimes gives only half the story. It is unwise to assume that only availability of information — as a result of designing information streams — is what future orientated planning will need.

In addition to this, a further integration step requires a planning methodology which is harmonised within its functions and carefully adapted to the user's tasks.

One possibility of doing this is 'prototypic' development. This approach was realised in a research project of the IAO together with four German companies and a software house, starting mid-1986. The ongoing project covers assembly process planning, especially for flow line and flexible assembly systems. Several laboratory prototypes for single planning steps were developed, and on this scientific base an industrial prototype was built. One further step was to assess and to evaluate this prototype and its functionality with respect to industrial conditions.

First, an overview of the functionality will be given, introducing to the interested reader other papers concerning system architecture and aspects of knowledge-based support (Seidel, 1987; Thaler, 1988).

39.2 System description and assessment

39.2.1 Overview

The assembly planning system aims at a closer link between product planning activities (construction, CAD) and assembly (assembly scheduling). The configuration covers a CIM process chain with different functionalities in a tool-box like system:

- transformation of construction BOM into assembly orientated product structure
- generation of precedence diagram
- description of operations and technology
- line balancing
- sequencing.

Input information is product descriptions like drawings and bills of material. At the moment the CAD interface used only allows the exchange of BOM structures, but not of the part data itself. The main output result is determined and documented in an assembly plan showing all activities

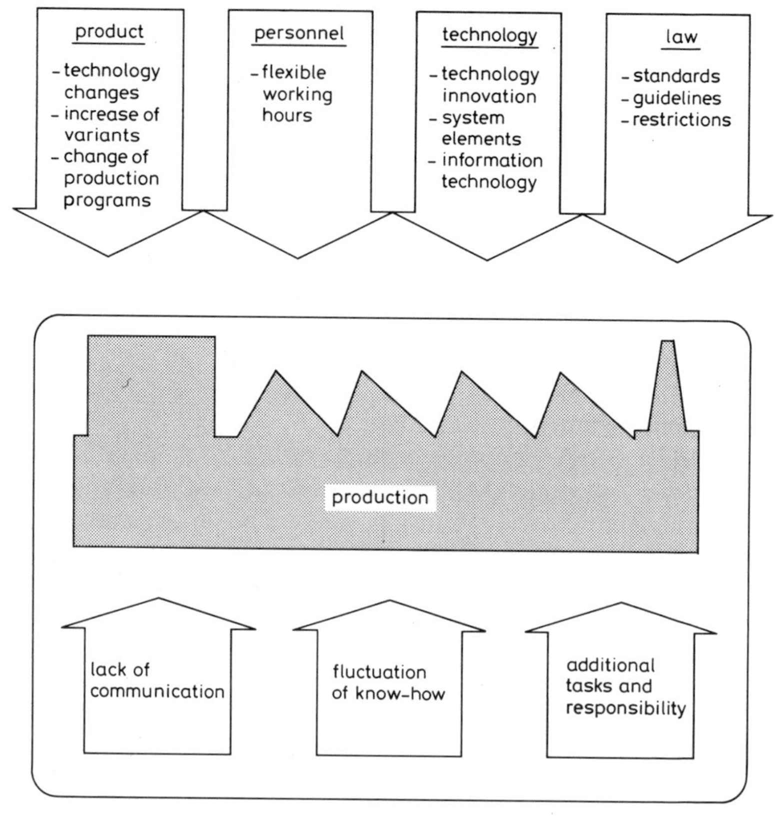

Fig. 39.1 Influences on assembly planning

which are relevant to produce the described product with its specific characteristics.

The evaluation was carried out by future users of the system, mainly experienced industrial engineers. During 1—2 day sessions working with the prototype, all relevant recommendations considering user interface as well as methodical aspects were protocolled. Each company took part in an own session with about 2—3 participants; so four sessions had to be protocolled on the whole. After completion of all protocols, a working group with all project participants assembled to put together common recommendations. All suggestions were weighed according to their realisation priority and their realisation effort. Altogether over 70 detailed suggestions were made, covering general suggestions, user interface and methodology.

The general opinion was that the prototype satisfied in most planning cases as far as planning functions were concerned. The user interface on the other hand was not considered sufficient and adequate for all occasions. The following summary differentiates between suggestions to improve the user interface and methodological recommendations.

39.2.2 User interface

The prototype is a graphically supported system running on a workstation in a LAN environment. Window-technique, mouse and icon/object oriented representation and direct manipulation technique are used. Relevant planning structures can be shown graphically to visualise planning steps and results. In particular, operation precedence diagram and work system structure are represented. Fig. 39.2 shows an example of this methodology. Work system and precedence diagram are represented as graphs.

As a result of the interface asessment, the use and handling of some screen functions were considered to be insufficient. Functions such as 'delete graphical element' were not standardised within the whole program. Secondly, visualisation of statistical planning information was considered poor. In particular, the planners missed bar and pie charts. Apart from these problems, it was confirmed in general that graphically visualised planning structures help planners do their tasks in a better and more efficient way.

39.2.3 Transformation of construction BOM

The structure of a product is generally described in a Bill of Materials (BOM), which consists of single parts and subassemblies. Usually structured according to product design function, the construction bill of material rarely meets assembly requirements.

The developed module supports the transformation of design orientated into an assembly orientated product structure. In practical use, the module is divided into the submodules 'BOM transformation', 'definition of component joints' and 'generation of component joint diagram'.

The BOM transformation is done interactively by the planner, who rearranges subassemblies and parts on the screen. The definition of component joints includes a procedure where all parts of the (assembly) BOM have to be systematically described. Related parts and component joints are defined interactively, specfiying their assembly sequence. The last step automatically generates a graph, which shows the assembly sequence of component joints as a precedence diagram. This so-called component joint graph will be the basis of systematic similarity planning which will be implemented in the next version of the system.

With regard to the assessment of this functionality, problems arose with the component information in the bill of material. There was no implemented function to show which kind of variants of a component existed. Secondly, there was a general need for a search filter to show

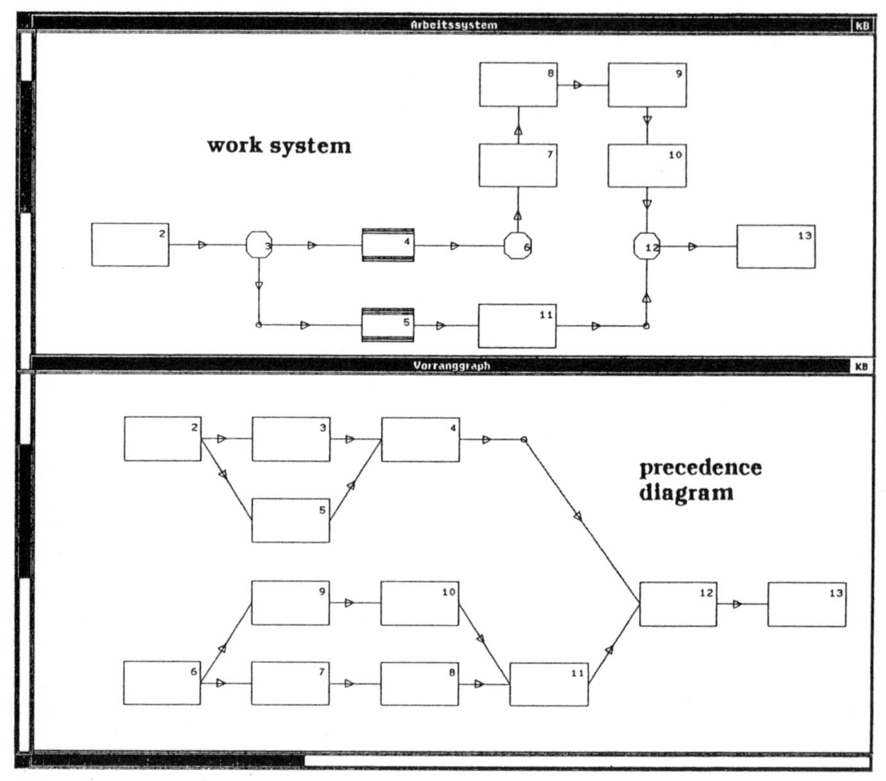

Fig. 39.2 Representation of assembly structures as graphs

specified parts on the screen. Another request was that specific references to drawings and to cutoff dates be made possible.

Another practical function planners desired was to change a subassembly (a subtree of a BOM) into a single part, in this way cutting away a subtree of the BOM. The reason for this was, in most cases, that a subassembly had to be 'changed' into a component which was produced or bought externally.

39.2.4 Generation of precedence diagram

Although a former study showed that on average more than 70% of the information of a precedence diagram is already aggregated within the bill of material (Bullinger, 1984), only a few companies use generic tools to reduce manual planning efforts.

The module described here permits the generation of the structure of a precedence diagram, based on the assembly bill of material and the component joint graph. Operational structures are derived and refined in several steps by the planner supported by the system. As a result of this procedure an assembly plan structure is gained, which still does not refer to a specific production system.

The operational structure is based on a concept which allows adaptability to different applications, especially cases where complex precedence diagrams have to be handled. The planner has the possibility of using a technique of gradual refinement, and in this way he is enabled to adapt the planning level to his own needs.

With regard to the assessment, a lack of this functionality meant that a reference to the components in the BOM was not possible any more after finishing this step. The reason was that the program 'lost' the generated references after completion of this step. Another point was the variant problem. Planners desired a function which, depending on a given variant, would show only the actual activities.

39.2.5 Operations and assembly technology

In many cases, planning and selecton of technology and, based on this, the optimal assembly process can be improved using the information from the precedence diagram (Fig. 39.3).

Using the prototype, the operation structure gained from the last step can be differentiated and refined, if not all operations are specified. In particular, activities which are not directly involved in the assembly operation can be determined and added to the precedence diagram. The precedence diagram structure can be extended by information about technology, operations and standard times.

In an interactive mode the planner is able to define or change operations, standard times and means of production. Catalogues of standard tools and standard operations are offered and their elements can be selected (Fig. 39.4).

In the assessment, handling of these catalogues and 'editing' operations were considered to be too complicated, compared to manual planning. Planners desired to use — in a more generic way — complete standardised assembly sequences, predefined and stored in a library. It showed that the planners would be able to define standardised operation sequences including rules, which reflect the inter-relations between component families, the assembly operations and desired assembly technology. In the next version of the system, these 'modules' will be implemented.

39.2.6 Assembly line balancing

Line balancing is known as the problem of assigning assembly tasks to capacity units, e.g. work stations, in an assembly line. This assignment can be carried out on several planning levels, for the whole work system as well as for subsystems.

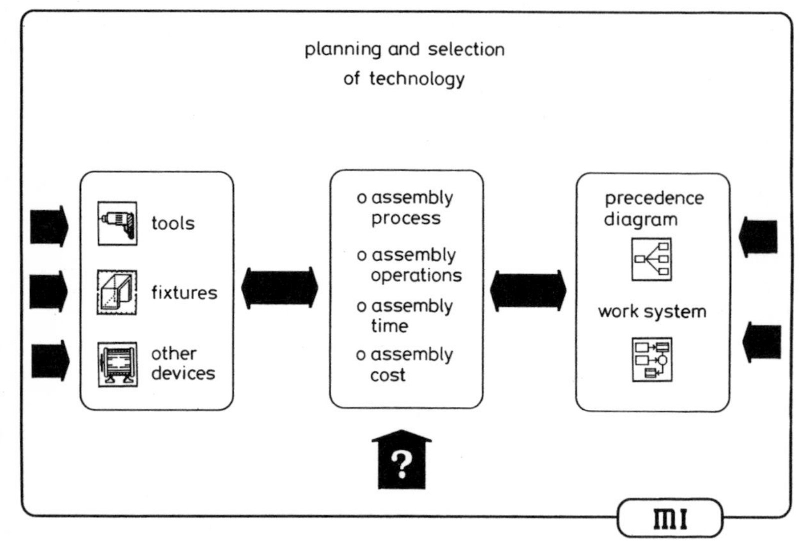

Fig. 39.3 Planning and selection technology

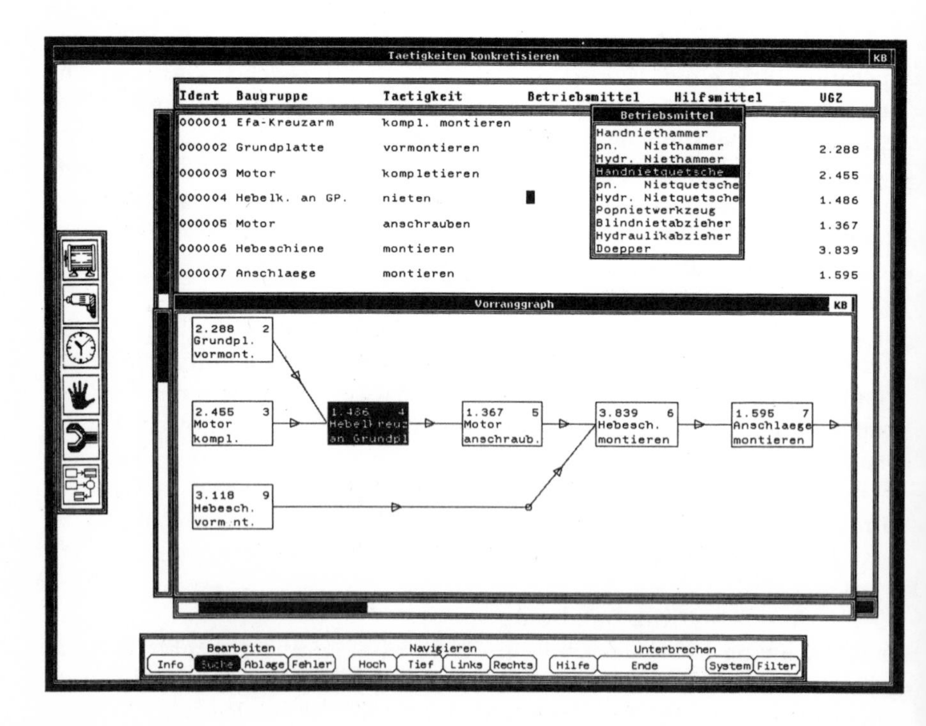

Fig. 39.4 Description of operations

The line balancing methodology allows assignment of operations to the work system according to technological restrictions. In practical use, the planner can interactively assign operations found in the precedence diagram to a graphically shown work system (Fig. 39.5). As a result, the line balancing module offers statistical information of how 'good' the assignment was.

The assessment showed that 'playing', which means to assign and reassign operations, is of great importance. The reassignment of operations at any chosen station has to be improved to get better support.

39.2.7 Sequencing (model-mix planning)

In mode-mix assemblies different standard times of one product (here called 'model') may lead to overload or to idle time at each station. Consequently, these time differences could cause extra costs. The total costs caused by this can be summarised as model-mix loss (Koether, 1987).

The system module 'sequence planning' offers the planner an evaluation tool to avoid and minimise model-mix loss. The planner can use three functions:

- *Generation of a cost-minimal sequence*: A cost-minimal sequence of products (models) is calculated for a given production program. Additionally, information is gained at which stations of the working system what costs are to be expected.
- *Evaluation of a given sequence*: By evaluating different (alternative) sequences, a production program might be evaluated according to its effect on the working system. This is an important completion of the evaluation of the line balancing methodology.
- *Design of the working system*: The work system can be defined, changed and extended interactively by the planner.

The evaluation showed that, owing to the statistical output of this function, a graphical representation of the results as pie and bar charts is of great value. Although the sequencing module gave planners a good insight in to how a production program would react dynamically, the representation of the statistical results could still be improved.

39.3 Conclusion

A report on an evaluation study of a planning system for assembly has been introduced.

It is interesting that the evaluation of the functionality showed a positive attitude towards using graphical and object based planning structures such

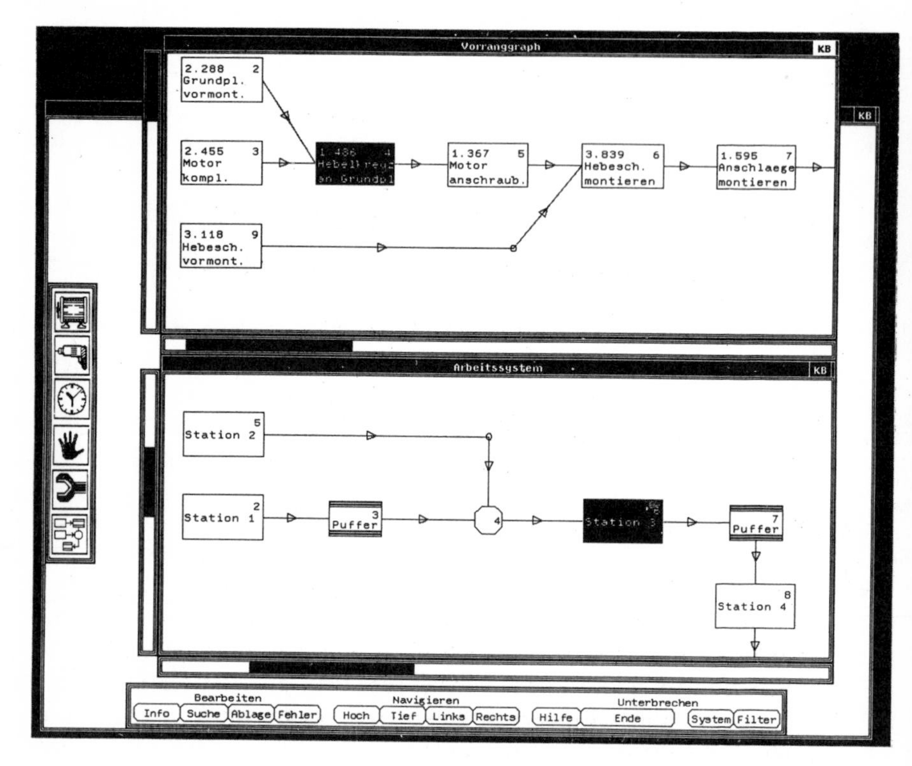

Fig. 39.5 Line balancing with precedence diagram and work system

as component joint graph, precedence diagram and work system. Although problems arose with special screen functions, representation of output data and other teething troubles of a prototype, the general assessment of the methodology was positive.

The specified planning system is a first step towards a CIM process chain between product planning and assembly activities (Hird *et al.*, 1988). It is intended to improve the functionality of this solution, from the experiences gained during the evaluation.

Chapter 40

A feature-based approach towards intelligent reasoning for process planning

A. H. Juri, A. Saia and A. de Pennington

University of Leeds, UK

40.1 Introduction

In the machining sector of the mechanical manufacturing industry, process planning can be defined as the function that establishes a set of manufacturing operations (e.g. machining, inspection and heat treatment), a suitable sequence and the appropriate tools necessary to convert a piece part from its initial state to its final pre-defined state. In addition, process planning also provides information to ensure the orderly running of an organisation. Attempts to decompose decision making in process planning (Weill, 1982) identify many activities which overlap and interact; planning is highly subjective, dealing with information from various inter-related functions to establish an orderly and efficient method of production. The use of current computer aids can assist planners in producing process plans; however, a more automated method, without heavy reliance on expert planners will bring the goal of CAD/CAM integration closer.

The success of future planning systems depends greatly on further advances in two areas:

(*a*) Enhanced product modelling systems capable of answering questions posed during the planning phase. A product model can be viewed as a structured 'computer consumable' representation of data. (Bloor, 1988; Step, 1988).

(*b*) Techniques for representing and manipulating planning knowledge. Process planning does not have an algorithmic solution. Emerging Intelligent Knowledge Based System (IKBS) techniques from Artificial Intelligence (AI) research offer an alternative approach.

This chapter discusses the use of a feature-based approach to the product data requirements of an intelligent planning system.

40.2 Knowledge representation in AI

The term 'knowledge' used by AI researches refers to the information a computer program needs before it can behave intelligently. It includes descriptions to identify and differentiate objects and classes, relationships expressing dependencies and associations between items, and procedures specifying operations to perform when attempting to reason or solve a problem. Various knowledge representation techniques have been developed (Rich, 1983); two of these are:

(*a*) *Frames*: Typically, a frame describes a class of objects, consisting of a collection of slots used to characterise a variety of elements associated with an object. These slots may be filled by attributes of the object, its realtionships to other and procedural information.

(*b*) *Production rules*: Two part statements that embody pieces of knowledge. Production rules take the format:

IF<conditions>THEN<actions or conclusion>

A collection of production rules forms the domain knowledge base (or rule base) of a production system.

Recognising the respective merits of different paradigms several knowledge engineering tools have been developed. This work employed a tool called Knowledge Craft (KC) (Carne, 1987), initially developed at Carnegie-Mellon University. KC provides a frame-based language called CRL and a forward-chaining production system called CRL-OPS.

40.3 Use of AI techniques in process planning

Research in CAPP (Wysk, 1985) has led to two approaches: variant and generative. The variant approach is based on group technology and is a computer-assisted extension of manual planning. The generative approach attempts to capture the knowledge used by planners in order to synthesise plans based on this logic along with information about the product, tooling and equipment. No fully generative planning systems exist, mainly because of problems with product representation and the complexity of identifying, extracting and understanding the human decisions in process planning.

The different areas of interest can be categorised as follows:

(*a*) the recognition, extraction and representation of form features from geometric models. Form features are high-level descriptions of a part that are more closely related to manufacturing compared to the low-level data defining the geometric model (Hende, 1984; Joshi, 1988)
(*b*) product modelling for process planning. Only geometric information may be extracted from a geometric model and this is only part of the complete product information required to support a planning system (Bloor, 1988)
(*c*) generation of high-level planning instructions (Wang, 1987)
(*d*) detailed planning in limited part domains, SIPS (Matsu, 1982).

40.4 Experimental software

Two modules of experimental software have been implemented. The first is a product modelling environment that investigates the use of a feature-based approach to modelling rotational parts in an environmental rule-based planning system whih forms the second module. The modelling environment is implemented partly in Common-Lisp and partly in CRL, whilst the planning system is implemented using a combination of Common-Lisp, CRL and CRL-OPS.

40.4.1 The product modelling environment

To support the product data requirements, this must model:

(*a*) the desired component
(*b*) the initial workpiece
(*c*) the intermediate stages of the workpiece.

These models are defined at two levels of abstraction: the components level and the form-features level. The components level is where general knowledge about the product as a whole is represented. The form-features level is used for representing more detailed knowledge about the product's geometry and technological requirements. At the top of the hierarchy is the COMPONENTS schema (the schema is the basic unit for representing knowledge in CRL). One example sub-class is called ROTATIONAL – PARTS:

ROTATIONAL—PARTS
ISA: components
PART – MAX – LENGTH: 0

PART – MAX – DIAMETER: 0
MIN – INT – DIAMETER: 0
PRIMARY – EXTERNAL – FEATURES:
PRIMARY – INTERNAL – FEATURES:

Detail geometric information is defined via pre-defined form-features linked together. There are two types of form-features; primary and secondary. Primary form-features define the basic shape of a component whilst secondary features are viewed as form-features that are superimposed on the basic shape which adapts it to meet certain prescribed functions. The external and internal primary form-features of a component are listed in the attributes of ROTATIONAL – PARTS. The sequence of form-features implicitly defines the spatial relationship between primary features (see Fig. 40.1).

ROTATIONAL – PARTS has two sub-classes, the FINAL – PARTS schema is used for modelling the desired component whilst the DYNAMIC – PARTS schema models the intermediate workpiece. Part of the schema for FINAL – PARTS is shown below:

FINAL – PARTS
IS – A: rotational – parts
PART – NO:
BATCH – SIZE
FUNCTIONAL – DESC:
MACHINE – ALL – OVER: 1.6
INITIAL – PART: INI – PART

The initial-part attribute holds a schema name for the initial workpiece from which the component is to be machined. Part of the DYNAMIC – PARTS schema is shown below:

DYNAMIC – PARTS
IS – A: rotational-parts
OF:
INITIAL – PART – TYPE:solid

The OF: attribute acts as a pointer to the name of the final part schema, to which the initial or intermediate workpiece is applicable. The initial-part-type attribute defines whether the initial workpiece is a solid bar or tubing. Both DYNAMIC – PARTS and FINAL – PARTS inherit all the non-relational attributes of ROTATIONAL – PARTS.

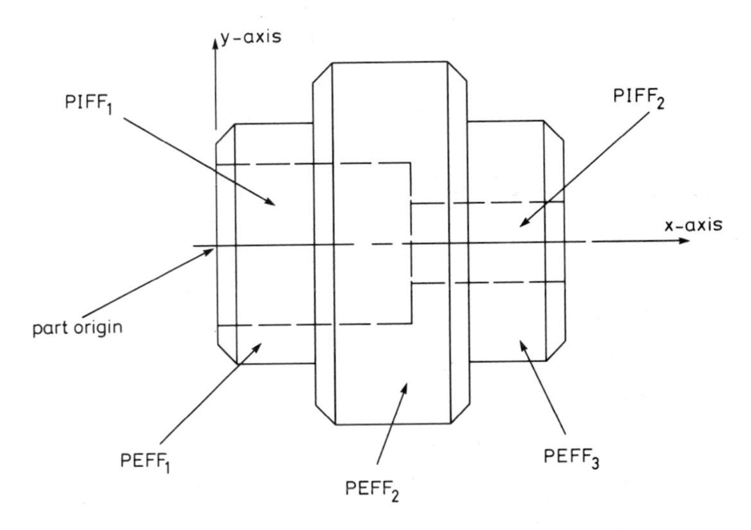

Fig. 40.1 An example component and its primary features

40.4.1.1 Form – features

Form-features are used to represent the shape and non-geometric characteristics of a component. The top level schema for form-features is:

FORM – FEATURES
IS – A: system-objects
FEATURE – TYPE
FEATURE – OF:
NOTES:

The feature-type attribute defines whether it is external or internal, the feature – of attribute points to the component level schema that uses the feature. For rotational parts, two primary form-features can be identified: cylinders and tapers. Because of their similarities, most of their attributes are located at the PRIMARY – FEATURES schema. These can then be inherited by the CYLINDERS and TAPERS schema which are related to the PRIMARY – FEATURES schema (see Fig. 40.2).

Secondary features are form-features often found on rotational components. Examples include chamfers, grooves and keyways. All secondary feature definitions are related to the SECONDARY – FEATURES schema, that contain attributes applicable to most secondary features.

The attributes for the form-features modelled in this work are by no

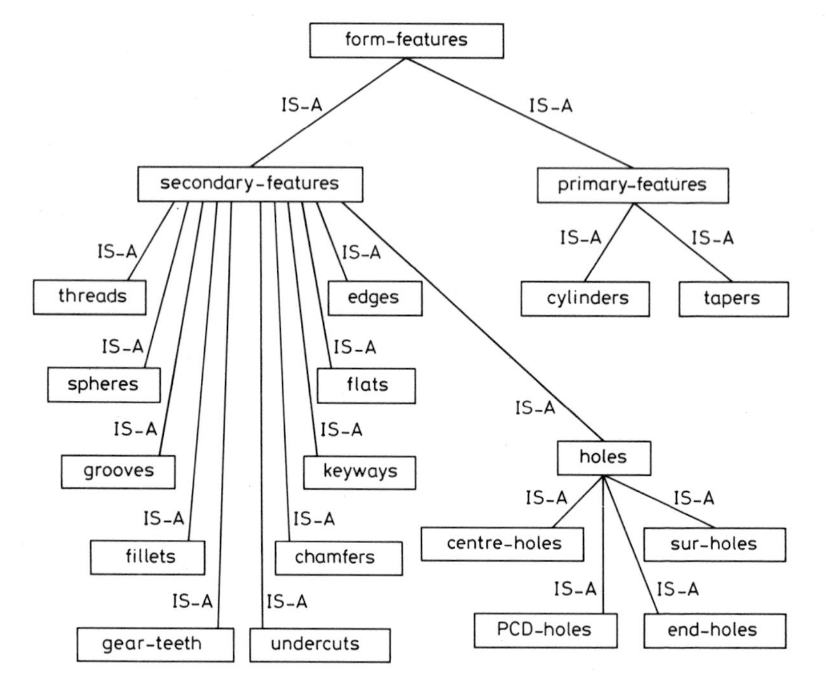

Fig. 40.2 The form-features hierarchy

means detailed or complete. They only identify the presence of a feature on a component and provide some knowledge of overall size and location to enable planning decisions to be made.

40.4.1.2 An example part

The modelling of an example part is given to illustrate the concepts introduced earlier. The engineering drawing of the part is shown in Fig. 40.3. A hierarchical breakdown of the from-features that make up the component is shown in Fig. 40.4. It has six cylindrical PEFFs (primary external form-features) called cylinder 1 to cylinder 6 and one cylindrical PIFFs (primary internal form-features) called cylinder 7.

When defining the PIFFs of a component, a dummy feature called the linker is used. The linker has base-position and length attributes to represent the amount of 'internal space' available for primary internal features. The length of the linker increases and decreases as PEFFs are defined and deleted, respectively. We can thus conceptualise the act of defining PIFFs as compressing the linker to make room for the new PIFF.

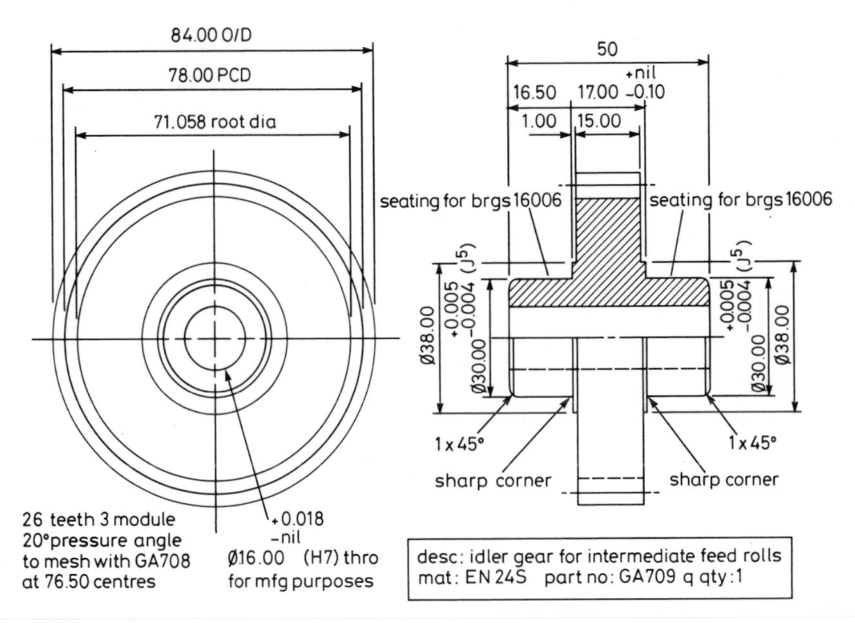

Fig. 40.3 The example part: idler gear

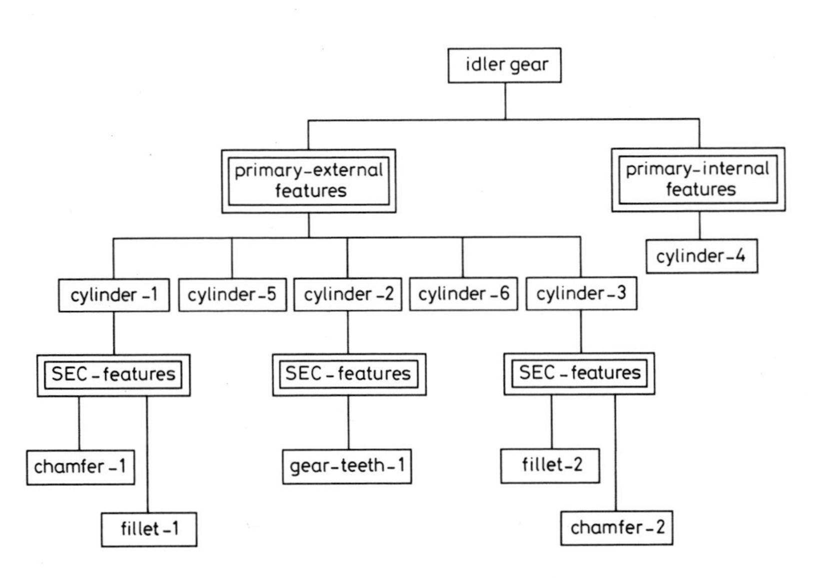

Fig. 40.4 Form-features for the idler gear

40.5 The experimental planning system

40.5.1 Overview

The system plans in a forward direction (i.e. from stock to component), and currently only follows one line of reasoning. However, the underlying data structures allow it to perform backtracking, although this facility has not been fully exploited. The overall goal is achieved through several states. The general strategy followed is to perform machining operations aimed at producing the currently accessible form-features. The external form-features are machined first followed by the internal form-features. There are special cases when an internal feature takes priority, such as drilling a through hole along the workpiece's axis. Secondary form-features are only considered after its primary form-feature has been formed. The precedence relationship between the various external and internal operations is defined implicitly through the condition elements on the left-hand side (LHS) of the rule representing the operation. For example, a rule representing a finishing operation will have a condition element to test that a roughing operation has been performed on the surface. There are, however, situations when it is not easy to define a precedence relationship through a rule. To overcome this, a facility has been developed that allows a CRL – OPS rule to be explicitly assigned a priority value.

The recursive planning concept is employed; each operation causes a new workpiece to be created. By comparing the current workpiece with the desired component, a decision about the next suitable operation is made.

A workpiece reversing operation is carried out when all accessible form-features that can be machined have been formed and the workpiece is still incomplete. When this is complete, the planning system shifts its attention to machining those form-features that cannot be created by turning processes. The end result of the planning system is a sequence of machining operations that can be used as a strategy for producing the component.

40.5.2 The dynamic-parts schema revisited

Apart from modelling the initial workpiece, the DYNAMIC – PARTS schema also provides a template for modelling intermediate workpieces. The schema has attributes used to maintain planning information generated during planning. The complete DYNAMIC – PARTS schema is shown below:

DYNAMIC – PARTS
IS – A: rotational-parts
OF:
OPERATION – DES:
FLAG:
WORK – HOLD:

The operation–des attribute maintains a description of the operation undertaken, resulting in the intermediate workpiece. A collection of these descriptions from the various intermediate workpieces will form the plan generated by the system. The flag attribute maintains an instance of the FLAGS schema, which is used for describing the completeness of turning and non-turning form-features on a workpiece.

40.5.3 Structuring the intermediate workpieces

The various states of the workpiece are structure hierarchically. The initial workpiece is at the top of the structure; when a new state is created, it is linked to the previous state through the IS–A relation. When a workpiece moves from one state to a new one, only information that changes is explicitly represented at the new state. Information that does not change can be inherited. Changes to the intermediate workpiece models are carried out by a combination of CRL and Common–Lisp functions that model the final effect of an operation on the workpiece. These functions form the action (or consequent) part of the planning rules.

40.5.4 Non-machining operations

Two methods of workpiece clamping have been considered, the workpiece can either be held between centres or in a chuck. If the component is long and slender, the workpiece is held between centres. This is quantified by the ratio of the component's overall length to its maximum external diameter.

Workpiece reversing is modelled by a single rule. A workpiece is reversed when all currently accessible external and internal turning form-features have been formed, but there are still turning form-features incomplete.

40.5.5 Machining operations

The machining operations currently modelled can be grouped into two categories:

(i) operations for machining primary form-features
(ii) operations for machining secondary form-features.

For PEFFs the main machining operations involve longitudinal turning and facing. For PIFFs, the main operations are drilling, boring and reaming. These are further classified as either roughing or finishing. Operations for machining secondary form-features are represented as a single operation capable of directly producing the required secondary features.

Apart from the machining operation rules, the planning system has several non-operation rules that generate conclusions to help control the execution of the machining rules.

40.5.6 Test results

The plan generated for the example part introduced earlier is shown in Fig. 40.5. Some of the workpiece geometric models generated by the planner at various stages are shown in Fig. 40.6.

Machining operations sequence	
Part name: idler gear	Part number: GA-709
Material: EN-24S	Drawing number: nil
Date: 25-5-1988	Prepared by: AHJ

No.	Operation description
10	Cut 90.0 mm diameter bar length 55.0 mm
20	Hold workpiece in chuck at position 40.0 mm from free-end
30	Rough and finish face free end length 2.5 mm
40	Centre drill and drill diameter 11.0 mm through
50	Rough turn diameter 84.0 + 1.0 mm length 32.5 mm
60	Rough turn diameter 38.0 + 1.0 mm length 17.5 mm
70	Rough turn diameter 30.0 + 1.0 mm length 16.5 mm
80	Finish turn diameter 84.0 mm length 15.0 mm
90	Finish turn diameter 38.0 mm length 1.0 mm and face shoulder
100	Finish turn diameter 30.0 + 0.5 mm length 16.5 mm, face shoulder and form sharp corner
110	Chamfer 1.0 mm by 45.0 on diameter 30.0 mm
120	Rough bore diameter 16.0 − 1.0 mm through
130	Finish bore diameter 16.0 − 0.5 mm
140	Ream diameter 16.0 + 0.018/0.0 mm H7 fit
150	Reverse workpiece
160	Hold workpiece in chuck at position 40.0 mm from free-end
170	Rough and finish face free end to length 50.0 mm
180	Rough turn diameter 38.0 + 1.0 mm length 17.5 mm
190	Rough turn diameter 30.0 + 1.0 mm length 16.5 mm
200	Finish turn diameter 38.0 mm length 1.0 mm and face shoulder
210	Finish turn diameter 30.0 + 0.5 mm length 16.5 mm, face shoulder and form sharp corner
220	Chamfer 1.0 mm by 45.0 on diameter 30.0 mm
230	Release workpiece from current clamping method
240	Finish grind diameter 30.0 + 0.005/ − 0.004 mm J5 fit length 16.5 mm
250	Finish grind diameter 30.0 + 0.005/ − 0.004 mm J5 fit length 16.5 mm
260	Plane 26.0 teeth on 78.0 PCD 26.0 teeth × 3.0 MOD × 20.0 PA

Fig. 40.5 Operating sequence generated for the idler gear

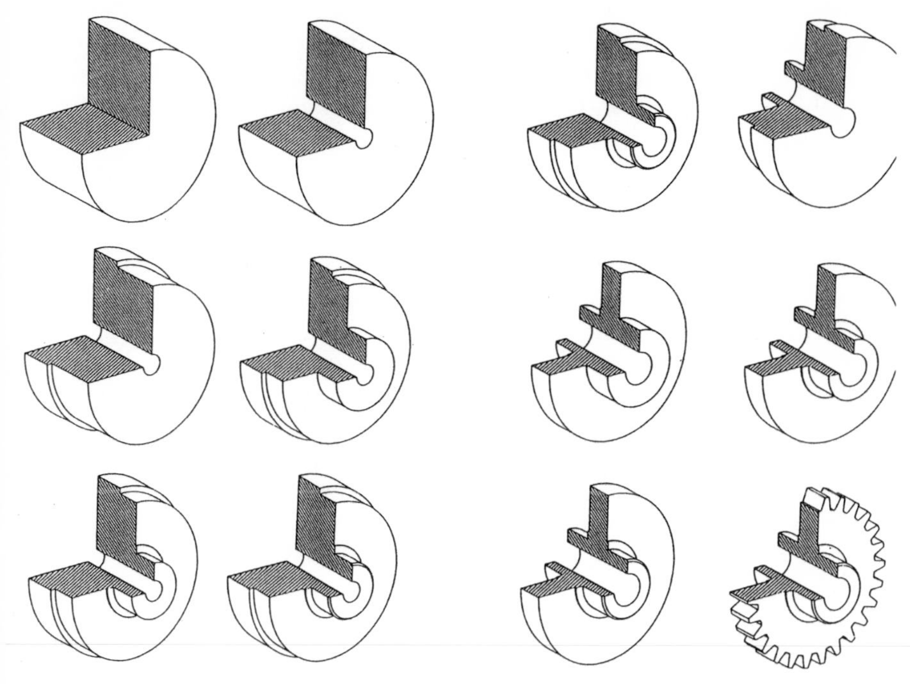

Fig. 40.6 Workpiece geometric models of the idler gear

40.6 Conclusion and discussion

This work has demonstrated that complex reasoning involved in process planning can be partially automated by modelling the product at a level appropriate for the task and by providing the system with the necessary knowledge. This is based on the ability of the experimental planning system to generate a practical machining sequence for a variety of engineering parts (Juri, 1988*a*; 1988*b*).

One of the advantages of modelling at feature-based level is that it can readily provide answers as to what features a component has and their characteristics. This is a question often posed to the model by the planning system through the condition elements of its rules. The relationship that exists between form-features and machining processes can also be exploited to assist the task of establishing the required operations. Information extracted from feature relationship data maintained implicitly by the model enables the system to make decisions about the producibility of a feature and its accessibility.

The production rule scheme appears to provide a format for

representing planning knowledge. For example, to represent the planning knowledge finish turn feature with larger diameter would first involve identifying that the feature is accessible, it has been rough turned and it is currently the feature with the largest diameter that has not been finish turned. This characteristic of the knowledge to be represented thus fits nicely into the format of production rules.

Areas can be identified for further work:

(i) *Modelling non-machining operations*: A complete plan for many engineering components can include non-machining operations such as heat-treatment and nitriding.

(ii) *Form-features hierarchical level*: Currently form-features are defined at two hierarchical levels. One approach where more flexibility could be introduced is to allow further levels of sub-features.

(iii) *Generation of part program*: Many machining operations today are performed on numerically controlled (NC) machine tools. It is thus important that the plans generated by a planning system can be transformed into a part program executable by NC machine tools.

Chapter 41

Experience with spreadsheets in production planning

C. Jones

University College of Swansea, UK

41.1 Introduction

Spreadsheets are one of the success stories in the short life of microcomputers. Since the appearance of VISICALC many others have comes and gone, leaving the now familiar names of the likes of Multiplan, Supercalc and Lotus 1–2–3. However, over the last few years spead-sheets have not changed much in their basic design, shape and fundamental functions. A spreadsheet in its simplest form can still be described as a matrix whose cells may contain text, numbers and formulae' (Anon., 1986). A cell is the basic unit of a spreadsheet and is defined by its row and column co-ordinates. A formula is a calculation rule consisting of an expression, usually involving a number of cells, which once evaluated provides a number (or possibly a string of text) to be displayed in a cell that owns the formula.

Most applications of spreadsheets have hitherto been in the related areas of accounting and finance, and include cashflow planning, investment appraisal and budgetary control. However, over the last few years a number of articles have appeared giving examples of the use of spreadsheets in production planning and control Although the author suspects many examples go unpublished there is still much scope for the use of spreadsheets in this area. This is particularly so in the smaller company because of the relatively modest outlays required and the attraction that the spreadsheet will also be of use to the accountant, financial controller or other functional manager. Indeed the accountant may already be using a spreadsheet. Thus the additional cost may be nil and some expertise will be present in the organisation.

Spreadsheets offer an 'easy entry point' for the beginner. Within hours if not minutes the user can produce simple models. The addition of extra facilities such as graphics, database management, 'programming' capabilities and powerful functions have enhanced the appeal of the spreadsheet and created a suitable environment for the serious modeller. Models can easily be modified as problems evolve and spreadsheets are excellent where 'what if' scenarios have to be evaluated.

A recent survey of articles on microcomputer planning and control for the small manufacturer found that applications were 'generally spreadsheet based' and many resulted from 'actual user applications' (Marucheck and Peterson, 1988). Spreadsheets have been used for safety stock calculation (Coleman 1987), master production scheduling and capacity planning (Hong and Maleyeff, 1987) warehouse costing (Cannella and Schuster, 1987), time-phased order planning (Peek and Blackstone, 1987)—the reader is warned of a typographical error in this paper (Erratum, 1988) and job-shop scheduling (Pendegraft, 1987). Also a series of six articles in the magazine *Lotus* give several examples (Jons, 1988*a*, 1988*b*, 1988*d*, 1988*e* and 1988*f*).

Mant (1988/1989), using Lotus 1 – 2 – 3, gives a useful introduction to the use of spreadsheets for inventory planning. Guidance on planning your spreadsheet and a machine loading example are included. Useful tips on the use of 'macros' to reduce the number keystrokes for repetitive tasks are given. The first macro the author usually adds to a large spreadsheet model produces a simple strip (similar to those used in the operation of Lotus 1 – 2 – 3 itself) GOTO menu so that the user can find his/her way round. The second is a similar macro to print out predefined areas of the spreadsheet.

The author has been involved in several spreadsheet applications in and around manufacturing and these are described in the next Section.

41.2 Some examples

41.2.1 Price lists

The earliest example of the use of a spreadsheet is in the production of price lists for 'flashings' (flat sheet formed in various shapes for gutterings and the like) and 'profile' sheet for a small company manufacturing lightweight insulated cladding systems. This material is typically used in the construction of factories and warehouses, and is sometimes used in the domestic sector. The company had a twin floppy disk IBM PC with 128k of RAM and, amongst other software, Multiplan 1.0. The maximum size of models in Multiplan was 63 columns by 225 rows, and by today's standards quite primitive.

The price of a flashing is a function of the basic material cost, length, wastage factor, girth, fixed cost, number of bends and the required profit margin. The user is permitted to change any of these but nothing else; all other cells are 'locked' (in Multiplan's Language). Thus the user cannot accidentally overwrite text or formulae. These values are used in formulae in the cells calculating the cost. Seven price lists of this form were created.

The second type of price list is for profile sheet. The price is a function of the 'SFH' price, transport cost and profit margin. On the face of it, this semed a much simpler problem until it was realised that the supplier usually informs the company of price changes (inevitable increases) by specifying an overall percentage change. One solution could have been to change the prices manually, but as there are over 60 this was not desirable. The difficulty arises because it is not possible for a cell to be a function of itself. The solution was to create a hidden column of prices. Let us call the cell of an arbitrary visible price A, the corresponding hidden price B and the input percentage increase C. The resulting relationships are $A = \mathrm{B}$ and $B = A(1 + \mathrm{c}/100)$. These are often referred to as circular formulae, and if Multiplan is allowed to 'recalculate' the table unchecked, the prices will continually be compounded at the specified percentage increase. No iteration limit could be set. The solution to this problem was to put the spreadsheet in 'manual' recalculation mode. The user input the percentage change, pressed a function key to start recalculation and then immediately pressed the escape key to stop recalculation — not a very elegant solution but it worked.

This work took less than a week to accomplish and is still being used several years later.

41.2.2 Stock calculation

Another early example is set in an animal food manufacturer. The company produces about 120 000 tons per annum at two mills. The daily output from both mills ranged between 400 and 600 tons. The bulky nature of the raw materials and finished products means that it is uneconomic to keep large stocks of either. Therefore, the finished goods are produced to order: to an advance notice of one or two days (and sometimes even shorter). Raw materials are called off against contracts again at very short notice. Thus, it is essential to have an up to date record of raw material stocks so that production can be planned for the start of the next day and the necessary raw materials ordered to complete the order book for the day.

One of the production mills is relatively modern and has a computerised mixing system with automatic stock recording while at the other mill mixing is essentially a manual operation. There were plans to completely modernise this latter mill. An interim solution was required. The company had an IBM PC Compatible 630k RAM and a copy of Lotus 1–2–3 Release 2.0. Lotus 1–2–3 was being used by the accountant for consolidating the accounts of the animal food manufacturer and the other companies in the group. There are some 72 products that can be produced at the mill from a selection of 77 raw materials. The particular formulations are determined by a service company using a least cost linear programming model. Formulations are usually changed monthly.

The basic problem was simple: to create a system whereby the quantities of raw materials received and finished products produced during the day could be input and the end of day stocks of raw materials calculated via the product formulations. However, extensive use was made of Lotus 1-2-3s Command Language (LCL) in macros to create a robust user-friendly system. The LCL allows the developed to blend together the convenience of the spreadsheet and those good old-fashioned skills of programming. The LCL contains many of the usual characteristics of high level programming languages such as looping, logical statements and branching, and statements specific to the use of Lotus 1-2-3, e.g. the creation of strip menus.

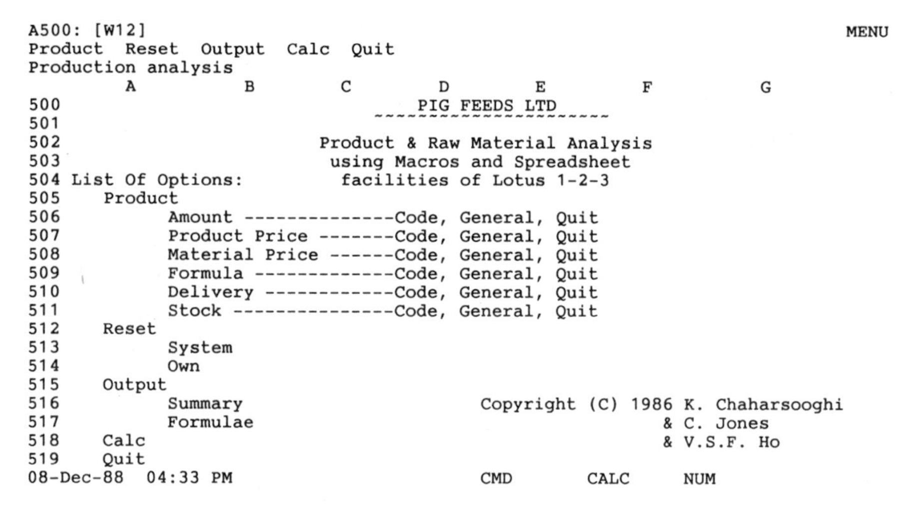

```
A500: [W12]                                                                  MENU
Product  Reset  Output  Calc  Quit
Production analysis
         A          B        C        D        E         F          G
500                                      PIG FEEDS LTD
501
502                                Product & Raw Material Analysis
503                                 using Macros and Spreadsheet
504 List Of Options:                 facilities of Lotus 1-2-3
505    Product
506          Amount --------------Code, General, Quit
507          Product Price -------Code, General, Quit
508          Material Price ------Code, General, Quit
509          Formula -------------Code, General, Quit
510          Delivery ------------Code, General, Quit
511          Stock ---------------Code, General, Quit
512    Reset
513          System
514          Own
515    Output
516          Summary                        Copyright (C) 1986 K. Chaharsooghi
517          Formulae                                         & C. Jones
518    Calc                                                   & V.S.F. Ho
519    Quit
08-Dec-88  04:33 PM                          CMD       CALC      NUM
```

Fig. 41.1 Opening screen of stock calculation system

When the model file is loaded the system starts up automatically (using a/0 macro) and the user is shown the range of options available (see Fig. 41.1). The highest level options are Product, Reset, Output, Calc, and Quit. The (MENUBRANCH ...) command is used to create all menus. The main option is Product. There are several options at this level namely Amount, P. Price, M. Price, Formula, Delivery, Stock, and Quit (see Fig. 41.2). Apart from Quit which returns the user to the higher level all the options work in the same way. They update cumulative production quantities, production prices, material prices, product formulations, deliveries of raw materials and beginning stocks, respectively. If, for example, Amount is selected the user has another menu: Code, General, Quit. Code allows the user to select a specific product by its factory code number so that its cumulative production quantity can be updated, whereas General permits many values to be changed. The combination of the Lotus instructions of /Range Input and /Range Unprotect within the LCL program prevents the user from straying onto forbidden territory in the spreadsheet. /Range Unprotect also has the effect of changing the colour of the contents of cells which enhances the display. Also, in Fig. 41.2 the values of '1' are the results of formulae that check that all the proportions of raw materials (or ingredients) add up to unity.

```
Z202: [W5]                                                                  MENU
Amount  P.price  M.price  Formula  Delivery  Stock  Quit
Quantity Produced
      Z          AA          AB     AC      AD        AE        AF        AG
202                                            PRODUCT ANALYSIS TABLE
203                                         ============================
204                      Product Code  0010     0011      0012      0013
205                                     Super   Sow Breed  Q Pig    N.P.D.
206                                    Starter 160 Rolls Weaner  Gilt rati
207                            CUM.QT      54        43        50        20
208                            P.PRIC     210       320       350       410
209 Today's Date:
210              29-Sep-86 BEGIN
211       RAW              STOCK DELIVE  Total:    Total:    Total:    Total:
212 CodesMATERIAL          (TON) (TON)       1         1         1         1
213
214 1005 Supp. Feed q       0.00 50.00
215
216 1006 Oat Groats         0.00  0.00
217
218 1007 Gross Meal         0.00  0.00
219
220 1008 Oats              0.00  0.00
221
12-Sep-88  09:23 AM                        CMD        CALC
```

Fig. 41.2 Screen when product option is selected

The Reset option makes all Amounts and Deliveries zero and copies ending stocks to beginning stocks to start another operating period. The Stock option may be used at this point if book and physical values differ. Output prints the current states of the spreadsheet, and Save saves the current state in a file.

It is worth reflecting at this stage on the extrapolation from small to larger versions of a model. One is tempted to produce small models to demonstrate ideas and methodologies as in Pendegraft (1987) and as was done in this particular case. A small version was built with only 10 or so products and 10 or so raw materials. Extending to the full version was not as simple as originally envisaged. The formulae became extremely long, too long for a single cell, and had to be split into two cells. Memory was nearly exhausted and calculation times increased greatly (but still far less than the manual method). So it is advisable to warn the client of the problem. The cause of this difficulty and a suggested solution are described in relation to the next example.

41.2.3 Material requirements planning

The author's first attempt at creating a material requirements planning (MRP) system used Multiplan 1.0. It was used in teaching to explan the methodology and demonstrate the effect of various lot sizing policies. A much more ambitious system was built in Lotus Symphony, an integrated software package based on/in a spreadsheet. This system uses the database facilities extensively and consists of a number of related database files: Item Master File, Subordinate Item File, Master Production Schedule and Bills of Material [see Vollmann *et.* (1984) for a comprehensive description of the components of an MRP system]. It was originally designed to help teaching, but as the system grew it was realised that, perhaps, there is scope for using the system in the real world. Over 700 records can be accommodated in a PC DOS based microcomputer with the maximum capacity of 640K. The breakthrough came when it was discovered that the linkages between the parts can be embedded within the Definition Range of the database. Symphony is essentially an all-in-memory integrated package and is limited by the maxiumum available memory. However, recent developments of enhanced memory support greatly increases the potential of the system. A fuller account is given by Jones and Chaharsooghi (1986).

```
L19: 'Save                                                              SHEET

                         SYMPHONY and
              Manufacturing Planning & Control Systems
              Copyright (c) 1986 Chaharsooghi & Jones.
                                                  COPYRIGHTY

                                        [( MAIN  MENU )]
                                 Please make your selection and
                                      press the Return key.

                                 MRP Database     Instructions
                                 Starting Date    Edit/Criteria
                                 B.O.M. DB        Transactions
                                 BOM Page Two     Reset MRP Links
                                 BOM MacExec      Print
                                 Update           Save
                                 MRP View         Load
                                 Graph View       Worksheet
                                                              MENU

08-Dec-88   04:39 PM                      Macro                  Num
```

Fig. 41.3 Opening screen of MRP system

The Main Menu to this system is shown in Fig. 41.3. The cursor is positioned on the required option and the enter key is pressed. The LCL does the rest. The system consists of data entry options to create the database files, options to view the MRP records and their graphical representation, options to do database queries and advance the system in time, and options for saving and loading models.

Let us now return to the memory hungry nature of spreadsheets. If a similar approach were used by someone using a high level language this would be regarded as very inefficient. If one wishes to perform the same calculation on different sets of data, one calculation routine (or set of formulae) would be written and each data set would pass through the routine. Arrays, looping and other facilities are available for the programmer to do this with little difficulty. In spreadsheets this is not the usual approach. Each set of data has its own set of formulae. However, the combination of the ICL and the database facilities in Symphony allows one to return to these old ways of doing things, and it is this that has made the MRP system a viable proposition for use in the real world, albeit the real world of a small company.

Another version has also been created in Supercalc3 for use in teaching (Supercalc3 being much cheaper for large classes). An MRP record is shown in Fig. 41.4*a*. The only values the students are allowed to change are shown in bold, as they would on the screen. All other cells are protected. Warning messages are displayed if invalid entries are made.

41.2.4 Scheduling

The Smart spreadsheet has been used to develop a scheduling system for a small contract furniture manufacturer (Jones and Chaharsooghi, 1988). The Smart System consists of three main modules: the spreadsheet, database management and word processing.The parent company had chosen Smart as its standard for use throughout the group for financial work. Smart is quite different to those above. It is not an all-in-memory system. Only parts of the spreadsheet need be in memory, and, in fact, parts of different spreadsheet models can be in memory at the same time; indeed cell formulae in one spreadsheet file can refer to cells in other spreadsheet files, effectively making Smart a three dimensional spreadsheet. (The next version of Lotus 1 – 2 – 3, version 3, will also have three dimensions.) The user has far greater control over the use of colours to enhance the appearance of systems and hence the friendliness.

The system allocates operators to jobs in each of four departments: machining(M), cabinet making(C), polishing(P) and dispatch(D). Fig. 41.4*b* shows a typical schedule for just two jobs (a more typical number would be about twelve), which is shown on the screen easier to read. The system tries to ensure all jobs are completed by their due dates, and if this is not possible it indicates where the problem lies. There are facilities built into the system for the manager to alter the schedule in any way he wishes.

Another scheduling example comes from a bakery. The problem was to schedule, for each dough type, two to four slicing and bagging machines, so that the total allocated time, including change over times, on each

4a

		Period		1	2	3	4	5	6	7	8	9	10	
010 IMP		G.R.		**100**	**150**	**150**	**80**	**70**	**200**	**100**	**100**	**150**	**100**	(Gross Requirements)
		S.R.		**240**	**0**	**0**	**0**	**0**	**0**	**0**	**0**	**0**	**0**	(Scheduled Receipts)
S.S.	**0**	P.I.	**10**	150	0	-150	-230	-300	-500	-600	-700	-850	-950	(Projected Inventory)
L.T.	1	A.I.	10	10	0	0	0	0	0	0	0	0	0	(Actual Inventory)
L.S.	**LFL**	N.R.		0	140	150	80	70	200	100	100	150	250	(Net Requirements)
		P.O.		140	150	80	70	200	100	100	150	250	0	(Planned Orders)

(S.S. = Safety Stock; L.T. = Lead Time; L.S. = Lot Size; LFL = Lot For Lot)

4b

```
PRODUCTION SCHEDULE · W.Days22Today:  02-Jan-80 Time: 00:25:04Total Jobs 2
  Days of Month:5 6 7 8 9 0 1 2 3 4 5 6 7 8 9 0 1 2 3 4 5 6 7 8 9 0 1 2 3 4 5
  25-Apr-87     S S M T W T F S S M T W T F S S M T W T F S S M T W T F S S M

Priority 1   M  · · · · · 7 7 7 ░ ▓ 7 6 6 6 ░ ░ 6 6 · · · ░ ░ · · · · · · ░ ░ ▓
 01-Jun-87   C  · · · · · · · · ░ ▓ · 101414 ░ ░ 1414141414 ░ ░ 141414· · ░ ░ ▓
PRODUCT ABC  P  · · · · · · · · ░ ▓ · · · · ░ ░ · · · · · ░ ░ · 5 9 9 9 ░ ░ ▓
   2031      D  · · · · · · · · ░ ▓ · · · · ░ ░ · · · · · ░ ░ · · · · · · ░ ░ ▓

Priority 2   M  4 4 4 4 4 3 3 3 ░ ▓ 3 4 4 4 ░ ░ · · · · · ░ ░ · · · · · · ░ ░ ▓
 15-May-87   C  · · · · · · · · ░ ▓ 8 7 7 7 ░ ░ 7 7 · · · ░ ░ · · · · · · ░ ░ ▓
PRODUCT XYZ  P  · · · · · · · · ░ ▓ · 1 3 3 ░ ░ 3 3 3 3 · ░ ░ · · · · · · ░ ░ ▓
   2015      D  · · · · · · · · ░ ▓ · · · · ░ ░ · · · 1 4 ░ ░ · · · · · · ░ ░ ▓

OPs Ava. in   M  6 6 6 6 6 0 0 0░ ▓  0 0 0 0░ ░  4 4101010░ ░ 1010101010░ ░ ▓
OPs Ava. in   C 2424242424242424░ ▓ 16 7 3 3░ ░  3 3101010░ ░ 1010102424░ ░ ▓
OPs Ava. in   P 1010101010101010░ ▓ 10 9 7 7░ ░  7 7 7 710░ ░ 10 5 1 1 1░ ░ ▓
OPs Ava. in   D  4 4 4 4 4 4 4 4░ ▓  4 4 4 4░ ░  4 4 4 3 0░ ░  4 4 4 4 4░ ░ ▓
              (░ = Non-working Sat. or Sun.; ▓ = Non-working weekday)
```

Fig. 41.4 a Example of MRP record from Supercalc3 system

b Schedule for two jobs showing unused operators in each department

machine was about the same. The lines would then be rebalanced for the next dough type. This is a fairly complex combinatorial problem. The formulation is given by Kemp and Jones (1988). The first approach was to use Lotus 1–2–3 linked to an integer programming system. However, this proved too slow. The next step was to use a macro to make obvious allocations and have the optimiser do the rest. The work is continuing.

41.3 Conclusion

All the above examples have used spreadsheet based software that was currently available within the organisations to design systems to improve the management of operations. This reduces the financial commitment that has to be made, and the time in learning different systems. In all examples reaction to the work has been very good. Even if some of the systems developed are not quite ready to run, they have demonstrated to the clients the potential for the software that they already possess.

The familiarity that managers now have with spreadsheets has led to the development of spreadsheet lookalikes as interfaces to other systems. Indeed, a new industry has developed around Lotus 1–2–3 since information on data structures etc. have been in the public domain. Software as varied as complex mathematical and statistical techniques to utilities for documenting spreadsheet models, printing sideways or seeing more on the screen have been written to link into the spreadsheet: 'add-ins' in the trade. Other spreadsheet writers are following.

Chapter 42

Hybrid simulation/knowledge-based systems for manufacturing

S. N. Farimani-Toroghi and S. N. Peck

Leeds Polytechnic, UK

42.1 Introduction

Modern developments in manufacturing and, in particular, flexible manufacturing systems (FMS) have been designed to achieve the efficiency and utilisation levels of mass production, while retaining the flexibility that job shops have in batch production. This recently developed technology is being adopted by many manufacturing industries in an attempt to overcome the current problems of high cost and low productivity in a highly competitive environment. The versatility of such systems allows considerable flexibility in assigning and scheduling operations, and so increasing output and utilisation. However, as a consequence of this, the management, control and efficient operation of such systems can be very difficult even for the most experienced supervisor. It is largely for this reason that the capacity of FMS is still underexploited (Brown *et al.*, 1984).

There is an unprecedented rise in the complexity and uncertainty of systems, and thus a corresponding need for sophisticated Decision Support Systems (DSS) in order to aid the making of more informed decisions. Such DSSs need to have an interactive capability that permits managers and operators to model their systems as completely and accurately as possible, and to evaluate the consequences of various alternative actions (Suri and Whitney, 1983). Of particular concern is the operational level of decision making in a discrete manufacturing environment, i.e. job scheduling.

Traditionally, simulation has been used as a stand alone decision support tool for studying the behaviour of manufacturing systems; either in proposing an entirely new design, or testing out various scheduling and sequencing rules. More recently, Intelligent Knowledge Based Systems

(IKBS) have also played an important role in all levels of decision activities within a manufacturing organisation. The close relationship between these two approaches suggests that it is logical to bring together the ideas and techniques of both to enhance decision support capabilities (O'Keefe, 1986).

Ultimately, our objective is to find improved control and scheduling rules in which the decision depends on the global and not some local state of the system. Whilst there are methods for developing optimal control rules for continuous systems, this is not so for automated discrete manufacturing systems because the number of states and state variables are very large (Buzacott, 1985). One approach in this case is to model the decision activities of the production supervisor and aim for sub-optimal control rules (Ben-Arieh, 1987). These are organised separately into a knowledge base which, together with simulation model, is interfaced to the simulator to form a hybrid simulation and knowledge based environment (Fig. 42.1).

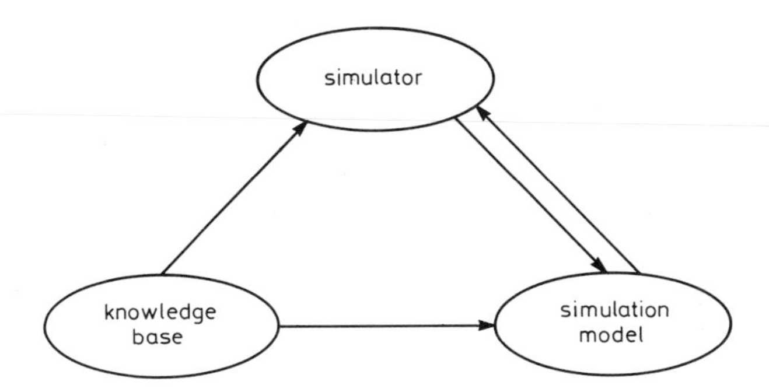

Fig. 42.1 Proposed structure of the system

This chapter outlines the preliminary work carried out in this area, starting from the simulation perspective and investigating how to incorporate Knowledge based rules at the operation and flow control level of decision making.

42.2 Scheduling, routing and allocation

In the manufacturing environment, the scheduling problem involves the simultaneous consideration of both the jobs to be performed and the resources which they require. The task is difficult because of the combinatorial complexity of the elements within the system (e.g. number of different ways with different resources that a job can be accomplished) and executional uncertainties of the exterior factors (e.g. breakdowns, urgent orders, absenteeism etc.). On the whole, the process of scheduling involves selecting a feasible sequence of operations (i.e. process routings) and assigning times and resources to each operation. From this it follows that, at the lowest level, the scheduling problem consists of two main decisions:

(i) the routing decision, which is the determination of a sequence of operations whose execution results in the complete production of a given order
(ii) the allocation decision, which is the assignment of the necessary resources and time intervals to the operations selected.

These two decisions are not entirely separable since the admissibility of a routing depends on the feasibility of each selected operation, and that in turn depends on the availability of its required resources (Smith *et al.*, 1986).

To fully utilise the flexibility available in today's manufacturing systems, it is essential that the plans for jobs are not limited to a single sequence of operations. However, this is not usually practical because of the difficulties mentioned above. Consequently, it would be more appropriate to have a scheduling system where the scheduling and routing decisions are made based on the actual state of the system. A knowledge-based environment provides a suitable framework to develop such real-time on-line scheduling system (Subramanyam and Askin, 1986), and hybrid simulation environments with such capabilities help in studying the behaviour of these systems before full implementation.

42.3 Example description

Discussion is based around a simple example which, while having nothing like the complexity of a FMS, will illustrate the various principles involved in developing discrete event simulation systems with knowledge base components.

Each of the 20 kilns in a tile factory are set by one of four available setting teams. Kilns are selected according to a priority rule. After setting, firing and cooling is started immediately and, following this, kilns are ready to be drawn by one of three available drawing teams. This sequence is then followed by another cycle of setting, firing and cooling, drawing and so on. From time to time it is necessary to schedule empty kilns for maintenance according to the state of the kiln and the workload.

Fig. 42.2 is a graphical representation of this example showing all the relevant events and the transitional states between these events.

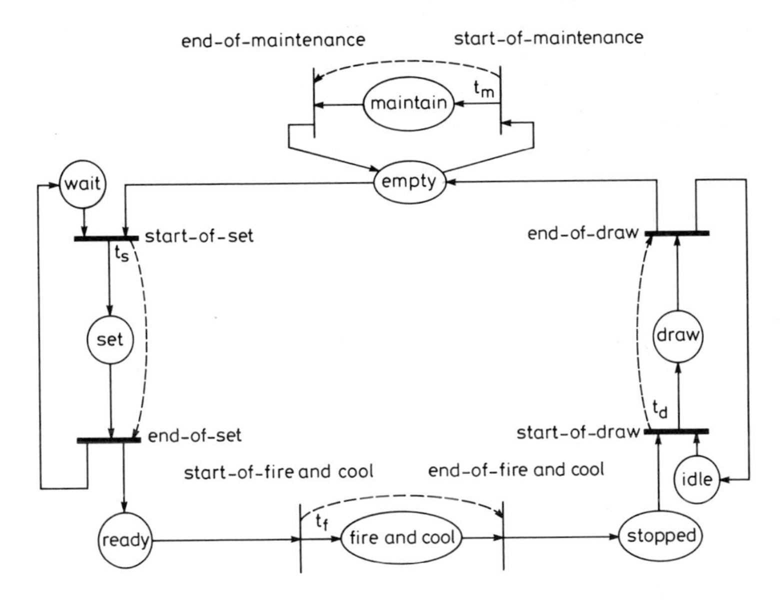

Fig. 42.2 Graphical representation of the kiln problem

42.4 Discrete event simulation and PROLOG

Like many researchers in this field we have chosen PROLOG as our implementation language. There are many different approaches to modelling discrete event systems, of which the most commonly used modelling world views are: event scheduling , activity scanning and process interaction (Carrie, 1988*a*). PROLOG has been used to model discrete event systems using all the modelling world views mentioned above. Futo and Gergely (1987) and Adelsberger (1984) have taken a process interaction approach in modelling and implementation of discrete event systems with PROLOG, Rodrigues (1986, 1987) has chosen activity scanning and Ben-Arieh (1986, 1987) and Fan and Sackett (1988) have adopted event scheduling. We have also chosen event scheduling as our modelling world view.

In an event scheduling world view, the modelling task consists of identifying all the possible event groups, defining the actions for each group including the scheduling by each event of future events through an event list. It should also be noted that there are two main categories of events (Overstreet and Nance, 1986):

(i) bound events whose condition(s) of occurrence can be expressed solely as a function of system time (e.g. end of setting, end of drawing etc.), and

(ii) contingent events whose condition(s) of occurrence contain(s) no attributes that are time-based signals (e.g. start of setting, start of drawing etc.). Thus each event routine describes related actions that may all occur at one instant.

The following shows, in PROLOG, the basis of a discrete-event simulation executive:

```
simulate :-
  initialisation,
  simulatel.

simulatel :-
  termination.

simulatel :-
  step,
  simulatel.
```

The simulation begins with initialisation, in which the system's status and the event list are initialised. Following this, the simulation is defined iteratively as a sequence of steps in each of which the sequence is to determine which event is most imminent, advance the simulation time to the time at which the event will occur and then execute the corresponding event routine. Typically, the event routine breaks down into four stages: (i) updating the system state to account for the fact that the bound event has occurred, (ii) generating the times of occurrence of future events and adding this information to the event list (bound events), (iii) testing the conditions on events which could occure at the same time (contingent events), and (iv) cancelling some of the future events on the list as a consequence. The cycle is continually repeated until the stopping condition of the termination routine is eventually satisfied.

The implementation of the model is described in the next Section.

42.5 Model implementation

As explained before, the simplest and the most basic form of representing the model is to first identify all the possible events and then define the conditions and actions for each event. Here, we use two examples to show how the model is implemented in PROLOG.

42.5.1 Example 1

Event: start of setting

```
condition (set_start,
           [A_kiln, A_setter_team]):-
  state ([A_setter_team], wait),
  state ([A_kiln],empty).
```

This means the condition for the setting to start with a kiln and a setter team is that

(i) there should be a setter team waiting
(ii) there should also be an empty kiln available.

As shown the status descriptive clauses are two argument clauses, i.e.

state ([Entities and their Attributes], Status).

The condition /2 predicate stated above is the simple case where there is no allocational decisions to be made. However, if we are to use rules (e.g. a priority rule for the kilns) in assigning resources to operations, the format of the predicate changes slightly:

```
condition(set_start,
                    [A_kiln, A_setter_team]):-
  state([A_setter_team], wait,
  find_best(A_kiln, empty, higher_priority).
```

This means the condition for the setting to start with a kiln and a setter team is:

(i) there should be a setter team waiting
(ii) out of all the empty kilns, if any, the one with the highest priority has to be chosen.

The find_best /3 predicate is defined as follows:

```
find_best(Resource, State, Selection_rule):-
  find_all(Resource, state([Resource],State),
                     List_of_Resources,
  sort_acctorule(List_of_Resources,
                 Selection_rule,[Resource:_]).
```

This clause uses the status descriptive clause explained above to find all the resources in a particular state; these are then sorted according to the rule and the required resource is chosen. It is important to note that the type of rule being considered at this time are simple heuristic rules; these rules are practical and easy to understand. They are, however, far from optimal for a dynamic system and are inappropriate to be used in such an environment because of their inability to exploit the flexibility in the system. This can in principle be overcome by defining meta rules that adaptively change the heuristic rule to suit the actual state of the system. A typical selection rule is defined as follows:

```
higher_priority(kiln(A,P1),kiln(B,P2)):-
  P1 > P2.
```

i.e. kiln A with priority P1 has a higher priority than kiln B with priority P2, if P1 is greater than P2:

```
actions(set_start,[kiln(Kn_no,Kpr_no),
                              setter(Sr_no)]):-
  remove(state([kiln(Kn_no,Kpr_no)],empty)),
  remove(state([setter(Sr_no)],wait)),
  insert(state([kiln(Kn_no,Kpr_no),
  duration(set,DT),
  system_time(T),
  Term_time is T + DT,
  schedule(set_end,[kiln(Kn_no,Kpr_no),
                      setter(Sr_no)],Term_time).
```

This means that, when the routine is activated, (i) the previous status of the kiln and the setter team are removed from the clause base, (ii) the present status of the kiln and the setter team is inserted into the clause base, (iii) the duration of the setting activity is generated and is added to the present system time to calculate the absolute termination time of the activity, and (iv) the end of setting is scheduled unconditionally (time is the only condition) and is added to the event list.

42.5.2 Example 2

Event: end of drawing

Since end of drawing is a bound event, no condition has to be defined for it:

```
actions(draw_end,[kiln(Kn_no,Kpr_no),
                    drawer(Dr_no)]):-
  remove(state([kiln(Kn_no,Kpr_no),
                 drawer(Dr_no)],draw)),
```

```
  insert(state([drawer(Dr_no)],idle)),
  insert(state([kiln(Kn_no,Kpr_no)],empty)),
  system_time(T),
  test_contingent([event(draw_start,
       [Another_kiln,drawer(Dr_no)])],_)
  test_contingent([event(start_maintenance,
       [kiln(Kn_no,Kpr_no)]),
                                    event(set_start,
       [kiln(Kn_no,Kpr_no),A_setter_team])],
       maintenance_rule).
```

The first four conditions are very similar to the previous example and there is no need for any explanation.

However, Unlike the schedule /3 predicate, the test_contingent /3 predicte activates only one event out of a number of possible events depending on the routing rules in the knowledge base (this is where the routing decisions are made). It is defined as follows:

```
test_contingent([Event],_):-
  check_conditions([Event],Invokable_event),
  invoke(Invokable_event),!.

test_contingent(List_of_events,
                 Routing rule):-
  check_conditions(List_of_events,
                      All_possible_e-
                      vents,
  decide_route(All_possible_events,
                 Routing_rule,
                 Invokable_event),
  invoke(Invokable_event),!.
```

The first clause is matched if there is only one route to be taken; for this, the conditions of the event are first checked and, if true, the event is invoked.

The second clause is matched if there is more than one possible route. In this case, the conditions of all the events are checked to produce the list of all possible events. From these events an event is produced which has had its routing conditions satisfied; this event is then invoked. A typical set of routing rules are illustrated as follows:

```
maintenance_rule(event(start_maintenance,
                      [A_kiln])):-
  queue(kiln,empty,Number),
```

```
    Number > 4,
    firing_time(A_kiln,TIme),
    Time > 1000.

maintenance_rule(event(start_setting,
                       [A_kiln,_])).
```

42.6 Conclusion

This chapter has presented an approach for building knowledge based aspects into simulation models as a stage in the development of Hybrid Decision Support Systems. The approach uses the PROLOG language and uses an event-oriented approach to the description of the simulation dynamics. Rules for Scheduling and Routing are separated from the main event routines in order to give a modularity to the total model and allow for the eventual incorporation of external knowledge bases.

The work is still in a preliminary phase and considerable work is expected in order to develop a consistent methodology with appropriate tools to support it. While the approach may seem rather theoretical, it is felt that only through the development of formal approaches will progress be made towards the goal of providing the Decision Support Systems necessary for modern manufacturing systems.

Chapter 43

Planning in FMS

V. V. Shafransky and A. P. Akimov
Research Computer Centre, Academy of Sciences, USSR

43.1 Introduction

The necessity to raise the efficiency and compatibility of machine building plants (and like enterprises) calls for the wide use of the automated manufacturing complexes, such as flexible manufacturing systems (FMS), and their further development — Computer Integrated Manufacturing Systems (CIM). And one of the most important issues is the structural balance of a manufacturing complex as a whole with the output production program not properly made.

Structural balance is the main topic of this chapter. Such balance should be considered as early as at the design stage of automated manufacturing complexes, together with the planning of their work. Generally, the problem is solved by means of mathematical modelling of a manufacturing complex and production programs. The most convenient way to undertake efficient mathematical modelling is to use special program support systems, which combine the means for the mathematical modelling, design and work planning of automated manufacturing complexes. The working out of such support systems, allowing one to analyse and synthesise many variants of production process and their organisation, is a difficult and time-consuming task. But without this process it is impossible to achieve a high level of project development. This can lead to over expenditure during the realisation of the production complex and low technological and economical efficiency.

This chapter considers how to work out intelligent program systems, aimed at the modelling, designing and work planning of automated manufacturing complexes.

43.2 General aims of the system development

Mathematical modelling used for this purpose allows one to consider in detail many different variants for the possible realisation of a production program on the given set of equipment (or the one under the design), to compare different variants of operations distributed on different kinds of machinery, to make a schedule for the order of the equipment to work on, to choose variants of the machinery sets and their layout.

The interelations between the above mentioned factors complicate the task still more. It means that a decision made on any one of them should be linked to the others and be mutually co-ordinated. This problem calls for a special set of user-oriented modelling procedures, which allow the designers to work out the decisions gradually, using the mechanism of iterative co-ordination of the decisions. This mechanism provides for backtracking to the previous stages of the modelling process (if necesary), and gives the user the possibility to change the parameters of the model.

That is the reason for building into the system effective means of organising complex procedures of multivariant calculations. It becomes necessary to provide co-ordination of the optimisation, calculations, information processing tasks, to give the user a free hand in choosing the calculation order, and to store the different variants of the resulting decisions.

Along with this, the system should provide a friendly interface, control over the calculations by means of menu, facilities for multiwindow work, convenient visualisation of all kinds of data (input data, intermediate and output results), paperless technology of design and planning, and support of the General Data Base.

It is also very useful for the system to have the means for modification of its configuration according to the needs of any particular user, taking into consideration his/her individual preferences and the peculiarities of the industrial objects he/she is designing. This concerns possible tuning of mathematical models, which are built in the system, and interface adapting. (It means adaption of dialogue support, results and data visualisation both in graphical and tabular form). That is why the system should incorporate the necessary spectrum of instrumental program tools supporting such adapting. The system should be able to have the ability to be modified and to make the process of step-by-step development more easy to fulfil.

43.3 Main tasks of the system

43.3.1 Modelling procedures

The range of questions which arise while working on the interrelated optimisation problems concerning the structure of manufacturing complexes is extremely wide. They cover:

- to decide on a rational set of equipment and its layout

- to evaluate the production capacity
- to form a production (output) program
- to pin point potential 'weak spots' and to devise steps for removing them
- increasing the utilisation coefficient for all kinds of equipment and speeding up the production rate of the complex as a whole
- to choose the optimal technologies for parts processing
- to co-ordinate technological operations for the given machinery
- to calculate the most rational loading of the transport and storage subsystems
- to find out and plan the need for tools and their fixture, particularly the rational parking of tools in machine magazines
- to formulate control rules for real-time management
- to calculate time scheduling of production equipment (for different time intervals)
- to provide efficient links of a production complex with other production environments.

The above list of problems is not exhaustive.

As stated above, the situation is aggravated still further by mutual interrelations of all the tasks. Optimal decisions in one set of procedures depend on what decisions have been made in others. It means that the decisions on all given issues must be mutually balanced. There are two alternative ways to make such a balance. One way is to consider simultaneously all important issues and circumstances. This leads to a single, but extremely complicated, mathematical model of a manufacturing complex and its production processes containing all the particular situations which can arise in production. The other way is to form a special system of procedures in the frame of which it is possible to co-ordinate the decisions made over separate groups of problems.

The first way, i.e. to make a single mathematical model, is theoretically possible, but in practice seems to be unacceptable because of its enormous dimensions and the great variability of the modelling techniques used (which arises because of modelling issues of different kinds). Such a situation does not allow one to make an efficient algorithm of the optimisation problem arising.

The second way is to design a system of modelling procedures, which allows one to work successively, step by step, at the problems and to co-ordinate them by means of an iterative mechanism of decision correlation (if necessary, one can return to any previous stage of the design process and change any parameters). In practice, this way seems more promising for the multidimentional problems.

43.3.2 Software means

The intelligent system for modelling, designing and planning should have the following components:

- *Optimisation models* for forming and choosing the technologies, for manufacturing processes, planning, equipment layout, optimisation of the equipment loading, optimisation of the structure of the instrumental kits, etc.
- *Manufacture knowledge* based on equipment and technologies, particularly:

    - equipment, its types, possibilities, the capacity of instrumental magazines, dimensions, precision specifications, the dimensions of the working areas and the weight limits of the parts to be processed
    - manufactured parts, alternative technologies for their production, geometrical parameters, processing precision requests
    - fixtures, their usability
    - instruments, their resources and usability according to the types of equipment and operations
    - typical times of equipment readjustment
    - statistics of equipment failures.

- *Knowledge bases* for the design techniques and methods of planning and management, particularly:

    - the preference for using certain kinds of equipment and their composition in different situations (that information is necessary for choosing the variants of the equipment set)
    - efficient techniques of equipment layout on the shop floor (including the library of standard composition of the equipment)
    - efficient rules for prescribing manufactured parts and operations to certain kinds of equipment (i.e. the rules for the distribution of elements of the output program on the executive equipment)
    - methods of dispatching production processes and means of adjustment for particular manufacturing systems

- *Built-in expert systems*, which form project variants and plans for the basis of expert knowledge
- *Simulation modelling system* for the detailed analysis of manufacturing processes under different management strategies and within different configurations of the manufacturing complex

- *Programs* for the analysis of the results of modelling and formulating suggestions for correction of versions of decisions
- *Instrumental tools* for configuration forming and system adapting, monitors, multiwindow work support, graphics and visualisation, user friendly interface, support systems for multivariant calculations
- *Monitoring dialogue system*, which supports the effective communication of users with the sets of modelling tools, and helps the user to influence the process of design and planning.

43.4 General characteristics and possibilities of the AMIGO system

At the Computer Center of the Academy of Sciences the new system AMIGO (Automated Modelling, Interactive Graphics, Optimisation) is now under development. The system is aimed at solving the above mentioned problems which may arise during the design and work planning of automated manufacturing complexes. The AMIGO system is developed for IBM PS XT/AT and the compatible hardware.

AMIGO is designed on the principle of an open integrated system and can suport the whole cycle of multistage modelling of an automated manufacturing complex. AMIGO allows us to input the data characterising the complex under design and to receive the results either in one session of work or several (depending on the dimensions of the complex under development and the quantity of data to be taken into account).

AMIGO is largely open-ended. It has the means for modification and extension, which allows it to receive versions of AMIGO with different functional possibilities according to the particular demands of different users. This is also very useful for designers who can make more and more powerful versions by simply adding new functional modules to the system.

The current version of AMIGO, which is already in use, suggests the following possibilities:

(i) dialogue input and editing of input data (i.e. description of composition and characteristics of machines, items of the production program and necessary technologies). This is done through the table interface
(ii) preliminary evaluation of the necessary quantity of the machinery
(iii) making a preliminary time schedule of a given production program for a more precise analysis of the possibilities of the machinery which was chosen in the process of solving the valuation problem
(iv) dialogue for forming the layout of the manufacturing complex, that is the transport scheme and positions of equipment

(v) calculating the shortest distances in a transport scheme and simulating transport movement in an animated cartoons mood
(vi) time scheduling of a production program for a particular set of machinery, within the current layout of the production complex. The results are visualised by means of Gantt diagrams and diagrams of machines loading
(vii) optimisation of the preliminary scheme of machine layout with the aim of minimising in a summary way the transport facilities
(viii) calculation and demonstration of the dynamics of parts queues at any machinery unit in the production process
(ix) distribution of all operations on different kinds of machinery
(x) calculation of the production of manufactured parts produced by different technologies
(xi) distribution of available tools between the given machinery
(xii) choosing the working tool and coolant for each operation
(xiii) forming the toolkit for machine magazines.

AMIGO has a set of mathematical models for solving these problems. Let us consider, in an illustrative manner, some of them.

We suggest that the ordered sequence of operations

$$\sigma_1^i, \ldots, \sigma_q^i, \ldots, \sigma_{Q(i)}^i \qquad (43.1)$$

for each manufacturing part i is given ($i = 1, 2, \ldots, I$).

Problem 1: Let $\{\overset{\circ}{x}_i\}_{i=1}^{I}$ be the manufacturing program for FMS, in which different kinds of equipment are disposed. Let A_n be the quantity of available n-kind equipment ($n = 1, 2, \ldots N$). The problem is to devise technological schemes

$$(\sigma_1^i, n_1^i), \ldots, (\sigma_q^i, n_q^i), \ldots, (\sigma_{Q(i)}^i, n_{Q(i)}^i) \qquad i = 1, 2, \ldots, I \quad (43.2)$$

for all given sequences (43.1) ($i = 1, 2, \ldots, I$), to minimise the top equipment loading in FMS. The model may be written in the form of linear Boolean programming:

$$\Delta \rightarrow \min,$$

$$\sum_{i=1}^{I} \sum_{q=1}^{Q(i)} \overset{\circ}{\chi}_i t_n(\sigma_q^i)(A_n)^{-1} z_{nq}^i \leqslant \Delta, n \in [1:N]$$

$$\sum_{n=1}^{N} z_{nq}^i = 1, i \in [1:I], q \in [1:Q(i)]$$

$$z^{i}_{nq} \in \{0,1\}, i \in [1:I], n \in [1:N], q \in [1:Q(i)]$$

where t_n is the time that is necessary to perform the operation on the equipment n.

Problem 1 may be generalised in the case, when alternative technologies for each manufacturing part are discussed.

By solving the Problem 2 we may answer the question: Equipments of what FMS subdivision should we use to realise technological schemes (43.2) if these equipments join several subdivisions?

Problem 2: The technological scheme (43.1) for manufacturing part i is given. In problem 2 we attempt to divide it into a set of fragments, so that all the operations of each fragment are done in the same subdivision, but the number of fragments is a minimum.

Let us fix manufacturing part i and consider the Boolean matrix $\boldsymbol{M}$ $[1{:}J], 1{:}\ Q(i)]$.

The element M_{jq} equals 1 if the n_q kind equipment is in subdivision j, and – 0 otherwise. Problem 2 may be formulated as in matrix $\boldsymbol{M}$ the 'stripes', consisting of nearly standing ones, so that

(a) the sum of stripe's length equals $Q(i)$
(b) each stripe intersects only one column of $\boldsymbol{M}$
(c) the number of stripes is a minimum.

This is the special case of the well known covering problem, the efficient solving algorithm of which was elaborated.

Problem 3: To form the manufacturing plan for each subdivision of FMS.

One of the most important features of the production process in FMS is the possibility of simultaneously making identical parts by means of different technologies. Consequently, we may variate the output of parts, made by each admissible technology, to improve such an important character of the production process as the evenness of equipment loading is. The relations for Problem 3 are:

$$\Theta \rightarrow \min,$$

$$(A^{j}_{n})^{-1} \sum_{i=1}^{I} \sum_{l=1}^{L_i} \sigma^{i}_{nie} \, \chi^{1}_{i} \leqslant \Theta, j \in [1{:}J], n \in [1{:}N],$$

$$\sigma^{j}_{nil} = \sum_{q} t_n(\sigma^{il}_{q})$$

and the summation is carried out on all operations to be done on equipment n in subdivision j:

$$\chi_i \geqslant \overset{\circ}{\chi}_i, \chi_i \leqslant \sum_{l=1}^{L_i} \chi^l_i, \quad i \in [1{:}I]$$

where L_i is the number of admissible technologies of part i, l is the index of technology. x^l_i is the number of part i, manufacturing by technology l, and the other designations are the same as above

The solution of Problem 3 is: $\overset{*}{\Theta}$, the optimal level of equipments loading, and $\{\overset{*}{\chi}{}^l_i\}$, the optimal production of each part i by each admissible technology.

Each of the above mentioned functional possibilities of AMIGO is supported by the convenient interface with a user at all levels of work. The set of functional models is constantly extended. Some of the mathematical models incorporated in a system are given in Shafransky *et al.* (1987) and Akimov and Shafransky (1987).

43.5 Logical structure of the AMIGO system

As we have stated above, the problems of planning and designing manufacturing systems are characterised by the multistage approach to problems solving; that is, step by step solving of subproblems with the co-ordination of mutually dependent parameters. In AMIGO the complex procedures for solving the above-mentioned tasks are presented as informational-logical structures, which reflect the possible sequences of actions (solving the subtasks) in given problem area.

Every informational-logical structure consists of a set of information dependent stages. At every concrete stage one of the subproblems of a general problem is solved. A user is requested to choose the order of those stages (within the limits denoted by the informational-logical structure of a complex procedure). The user can:

- analyse the intermediate results, received at any stage
- pass to the next stages, logically permitted
- backtrack to any former stage
- correct the input parameters of a subproblem
- store the intermediate decision results (it is necessary to compare them with the results obtained with different input data).

- interrupt the process of problem solving at any stage, with the possibility of continuing in future.

The system may incorporate so-called 'parallel' stages; that is, stages processing independent data. Of course, they can be fulfilled in an arbitrary order. In the process the designer may come across different situations, which influence his future actions and calculations. In that case the following may become necessary:

- redoing certain calculations, or part of them, beginning from a certain point, and using results gained from succeeding stages (iteration)
- choosing one of the alternative paths of continuation of calculations (alternative)
- return to a former stage, with refusal from a part of the calculations done already, and re-do the calculations based on corrected input data (return).

For all the above mentioned situations the AMIGO system provides special means which allow intelligent and effective control of the process of gaining the final results.

The ideology and principles of the AMIGO logical structure are described more fully by Orlova and Shafransky (1984) and Orlova and Terentyev (1986).

It is necessary to point out that the approach suggested allows us to realise the principles of an open ended and adaptable system, because the most labour-consuming monitoring module of a system stays unchanged through all the modifications of AMIGO. Only the set of closed modules and the logical scheme of their interface should be changed. The description of the logical schemes, in its turn, is also automated and is designed during the interface of the system administrator with the special subsystem, called 'the designer'.

Chapter 44

Implementation of MRP II production planning and inventory control methodologies using expert systems and computer simulation techniques

W. Walker and K. Maughan

Sunderland Polytechnic, UK

44.1 Introduction

In recent years considerable effort has been devoted to revising formal methodologies for improving the effectiveness of the production planning and inventory control procedures used within manufacturing organisations. In particular, Just-in-Time (JIT) and Manufacturing Resources Planning (MRP II) systems have received a great deal of attention (Voss, 1987; Ho & Dilts, 1988). However, despite this high level of interest, practical implementations of JIT and MRP II systems are not common (New & Myers, 1985). This low level of usage can be attributed to the complexity of the problems associated with the generation of material and work flow schedules within the manufacturing environment. Within this chapter we discuss the nature of these problems and propose a methodology for applying expert systems and computer simulation techniques to their solution.

Within any manufacturing organisation the production management function must attempt to meet the following objectives:

(a) guaranteed meeting of the due dates associated with each customer order
(b) efficient utilisation of manufacturing resources
(c) minimisation of manufacturing lead times
(d) minimisation of raw material and bought in component inventory levels
(e) minimisation of work in progress (WIP) inventory levels

(f) minimisation of intermediate finished goods inventory levels (i.e. component and sub-assemblies which have been manufactured and are waiting to be included in higher level sub-assemblies or finished goods)
(g) minimisation of finished goods inventory levels.

Within an MRP II based production management system the above objectives are achieved by resolving the production management function into a hierarchical sequence of production planning and inventory control procedures as shown in Fig. 44.1. The MRP II system shown in Fig. 44.1 generates two distinct schedules:

(a) a production schedule which provides the dates and machinery on which basic components and sub-assemblies should be manufactured
(b) an MRP schedule which provides the dates on which orders for raw materials and bought in components should be dispatched in order to ensure that these materials arrive on the shop floor in time to meet the requirements of the production schedule.

In an ideal MRP II system the production schedule and the MRP schedule should combine to provide a JIT production plan, i.e.:

(*a*) Raw materials and bought-in components are scheduled to arrive from external sources onto the shop floor just in time to be processed into product parts and/or included in higher level and sub-assemblies.
(*b*) The manufacture of components and sub-assemblies is scheduled to be completed just in time for these parts to be included in higher level sub-assemblies.
(*c*) The manufacture of all the individual components and sub-assemblies associated with specific orders are scheduled to be *simultaneously* completed just in time for these components and sub-assemblies to be assembled into finished goods.
(*d*) The assembly and final processing of finished products are scheduled to be completed just in time to meet the due dates of customers orders.

The MRP II architecture shown in Fig. 44.1 has been most successfully applied within mass production industries which use large batch sizes and flow line manufacturing techniques. In these manufacturing environments it is fairly easy to calculate future material requirements, and hence generate an MRP schedule. Furthermore, the sequential nature of the production process makes it possible to manually calculate the future

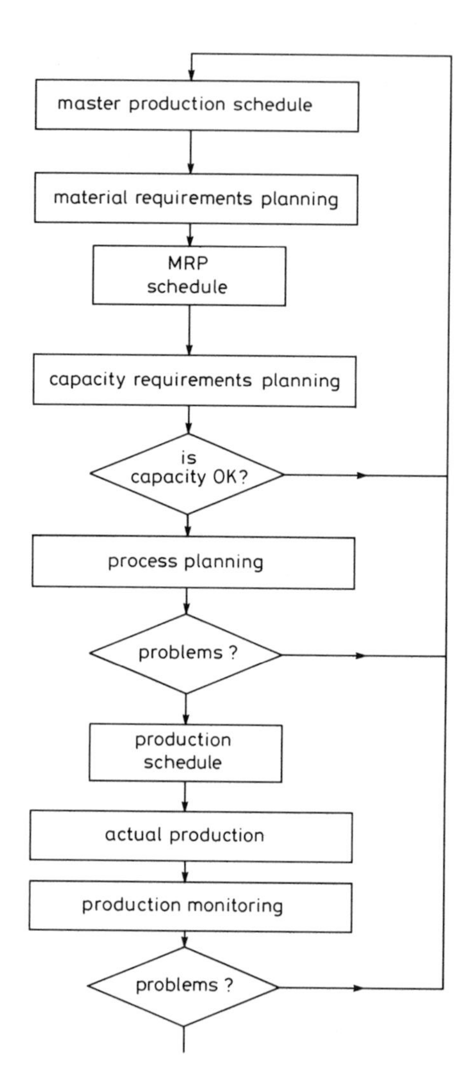

Fig. 44.1 The MRP II architecture

loading of manufacturing resources for any given master production schedule. These factors, together with the generally constant nature of the flow-line manufacturing environment, have made it possible to implement flow-line MRP II production management systems using commercially available MRP packages and manual or computer-aided production scheduling techniques (Harrison, 1988; McAreavey *et al.*, 1988).

44.2 Application of MRP II methodologies to non-flow-line manufacturing systems

Attempts have been made to apply the MRP II architecture shown in Fig. 44.1 to manufacturing systems which use non-flow-line production systems and small-medium batch sizes. However, the production planning and inventory control problems found within this manufacturing environment are considerably more complex than those found within the flow-line manufacturing systems where most MRP II systems have been implemented (Monniot, 1987). This higher degree of complexity can be attributed to the following factors:

(*a*) the large number of different products, components and sub-assemblies which must be manufactured on the same production equipment
(*b*) the complex machine routings associated with the processing of each individual component and sub-assembly
(*c*) the need to co-ordinate the manufacturing times of each of the individual components associated with each individual customer order for a specific product
(*d*) the need to meet the due date requirements of orders from a large customer base with each customer possibly placing multiple time-phased orders for a range of different products.

If this form of manufacturing environment is to operate efficiently, it is essential that the production management procedures take into account the complex interactions between each individual aspect of the production planning and inventory control problem. In particular, it is necessary to consider the constraints which the availability of material and manufacturing resources impose on the feasibility of the production schedules generated by an MRP II system (Smith, 1988). For example, during the generation of the MRP schedule it is necessary to take into account any manufacturing resource constraints which may result in delays in the manufacture of basic components and sub-assemblies (these delays affect the times at which materials will actually be required). Correspondingly, during the generation of the production schedule it is necessary to take into account any material requirements constraints which may affect the feasibility of the production schedules generated by the system (i.e. the scheduling process must avoid scheduling work operations if the materials required to carry out those operations will not be available at the appropriate time). This suggests that, within any effective production management system, the Materials Requirements Planning (MRP), Capacity Requirements Planning (CRP) and Process Planning (PP) functions should be integrated in order to ensure that realistic material requirements and production schedules are generated. However, as can be seen from Fig. 44.1, current practical implementations of MRP II systems use architectures in which the MRP, CRP and PP functions are implemented independently (although in practice there is usually some form of informal interaction between each of these functions).

Alternative MRP II architectures have been proposed in which production planning is automated using conventional algorithmic data processing techniques. However, within most production systems, the large number of system parameters and potential system configurations makes a deterministic search for a feasible production plan computationally unfeasible even if an algorithmic approach could be defined. Hence many commercially available MRP II systems must make simplifying assumptions which render much of their output of limited use.

Expert systems and computer simulation modelling techniques potentially provide a means of overcoming many of the deficiencies of more conventional production planning tools (Moser, 1986; Flitman & Hurrion, 1987). Expert systems techniques can be used to generate production plans automatically using methodologies similar to those employed by human experts. Simulation tools provide a means of evaluating the production schedules generated by an expert production scheduling system. Hence, by integrating expert systems and simulation of modelling techniques, it should be possible to both generate and validate production plans. Furthermore, the results generated by the simulation tool can be analysed by the expert system, which may then decide to modify the production plan in order to avoid any problems identified during the simulation phase. This feedback and iteration characteristic reflects the behaviour of closed loop MRP II production management systems.

44.3 A modified MRP II architecture using expert systems and computer simulation techniques

It is clear from the above discussions that the MRP II architecture of Fig. 44.1 is not suitable for use within manufacturing environments which use non-flow-line production systems and small—medium batch sizes. Fig. 44.2 shows an alternative MRP II architecture based on the use of expert systems and computer simulation techniques. As in the case of Fig. 44.1, the MRP II architecture of Fig. 44.2 generates a materials requirements schedule and a production schedule. However, in the case of Fig. 44.2, these schedules are generated in parallel in order to ensure that they are both mutually consistent. Furthermore, the validity of these material

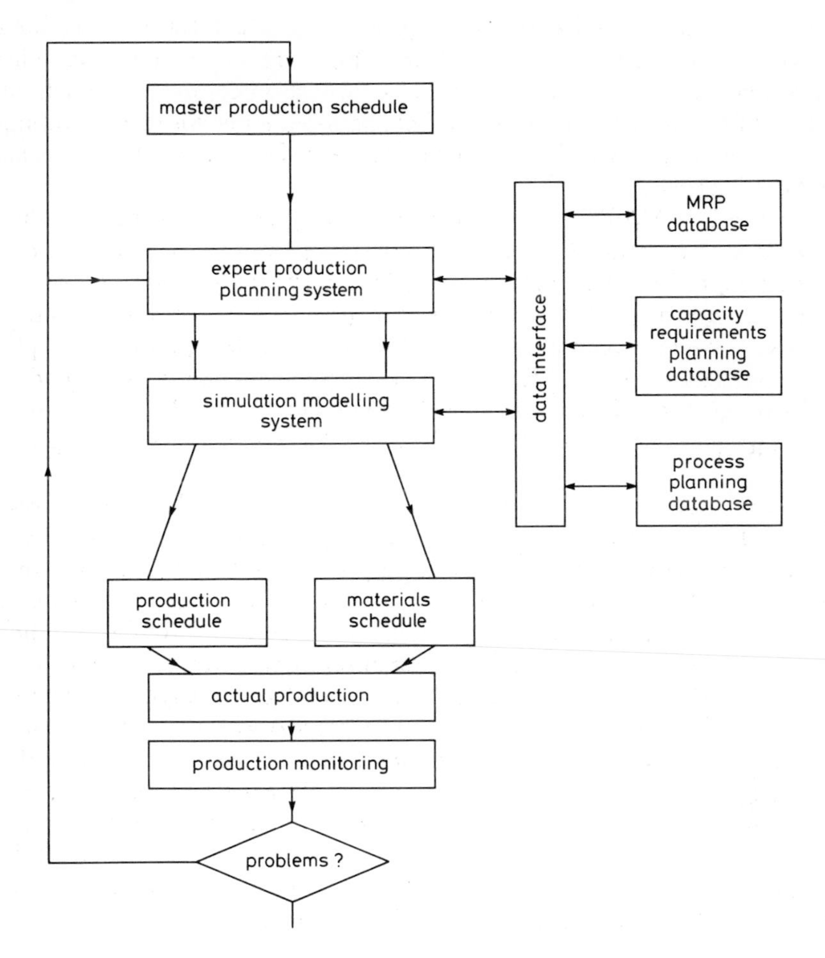

Fig. 44.2 Modified MRP II architecture using expert systems and simulation modelling techniques

and production schedules is investigated using the simulation facilities built into the modified MRP II architectire. Iterative schedule generation – simulation cycles are then used in order to ensure that the MRP and production schedules meet the requirements of the Master Production Schedule without violating the constraints imposed by the production system or the lead times on the acquisition of externally sourced components and raw materials.

Fig. 44.3 provides a more detailed illustration of how the expert systems and simulation modelling components of the modified MRP II architecture are implemented. Within conventional production management systems, three separate data bases are usually used in order to

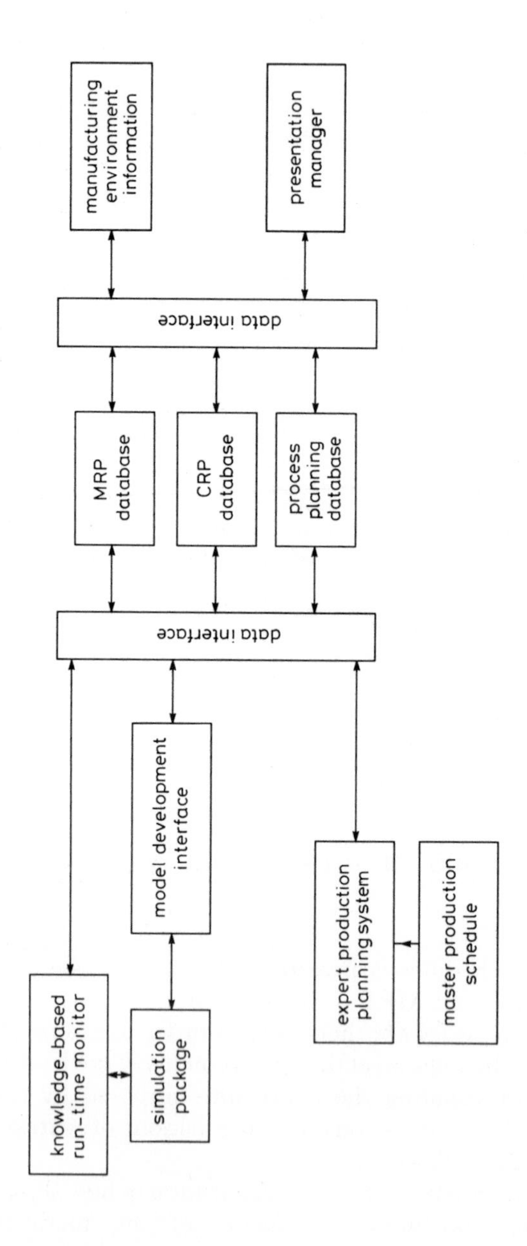

Fig. 44.3 Integration of expert systems and computer simulation into an MRP II system

implement the production planning and inventory control functions. These data bases are:

(*a*) *The capacity requirements planning data base*: this data base contains information on the manufacturing resources which are available on the shop floor.
(*b*) *The materials requirements planning data base*: This data base contains a description of how complete products are built up from basic components and sub-assemblies. This description is normally implemented as a hierarchical Bill of Materials data structure. The MRP data base also contains information on current and projected inventory levels for raw materials, components and sub-assemblies.
(*c*) *The process planning data base*: This data base contains a description of the manufacturing operations (machine routings, timings etc.) required to make each individual product.

Taken collectively, these data bases provide a description of the production system and the constraints under which the material and production schedules must be developed. The MRP II architecture of Fig. 44.2 uses an enhanced version of these data bases: very briefly, the information contained within the modified data bases can be divided into three categories:

(*a*) *Manufacturing environment data*: Essentially this information is identical to that held on the conventional CRP, MRP and process planning data bases. This information is regarded as constant and is used to define the manufacturing environment in terms of available manufacturing resources and a description of the products which are required to be manufactured using these resources.
(*b*) *Expert systems related data*: This information largely consists of the knowledge bases which the expert production planning system uses when generating the MRP and production schedules.
(*c*) *Simulation modelling related data*: This information relates to the intermediate and final results generated by the simulation model.

The expert production planning system and the simulation modelling system are both implemented independently. Interaction between the two is achieved via the data bases described above. In operation, the behaviour of the MRP II architecture shown in Fig. 44.2 is as follows:

(*a*) *The Master Production Schedule* is input into the expert production planning system.
(*b*) *The expert production planning system* then uses expert systems techniques and the manufacturing environment data to construct an initial version of

the MRP and productions schedules (see Section 44.4). These schedules are then asserted into the data bases shown in Fig. 44.3. Control is then passed to the simulation modelling system.

(*c*) *The simulation modelling system* then uses the manufacturing environment data and the schedules generated by the expert production planning system in order to construct a simulation model of the production system. This model is then executed and the results asserted into the data bases of Fig. 44.3. After termination of the simulation phase, control is returned to the expert production planning system.

d) *The expert production planning system* then examines the results generated by the simulation modelling systems and compares these results with the requirements of the Master Production Schedule. In the event of any major discrepancies occurring, the expert production planning system uses the results generated by the simulation system in order to identify the relevant problem areas. The expert production planning system then attempts to regenerate the parts of the schedules associated with these problem areas. Control is then passed once again to the simulation modelling system which re-models the production system using the amended schedules.

(*e*) *The schedule generation-simulation cycle* is then repeated until an MRP and a production schedule which meet the requirements of the Master Production Schedule are achieved. These schedules are then presented to the user via the presentation management system shown in Fig. 44.3.

44.4 Implementation of an expert production planning system

As indicated earlier, the objectives of the expert production planning system are to generate an MRP and a production schedule which meet the objectives of the Master Production Schedule (MPS) without violating the constraints of the production system. The MPS itself consists of a set of time-phased orders from individual customers for products from the company's product range. A brief overview of the production planning process associated with each other is as follows.:

(*a*) Each customer order is analysed and broken down into its component level and sub-assembly level material and manufacturing resource requirements.

(*b*) A process plan (machine routing) is worked out for each component and sub-assembly.

(*c*) A decision is made on which machines to carry out the processing of each part and sub-assembly.

(*d*) A decision is made on when to start the manufacturing of each component.

(*e*) A decision is made on when to place orders for externally sourced raw materials, components and sub-assemblies.

The above production planning process must be carried out for each individual order in the MPS. If the expert production planning system is to overcome the deficiencies of existing MRP driven production management systems, two factors need to be taken into account when generating the MRP and production schedules:

(*a*) *The lead times on acquiring externally sourced materials*: Many commercially available production scheduling systems assume zero lead times or infinite material resources.
(*b*) *The finite nature of the manufacturing rsources*: Commercially available MRP and production scheduling packages often assume infinite manufacturing resources; hence the queueing times incurred by parts as they pass through the production system are ignored.

Incorporation of material acquisition lead times into an expert production planning system is reasonably straightforward since these lead times are relatively constant. However, incorporating the finite nature of the manufacturing resources into the production planning system is made difficult because of the complex interactions between the work queues at each individual work centre. During the generation of the production schedule, individual orders are broken down into their component level requirements, and the production of these components is scheduled onto the available manufacturing resources. However, each component associated with an individual product may be manufactured on a different machine. Consequently, the component parts of individual products may be widely distributed throughout the various work centres on the shop floor. Furthermore, since the production equipment is usually general purpose, the work queues at each individual machine will contain components belonging to a range of unrelated products. Hence, in order to co-ordinate the manufacture of complete products, the production scheduling system must operate at two levels:

(*a*) *Local level*: At a local level the scheduling system must decide in which order to process the various parts which are waiting at a particular work centre.
(*b*) *Global level*: At a global level the scheduling system must consider how the order in which parts are processed at individual machines affects the completion dates of finished products and sub-assemblies.

The expert production management system of Fig. 44.3 is intended to be implemented in a modular fashion with an individual local expert scheduling module being associated with each work centre. A global scheduling module is then used to co-ordinate the scheduling activities carried out at a local level. The global scheduling module interacts with the localised scheduling modules by providing these localised scheduling modules with information relating to the future requirements for parts processed at the local work centre. For example, at a local level, the production of certain parts can be delayed if these parts are intended to be included in a higher level sub-assembly which has yet to be completed at another work centre. Alternatively, the manufacture of some parts can be given a high priority if this would contribute to a reduction of WIP at other work centres by allowing the completion of a finished product or sub-asembly currently being processed at these work centres. The global scheduling module also seeks to load the localised work centres in such a way as to avoid bottlenecks at heavily loaded work centres. This is achieved by analysing the flow of work through these heavily loaded work centres and attempting to re-route the work scheduled at the work centres which feed the heavily loaded work centre in such a way as to obtain a more uniform non-critical load at the potential bottleneck resources.

The resolution of the scheduling problem into interacting local and global scheduling functions allows the production planning process to be carried out at a product level rather than a component level. In addition, it is possible to incorporate a wide range of both domain independent knowledge into the expert production management system. The knowledge incorporated into the production management system can be roughly divided into four categories:-

(*a*) *Environmental knowledge*: This knowledge relates to the efficient and effective use of the production technology and may include, for example, information on which machines are most suitable for manufacturing specific components. Information on optimum tool set up times and job sequencing can also be exploited.
(*b*) *Logistical queue management knowledge*: This knowledge relates to the various strategies used to analyse the flow of work both at a machine level and at a global plant-wide level.
(*c*) *Managerial knowledge*: This knowledge relates to the various decision making processes which production managers use in order to resolve conflicts arising from the limitations of the production technology; for example, the prioritisation of specific orders, relaxation of due date and WIP inventory level constraints and the re-allocation of manufacturing resources.
(*d*) *Heuristic knowledge*: This knowledge relates to information which has been learnt about the production system either from past experience or from the simulation modelling results. This knowledge can be used, for example, to indentify potential problem areas such as bottleneck resources.

Using the MRP II architecture technology of Fig. 44.2, the above knowledge can be exploited during the production planning process in order to achieve feasible production plans.

44.5 Integration of simulation technology within an expert-system driven MRP II package

Complementary to the expert scheduling package, the second component of the iterative schedule generation-simulation cycle mentioned earlier is the application of computerised simulation technology. Such a facility could be accessed by the expert system as a tool to investigate the complex queue interactions inherent within production systems, particularly when addressing a non-flow-line environment. Fig. 44.3 shows how a simulation package would be integrated into the MRP II system. In addition to accentuating the functionality of the expert scheduler component of the system, by providing a method of assessing the impact of proposed production and materials schedules on the production process under investigation, such an application of the technique could overcome many existing problems with the implementation of simulation technology within manufacturing industry (Maughan, 1988; Murray & Sheppard, 1988; Shannon *et al.*, 1985) and help to improve the currently low take-up rate of this powerful decision support tool (Walker, to appear).

44.5.1 Implementation of modelling component

There are a number of stages in a single iteration of the simulation cycle:

(*a*) Once the expert scheduler has generated the initial product and materials schedules, the resultant data is passed to the MRP, CRP and PP data bases. Once triggered, the simulation package interrogates the data bases and extracts the information required to build that simulation model.
(*b*) The Model Development Interface (MDI) uses this data, in conjunction with domain-specific knowledge on model construction rules, to create a computer representation of the production system and strategem implied by the schedules. This automatic model generation and execution facility directly addresses many of the outstanding difficulties in simulation implementation.
(*c*) The simulation package executes the model created by the MDI, under the influence of a knowledge-based run-time monitor. This software continually checks the on-going simulation results against parameters, (e.g. maximum queue lengths, required machine and operator utilisation statistics etc.) derived from the data bases and the expert scheduler. Should any modelled conditions exceed these constraints, control can be returned to the expert system immediately.

(*d*) On completion of the simulation, the model returns data, via the MDI, to the system data bases, and passes control back to the scheduler as described in Section 44.3.

This process enables the system to derive and present precise details on the implications for the production process of the original MPS. The parallel, self-consistent production and materials schedules can be implemented on the model for comprehensive validation before they are applied to the shop floor. Graphics-based presentation management techniques can help to ensure that these results are fully appreciated by users. The integration of simulation modelling within an integrated, expert-system driven MRP II package provides the system with an accurate, rapid method of assessing, manipulating and presenting the results of the schedule generation exercise.

44.6 Conclusion

Practical implementations of MRP II production management systems have largely been confined to manufacturing organisations which use large batch sizes and flow-line production technologies. Within these manufacturing environments the linearity and stability of the production system greatly simplifies the task of implementing each individual component of the MRP II system; in particular it is possible to implement the MRP and the production scheduling functions as separate stand-alone system which are both driven by the same master production schedule.

In this chapter we have discussed some of the problems associated with the application of MRP II methodologies to manufacturing organisations which use small-medium batch sizes and non-flow-line production technologies. It has been suggested that conventional MRP II architectures are not suitable for use within this type of manufacturing environment because of the complex interactions between the requirements of the MRP and production scheduling systems. Hence an alternative MRP II architecture has been proposed in which expert systems and computer simulation techniques are used in order to integrate the MRP and production scheduling functions. This alternative MRP II architecture uses constraints based production planning methods in order to ensure that the MRP and production schedules generated by the system are both mutually consistent and feasible within the constraints imposed by the production system. The constraints under which the production planning process must be carried out are derived from the MRP, CRP and process planning data bases used within conventional manual or computer-based production management systems.

As well as providing production schedules for existing shop-floor layouts, the MRP II system described above would also be of great value in the design and implementation of new or adjusted production facilities. In such a project, the emphasis is subtly changed. Instead of inputting the known MRP, CRP and PP data, and asking the MRP II system for the most efficient way of implementing a particular MPS, the user creates an artificial MPS, which is to be used to test the new production facility. This MPS would be based on the known requirements for productivity from the new system. The user would then try a number of potential strategies for layout, machine utilisations, product mix etc., by inputting the required data into the MRP, CRP and PP data bases, and the MRP II system, in its new role, would provide information on the operation of the hypothetical system under known MPS conditions.

References

ACARD Report (1983): 'New opportunities in manufacturing: the management of technology', October

ADELSBERGER, H. H. (1984): 'PROLOG as a simulation language', Proc. of the 1984 Winter Simulation Conference, SCS, pp.501 – 504

AKIMOV, A., and SHAFRANSKY, V. (1987): 'Planning of flexible manufacturing systems', *Technical Cybernetics*, (4)

AMMER, D., DUNGS, K., SEIDEL, U.A., and WELLER, B. (1986): 'Systematische Montage-planung' *in* BULLINGER, J. (Ed.) 'Handbuch fur die Praxis', (Carl Hanser Verlag, Munchen, FRG)

ANDREASEN, M. M., and HEIN, L. (1987): 'Integrated product development', (IFS Publications Ltd., Bedford, UK)

ANON (1986): 'Lucas cuts 40% of labour through Kanban to boost business', *Auto Industry*'

ANON. (1986): *PC Week*, October, pp. 19 – 22

APICS Report (1984): 'APICS JIT and MRP II: Partners in manufacturing strategy'. Report on 25th APICS Conference, Modern Materials Handling, APICS, December, pp. 58 – 60

ATKINSON, J. (1984): 'Manpower strategies for flexible organisations', *Personnel Management*, **16**, pp. 28 – 31

BANERJI, S. K. (1986): 'Information systems for computer integrated manufacture-methodology'. Computer-Aided Production Engineering Conference Proceedings, Edinburgh

BARKMEYER, E. *et al.* (1986): 'An architecture for distributed data management in CIM'. National Bureau of Standards, USA (NBSIR 86-3312)

BARRAR, P. (1988): 'Organising for systems integration'. BPICS Euro Conf. Proc. 'Integration for success', Birmingham, England

BAXTER, L. F., FERGUSON, N., MACBETH, D. K., and NEIL, G. C. (1988): 'Materials and information flow: Managing the supply chain', Factory 2000 International Conference, IERE, Cambridge, pp. 81 – 86

BEN-ARIEH, D. (1986): 'Manufacturing system application of a knowledge-based simulation', Proc. 8th Annual Conf. on Computers and Industrial Engineering, Vol. **1**, 1.4, pp. 459 – 63

BEN-ARIEH, D., and MOODIE, C. L. (1987): 'Knowledge-based routing and sequencing for discrete part production', *J. Manufacturing Systems*, **4**, pp. 287 – 297

BENNETT, D. J., and FORRESTER, P. L. (1988*a*): 'System integration of modularised assembly FMS', Factory 2000 Conference, Cambridge, August/September, IERE Publication 80

BENNETT, D. J., FORRESTER, P. L., RAJPUT, S., and HASSARD, J. (1988*b*): 'The effect of market driven business policies on the design and implementation of production systems'. British Academy of Management Conference, Cardiff, September

BESSANT, J.R., and HAYWOOD, B.W. (1985): 'The introduction of flexible manufacturing systems as an example of computer integrated manufacturing'. Brighton Polytechnic

BITITCI, U. S. (1988*a*): Internal report, University of Stratchclyde, MEM Division, Glasgow, UK

BITITCI, U. S., and ROSS, A. (1988*b*): 'PLC based diagnostic systems for FMS'. FMS-7 Conference Proceedings, Stuttgart, West Germany

BLOOR, M. S. *et al.* (1988): 'Towards integrated design and manufacturing systems'. Paper submitted to the Factory 2000 Conference

BLUE, R. (1985): 'Developing an Executive Early Warning System', Proc. of CAMP, Berlin, pp. 308 – 310

BOADEN, R. J., and DALE, B. G. (1986): 'What is computer integrating manufacturing?', *Int. J. Operations and Production Management,* **6**, pp. 30 – 37

BOOTHROYD, G., and DEWHURST, P. (1983): 'Design for assembly – A designer's handbook'. Department of Mechanical Engineering, University of Massachusetts Amherst, USA

BOURELY (1987): 'Application of the GRAI method by the company game Ingeniere at the transfusion centre Bordeaux'. GRAI User Conference, Bordeaux

BRAMER, M. A. (1985): 'Expert systems – The visions and the reality'. Proceedings of the 4th Technical Conference of the British Computer Society Specialist Group on Expert Systems, Cambridge, (Cambridge University Press)

BRAVOCO, R. R., and YADAV, S. B. (1985): *Computers in Industry*, **6**, pp. 345 – 361

BRIGHAM, E. O. (1974): 'The Fast Fourier Transform' (Prentice-Hall)

BROWNE, J., DUBOIS, D., RATHMILL, K., SETHI, S., and STECK, K. (1984): 'Classification of flexible manufacturing systems', *The FMS Magazine,* (IFS Publications)

BROWNE, J., HARHEN, J., and SHIVNAN, J. (1988): 'Production management systems: a CIM perspective' (Addison Wesley)

BULLINGER, H. – J. (1984): 'Computer-aided depicting of precedence diagrams – A step towards efficient planning in assembly', *Computer and Industrial Engineering*, **8**, (3/4)

BULLINGER, H. – J. (1988): 'The Management of information technology' Management Seminar at 1st EURINFO, Athens

BUNN, P. R., and PANDYA, K. V. (1988): 'Requirements to interface MMS protocol to production control systems'. Proc. 4th National Conference on Production Research, UK, pp. 179 – 182

BURBRIDGE, J. L. (1988): 'IM Before CIM'. 27th International Matador Conference, Manchester, UK

BUZACOTT, J. A. (1985): 'Modelling Manufacturing Systems', *Robotics & CIM,* **2**,(1), pp. 25 – 32

CAM – I (1980): Computer-aided Manufacturing International Inc., Architects Manual I-CAM Definition Method

CAM – I (1984): Computer-Aided Manufacturing International Inc., Study of the Usefulness of IDEFo, Report No. R – 84 – CF – 01

CANELLA, T., and SCHUSTER, (1987): *Production and Inventory Management,* **28**, (4), pp. 1 – 5

CARNEGIE GROUP (1987): Knowledge craft Reference Manual, Version 3.1, Carnegie Group Inc., Pittsburgh, PA

CARRIE, A. S., ADHAMI, E., STEPHENS, A., and MURDOCH, I. C. (1984): 'Introducing a flexible manufacturing system', Int. J of Production Research, **22**, (6) pp. 907 – 916

CARRIE, A. (1988*a*): 'Simulation of manufacturing systems', (John Wiley, Chichester, England)

CARRIE, A. S., and BITITCI, U. S. (1988*b*): 'Tool management in FMS for integrating information and material flow', Proc. International Conf. on Factory 2000, IERE Aug. – Sept., Cambridge, IERE Publication 80

CAULKIN, S. (1987): *Management Today*, January

CHAMBERS, J. C., MULLICK, S. K., and SMITH, D. D. (1971): 'How to choose the right forecasting technique', *Harvard Business Rev.,* July – August, (49) pp. 45 – 74

CHANG, T. C., and WYSK, R. A. (1985): 'An introduction to automated process planning systems', (Prentice-Hall Inc.)

CHECKLAND, P. (1984): 'Systems thinking, systems practice', (John Wiley UK)

COLEMAN, H. W. (1987): *Production and Inventory Management*, **28**, (2) pp. 110 – 116

COLQUHOUN, G. J., GAMBLE, J., and BAINES R. W. (1989): 'The use of IDEFo to link design and manufacture in a CIM environment', *Int. J. Oper. Prod. Manag.*, **9**, (4)

CONRAD, S., and PUKANIC, R. (1986): *Ind. Engineering*, **18**, (2)

CROSSFIELD, R. T., TAYLOR, J., DALE, B. G., and PLUNKETT, J. J. (1988): 'The development of IDEFo as an effective tool for mapping quality management systems'.

DALE, M. W., and JOHNSON, P. (1986): 'The redesign of a manufacturing business', Research in advanced manufacture, *Proc., I. Mech. E.*

DE PIETRO, R. A., and SCHREMSER, G. M. (1987): 'The introduction of advanced manufacturing technology and its impact on skilled workers; perceptions and communication, interaction, and other job outcomes at a large manufacturing plant', *IEEE Trans. Engng. Management,* **EM-34**, 1

DEERY, F. C., and CHAMBERLAIN, N. H. (1964): 'A study of thread tension variations during the working cycle in a lockstitch sewing machine'. Clothing Institute Technological Report No. 15, July 1964

DIS 9506: Parts 1 and 2, available from BSI

DOHERTY, J. G., LAIRD, R. J., MILLER, W., and HERDMAN, T. J. (1987): 'Skill requirements for new technology in Northern Ireland'. Northern Ireland Training Authority, Newtownabbey, Northern Ireland

DOUMEINGTS, G. (1984): 'Methodology to design computer integrated manufacturing and control of manufacturing unit' *in* 'Methods and tools for computer integrated Manufacture', (Springer – Verlag)
DOUMEINGTS, G. (1985): 'How to decentralise decisions through GRAI model in production management', *Computers in Industry* (6), North Holland Publishing
DOUMEINGTS, G. *et al.* (1986): 'Design methodology of computer integrated manufacturing and control of manufacturing units' *in* RUMBOLD and DILLMAN (Eds.): 'Computer-aided design and manufacturing: Methods and tools' (Springer-Verlag) Chap. 5

EDWARDS, G. (1988): *J. Inst. Prod. Engs.*, **5**, p. 32–35
ERRATUM (1988): *Production and Inventory Management*, **29**, pp. 89

FAN, I. S., and SACKETT, P. J. (1988): 'A PROLOG simulator for interactive flexible manufacturing systems control', *Simulation*, **50**, (6), pp. 239–247
FAVERO, R.: 'Graphics, a direct and universal language'. *Management and Information*, **22**, (10), pp. 625–30
FEIGENBAUM, A. V. (1983): 'Total quality control', (McGraw Hill, 3rd ed.,) p.6
FERDOWS, K. (1985): 'The state of large manufacturers in Europe'. Results of the 1985 European Manufacturing Futures Survey, INSEAD
FERGUSON, N., BAXTER, L. F., NEIL, G. C., and MACBETH, D. K. (1988): 'Better management of the supply chain in support of JIT manufacturing operations'. IIE International Systems Conference, St. Louis.
FLITMAN, A. M., and HURRION, R. D., (1987): 'Linking discrete-event simulation models with expert systems', *J. Opl. Res. Soc.*, **38**, (8), pp. 723–733
FORD, E.: 'Pump Just in Time at ITT Jabsco', (to be published)
FORSYTH, R. (Ed.) (1984): 'Expert systems, principles and case studies' (Chapman & Hall, London)
FUTO, I., and GERGELY, T. (1987): 'Logic programming in simulation', *Trans. SCS*, **3**, (3), pp. 195–216

GALBRAITH, J. R. (1971): *Bus. Horizons*, February.
GERRARD, W. (1988*a*): 'A survey of selection methodologies adopted by industry for the selection of new technology/machine tools'. Proceedings of the 27th Matador Conference, UMIST, April
GERRARD, W. (1988*b*): 'A Strategy for selecting and introducing new technology/machine tools'. Proceedings of the 4th National Conference of Production Research, City Polytechnic of Sheffield, September
GLEN, R. H. (1985): 'The CIM revolution: A Delphi study of the strategic and structural implications of computer integrated manufacturing'. Proc., IEEE Int. Conf. on Cybernetics and Soc., Tucson, Arizona
GODDARD, W. E. (1982): *in* 'Modern materials handling (Cahners Publishing Co.)
GODWIN, A. N., GLEESON, J. W., and GWILLIAM, D. (1989): *Info. systems,* **14**, (1), TBA
GOLD, B., and RADAR, C. M. (1969): 'Digital processing of signals' (McGraw-Hill)

GOLDHAR, J. D., and JELINEK, M. (1983): 'Plan for economies of scope', *Harvard Business Review*, November, pp. 141 – 148
GOLDRATT, E. M., and COX, J. (1986): 'The goal', (Creative Output Books)
GREEN, N. L. (1984): 'Interfacing interactive graphics with fuzzy production rules' Proc. IEE Conference on Cybernetics, Halifax, October, pp. 226 – 228
GREENWALD, R. (1985): *Ind. Engineering,* **17**, (4)
GUSTAVSEN, B. (1986): 'Evolving patterns of enterprise organisation: The move towards greater flexibility', *Int. Labour Rev.,* **125**, pp. 367 – 382

HARRIS, F. J. (1978): 'On the use of windows for harmonic analysis with the Discrete Fourier Transform', *Proc. IEEE,* **66**, pp. 51 – 83
HARRISON, M. (1988): 'MRP delivers just in time', *Industrial Computing,* December, pp. 26 – 28
HARTLAND-SWANN, J. (1987): '3 Steps to CIM', *Industrial Computing*, (UK)
HARTLAND-SWANN, J. (1988): 'The CIM fix: From a strategy into reality', *Industrial Computing*, February
HAYES, R. H., and WHEELRIGHT, S. C. (1979): 'Linked manufacturing process and product life cycles', *Harvard Business Review*, January, pp. 133 – 40.
HENDERSON, M. R. (1984): 'Extraction of feature information from three-dimensional CAD data'. Ph.D. Thesis, Purdue University
HILL, T. (1985): 'Manufacturing Strategy', (Macmillan Education)
HIRD, G., *et al.* (1988): 'Possibilities for integrated design and assembly planning'. Product Design for Assembly Seminar Notes, IFS Conferences, Kempston, Bedford
HO, C., and DILTS, D. (1988): 'Integrating MRP with information systems through the MEISE Grid'. *Int. J. of Operations & Production Management,* **9**, pp. 36 – 47
HONG, S., and MALEYEFF, J. (1987): *Production and Inventory Management*, 28(1) pp. 46 – 54
HOUSE OF LORDS (1985): Select Committee Report on Overseas Trade, (HMSO.)
HRUZ, B. (1984): 'Prozessgerechter Entwurf von Monitoren in Echtzeitbetriebssystemen', *Elektronische Rechenanlagen,* **26**, (1)
HUGHES, D. R., and BAINES, R. W., (1985): 'Evolutionary design of computer integrated manufacturing systems'. Proc. 1st National Conference on Production Research, University of Nottingham, September
HUGHES, J. (1987): *FMS Magazine*, **5**, (2)

ISO (1986): 'The Ottawa report on reference models for manufacturing standards'. ISO Publications Oct., TC184/SC5/WG1 N51
Industrial Computing (1987): 'MRP II/production management systems', November pp. 35 – 41
INGERSOLL ENGINEERS (1985*a*): 'Integrated manufacture', MORTIMER, J. (Ed.) (IFS Publications Ltd.)
INGERSOLL ENGINEERS (1985*b*): Hitech can flop in factories'. Report of a poll commissioned by Ingersoll Engineers, *The Guardian*, 12th May
INGERSOLL ENGINEERS (1986): 'Integrated manufacture' (IFS Publication Ltd.)
INSTITUTION OF CHEMICAL AND ASSOCIATION OF COST ENGINEERS (1988): 'A guide to capital cost estimating'. 3rd edn.

JARVIS, J. W. (1986): 'Computer integrated manufacturing technology and status – An overview for management'. IEE Advanced Manufacturing Information
JONES, A., and WEBB, T. (1987): *General Management*, **12**, (4), pp. 60 – 74
JONES, C., and CHAHARSOOGHI, K. (1986): 'Symphony and manufacturing planning and control systems'. Proc. of Second National Conference on Production Research, Napier College, Edinburgh
JONES, C., and CHAHARSOOGHI, S. K. (1988): 'A "smart" production Planning System' *in* WORTHINGTON, B. (Ed.): 'Advances in manufacturing technology': III', (Kogan Page, London) pp. 310 – 314
JONS, A. (1988*a*): *Lotus*, **2**, (1) pp. 52 – 54
JONS, A. (1988*b*): *Lotus*, **2**, (2), pp. 49 – 51
JONS, A. (1988*c*): *Lotus*, **2**, (3), pp. 54 – 58
JONS, A. (1988*d*): *Lotus*, **2**, (4) pp. 29 – 32
JONS, A. (1988*e*): *Lotus*, 2, (5), pp. 26 – 28
JONS, A. (1988*f*): *Lotus*, **2**, (6), pp. 21 – 24
JOSHI, S., VISSA, N. N., and CHANG, T.C. (1988): 'Expert process planning with solid model interface', *Int. Journal Production Research*, **26**, (5), pp. 863 – 885
JURI, A. H. (1988*a*): 'Reasoning about machining operations to aid process planning – a case study using components from GEC Turbine Generators Ltd'. GMP Internal Report, CAE Series, University of Leeds, June
JURI, A. H. (1988*b*): 'Reasoning about machining operations for rotational parts using feature-based models'. Ph.D. Thesis, University of Leeds, September

KALBFLEISCH, J. G. (1979): 'Probability and statistical inference: II', (Springer-Verlag, New York, USA.)
KAPLAN, R. S. (1986): 'Must CIM be justified by faith alone'. CIMTECH Conference Proc., Chicago, Ill., USA
KEMP, J., and JONES, C. (1988): 'production planning is as easy as 1 – 2 – 3' *in* WORTHINGTON, B. (Ed.): 'Advances in manufacturing Technology: III' (Kogan Page, London) pp. 304 – 309
KIDD, A. L. (Ed.) (1987): 'Knowledge acquisition for expert systems – A practical handbook', (Plenum Press, London)
KIMBLE, C., and PRABHU, V. B. (1988): 'CIM and manufacturing industry in the north east of England: A survey of some current issues' *in* KARWOWSKI, W., PARSAEI, H. R., and WILHELM, M.R. (Eds.) 'Ergonomics of hybrid automated systems, (Elsevier) pp. 133 – 140
KNOBEL, L. (1988): *Management Today*, June
KOETHER, R. (1987): 'Design of a sequencing buffer for model-mix assembly in the automotive industry', Proc. IXth ICPR, Cincinatti
KRIEG, T. E. (1979): 'An introduction to parametric cost estimating', *Trans. Amer. Association of Cost Engineers*, pp. F.3.1 – F.3.6
KUMPE, T., and BOLWJN, P. T. (1988): *HBR*, **88**, p.2
KUVIK, J., HRUZ, B., and TEKUS, S. (1983): 'Automatizovany system riadenia manipulacie gulatiny (Automated control system of log handling)'. DREVO 38, Bratislava, No. 4.

LAWRANCE, A. (1987): 'MRP, OPT, JIT: The Facts', *Industrial Computing*, September pp. 16 – 21

LEVENBACH, H., and CLEARY, J. P. (1982): 'The professional forecaster – The forecasting process through data analysis', (Belmont: Lifetime Learning Publications)

LEWIS, R., and TAGG, E. D. (1987): 'Trends in computer assisted education' (Blackwell Scientific Publications, Oxford)

LOWE, J. (1987): 'Managing the change to new methods of working'. Manufacturing Technology International, Europe, Sterling Publications, London

MACBETH, D. K., *et al.*: 'Managing Suppliers in an AMT Environment'. Grant Number GR/E/12337, ACME Directorate of SERC, Swindon, England

MACBETH, D. K. (1985): *Int. J. Operations and Production Management*, **5**, (1)

MACBETH, D. K., BAXTER, L. F., FERGUSON, N., and NEIL, G. C. (1988): 'Buyer – vendor relationships with Just-in-Time: Lessons from US multinationals', *IE*, September, pp. 34 – 41

MAISONNEUVE (1987): 'Application of the GRAI Method by the Company GRAI Productique'. Company SMG/Cegedor Pechineg GRAI User Group Conference, Bordeaux

MAJI, R. K. (1988): 'Tools for development of information systems in CIM'. *AME*, (1), **1**, pp. 26 – 34

MAKHOUL, J. (1975): 'Linear prediction: A tutorial review', *Proc. IEEE*, **63**, pp. 561 – 580

MAKRIFAKIS, S., and HILON, M. (1979): 'Accuracy of forecasting: An empirical investigation', *J Roy. Statistical Soc.*, **A142**, pp. 97 – 104

MALOUBIER, H., BREUIL, D., DOUMEINGTES, G. and GAVARD, J. (1984): 'Use of GRAI methods to analyse and design a production management system'. *Proc. of IFIP*, Work Gp S7 (North Holland) pp. 127 – 140

MANT, J. (1988/89): *Control*, pp. 30 – 38

MARUCHEK, A. S., and PETERSON, D. K. (1988): *Production and Inventory Management*, **29**, (1), pp. 34 – 38

MATSUSHIMA, K., OKADA, N., and SATA, T., (1982): 'The integration of CAD and CAM by application of artificial intelligence techniques', *Ann CIPP*, **31**, (1), pp. 329 – 332

MAUGHAN, K. (1988): 'Influence of user interfaces on implementation of simulation technology'. Internal Report, Sunderland Polytechnic

MCAREAVEY, D., HOEY, J., and LEONARD, R. (1988): 'Designing the closed loop elements of a material requirements system in a low volume, make-to-order company', *Int. J. Prod. Res.*, **26**, (7), pp. 1141 – 1159

MEREDITH, J. (1986): 'Automation strategy must give careful attention to the firm's ''Infrastructure'' ', *Industrial Engineering*, **18**, pp. 68 – 73

MING WANG, and SMITH, G. W. (1988): 'Modelling CIM systems'. Part 1: Methodologies computer integrated manufacturing systems', **1**, (1), (Butterworth & Co. Ltd)

MINTZBERG, H., RAISINGHANI, D., and THEORET, A. (1976): *Administrative Science*, **21**

MIYAKAWA, S., and TOSHIJIRO, O. (1986): 'The Hitachi assemblability method (AEM)', Int. Conference on Product Design for Assembly, 15 – 17 April, Newport, Rhode Island, USA

MONDEN, Y., (1983): 'Toyota production system'. AMIE, Atlanta, America
MONNIOT, J. P., *et al.* (1987): 'Report of a study of computer-aided production management in UK batch manufacturing', *Int. J. of Production Management*, **7**, (2) pp. 7 – 32
MOSER, J. G. (1986): 'Integration of artificial intelligence and simulation in a comprehensive decision-support system', *Simulation*, December, pp. 223 – 229.
MURRAY, K. J., and SHEPPARD, S. V. (1988) 'Knowledge-based simulation model specification', *Simulation*, **50** (3) pp. 112 – 119

NAGARKER, C. V., and BENNETT, D. J. (1988): *Ind. Engineering*, **20**, p.11
NEDC Report (1984): 'Manpower aspects of manufacturing change'
NEIL, G. C., BAXTER, L. F. FERGUSON, N., and MACBETH, D. K. (1988): 'The continuing importance of cost in supplier selection in support of JIT', Proc. 3rd International Conference on JIT, (IFS Publications Ltd., Kensington, England)
NEW, C. C. (1986): 'Managing manufacturing operations in the UK'
NEW, C. C., and MYERS, A. (1986): 'Managing manufacturing operations in the UK 1975 – 1985'. British Institute of Management

O'GRADY, P. (1986): 'Control of automated systems', (Kogan Page Ltd., UK)
O'KEEFE, R. M. (1986): 'Simulation and expert systems: A taxonomy and some examples', *Simulation*, 46, (1) pp. 10 – 16
OKAWA, Y. (1984): *Comp. Vision, Graphics and Image processing*, **25**, pp. 89 – 112
OLDHAM, K. (1988): 'Computer integrated manufacturing guidelines to good practice'. Lucas Engineering and Systems, Birmingham, England
ORLOVA, G., and SHAFRANSKY, V. (1984): 'Flexible Dialogue planning system for collective work of users DISKOR-2' *in* 'Knowledge representation in man – machine systems and robotics.' Moscow, Computer Center Academy of Sciences (in Russian)
ORLOVA, G., and TERENTYEV, A. (1986): 'The structure of a flexible dialogue planning system (logical description)'. Moscow, Computer Center, Academy of Sciencies (in Russian)
OVENDEN, A. (1986): 'Competitiveness in UK manufacturing industry'. British Institute of Management
OVERSTREET, C. M., and NANCE, R. E. (1986): 'World view based discrete-event model simplification', *in* ELZAS, M. S., *et al.* (Eds.): 'Modelling and simulation methodology in the artificial intelligence era' (North Holland) pp. 165 – 179

PARNABY, J. (1986): *Int. J. of Tech. Manag.*, **1**, (3/4) pp. 385 – 396
PARNABY, J., and JOHNSON, P. (1987*a*): 'Development of the JIT-MRP Factory Control System'. Proceedings 2nd International Conference on CAE, I. Mech. E.
PARNABY, J. (1987*b*): *Prog. in Rub. and Plas. Tech.*, **5**, pp.42 – 50
PARNABY, J. (1988*a*): *J. Inst. of Prod. Engs.*, **7**, pp.24 – 28
PARNABY, J. (1988*b*): *Int. J. Prod. Res.*, **26**, pp.483 – 492
PEEK, L. E., and BLACKSTONE, J. H. (1987): *Production and Inventory Management*, **28**, (4) pp.6 – 10

PENDEGRAFT, N. (1987): *Production and Inventory Management*, **28**, (4) pp.11 – 14
PERERA, D. T. S., and CARRIE, A.S. (1987): 'Simulation of tool flow within a flexible manufacuring system'. Proceedings 6th International Conference on Flexible Manufacturing systems, 4 – 6 November, Turin, Italy
PERERA, D. T. S. (1988*a*): 'The production planning problems of flexible manufacturing systems with high tool variety'. Ph.D. Thesis, University of Strathclyde, Glasgow
PERERA, D. T. S. (1988*b*): 'Solving production planning problems of FMS with computer simulation'. Proceedings 4th International Conference on Simulation in Manufacturing, 2 – 3 November, London
PERERA, D. T., S. (1988*c*): 'The development of operational strategies for an FMS using graphical animation'. Proc. 4th National Conference on Production Research, Sheffield, UK
PIPER, A. (1985): *Int. Management*, November
PORTER, G. B., and MUNDY, J. L. (1982): 'Visual inspection of metal surfaces'. 5th Int. Joint Conf. Pattern Recognition pp.149 – 151
POWELL, A. (1986): *The Production Engineer*, **65**, (5)
PRICE, C. D., and J. E. B. (1982*a*): 'Velocities and accelerations of moving parts of sewing machines'. Wira Report Ref. 1/399, May
PRICE, C. D., and RAE, A. (1982*b*): 'Variability in sewing'. British Clothing Centre Report, 21st October
PROAKIS, J. G., and MANOLAKIS, D. G. (1988): 'Introduction to digital signal processing', (Macmillan)
PUN, L., *et al.* (1986): 'The GRAI approach to the structural design of flexible manufacturing systems', *Int. J. Production Research*, **23**, (6)

RANKY, P. G. (1985): 'Computer integrated manufacturing'. ISBN 0 – 13 – 165655 – 4
RANLEFS, P. (1984): 'Knowledge processing expert systems'. Proceedings Joint Technology Assessment Conference of the Gottlief Duttweiller Institute and the European Co-ordinating Committee for Artificial Intelligence, Amsterdam, (North Holland)
REDFORD, A. H., and SWIFT, K. G. (1980): 'Design for mechanised assembly'. MTDR, 20 September, pp.619 – 626
RICH, E. (1983): 'Artificial intelligence' (McGraw-Hill International)
ROBEY, D. (1979): 'User attitudes and management information system use', *Academy of Management J.*, **22**, pp.527 – 538
RODRIGUES, A. J. L. (1986): 'Knowledge based modelling and simulation', *in* COELHO, J. D., *et al.* (Eds.): 'OR models on microcomputers', (North Holland) pp.161 – 169
RODRIGUES, A. J. L. (1987): 'Structure and logic in knowledge-based representations of system simulation models'. 4th EURO Summer Institute on Systems Science, Turku-Finland, pp.1 – 16
ROSS, D. T. (1977): *IEEE Trans.*, **SE-3**, (1), pp.16 – 34
RZEVSKI, G. (1988): 'From business plan to CIM strategy', *Industrial Computing*, February

SANDERSON, A. C., WEISS, L. E., and NAYAR, S. J. (1988): *IEEE Trans. Patt. Anal. and Machine Int.*, **10**, pp.44 – 55

SCHEER, A. – W. (1988): 'CIM — Computer steered industry' (Springer, Berlin FRG)
SCHONBERGER, R. J. (1982): 'Japanese manufacturing techniques: Nine hidden lessons in simplicity (Free Press, NY)
SCHONBERGER, R. J. (1984): 'Japanese manufacturing techniques', (Free Press, NY)
SCHONBERGER, R. J. (1986): 'World class manufacturing: The lessons of simplicity applied' (Free Press, NY)
SEIDEL, U. A. (1987): 'Systematic design of an integrated assembly planning toolbox'. Proc. IXth ICPR, Cincinatti
SERLIN, O. (1972): 'Scheduling of time critical processes'. Preprints of Spring Joint Computer Conference
SHAFRANSKY, V., AKIMOV, A., and TSAREVSKY, N. (1987): 'The structure of a system of planning models for flexible manufacturing complexes' *in* 'State of art and future development of flexible manufacturing systems'. Moscow (in Russian)
SHAH, J. J., and ROGERS, M. T. (1988): 'Functional requirements and conceptual design of the feature-based modelling system', *Computer-Aided Engineering J.*, February, pp.9 – 15
SHANNON, R. E., MAYER, R., and ADELSBERGER, H. H. (1985): 'Expert system and simulation', *Simulation*, **44**, (6), pp.275 – 284
SHAPIRO, R. D.: 'Towards effective supplier management: International comparisons'. Working paper 9 – 785 – 062, Harvard Business School, Boston
SHINGO, S. (1981): 'Study of Toyota production system from industrial engineering viewpoint'. Japan Management Association
SID-AHMED, M. A., SOLTIS, J. J. and RAJENDRAN, N. (1986): *Computers in Industry*, **7**, pp.131 – 143
SKINNER, W. (1985): 'Manufacturing: the formidable competitive weapon', (Wiley)
SLACK, N. (1987*a*): 'The flexibility of manufacuring systems', *Int. J. Operations and Production Management*, **7**, pp.35 – 45
SLACK, N. (1987*b*): *Int. J. Operations and Production Management*, **7**, p.4
SLAUTTERBACK, W. H. (1984): 'Manufacturing environment in the Year 2000'. Proc. of Synergy, Chicago, November, American Production and Inventory Control Society, pp.141 – 143
SMITH, S. F., FOX, M. S., and OW, P. S. (1986): 'Constructing and maintaining detailed production plans: Investigations into the development of knowledge based factory scheduling systems', *AI Magazine*, pp.45 – 61
SMITH, S. F. (1988): 'A constraint-based framework for reactive management of factory schedules'. Intelligent Manufacturing Proceedings from the First International Conference on Expert Systems and the Leading Edge in Production Planning and Control (Benjamin Press), pp. 113 – 130
SNADER, K. R. (1986): *Production and Inventory Management*, **27**, (4)
STAVELEY, J. C., and DALE, B. G. (1987): *Quality and Rel. Eng. Int.*, **3**, pp.265 – 271
STEP/PDES (1988): 'Conceptual schema (Applications). ISO TC184/SC/WG1, March
STOLL, H. W. (1988): 'Design for manufacture', *Manufacturing Engineering*, January, pp.67 – 73

STORJOHANN, J. (1986): *Ind. Engineering*, **18**, (2)
STRAUTS, E. J., and FLAHERTY, J. J. (1981): *Proc. SPIE*, (281), pp.176 – 181
SUBRAMANYAM, S., and ASKIN, R. G. (1986): 'An expert system approach to scheduling in flexible manufacturing systems', *in* KUSIAK, A. (Ed.): 'Flexible manufacturing systems: Methods and studies' (North Holland) pp.243 – 256
SURI, R., and WHITNEY, C. K. (1983): 'Decision support requirements in flexible manufacturing', *J. Manufacturing Systems*, **3**, (1), pp.314 – 326
SUZUKI, K. (1985): *Technometrics*, **10**, pp.263 – 271

TANNOCK, J. D. T., and WORT, R. G. (1988): 'Quality system design using IDEFo with a software tool', Seminar on 'Using IDEFo for Mapping Quality Systems, Bristol Polytechnic
THALER, K. (1988). 'Concepts of knowledge based support for assembly process planning'. Proc. 1st EURINFO, Athens
THOMPSON, J.,: 'CIM — Bringing the islands together', *Chartered Mech. Eng.*, March
THURWACHTER, W. A. (1986), *Ind. Eng.*, November
TRANFIELD, D., and SMITH, S. (1987): 'A strategic methodology for implementing technical change in manufacturing'. Paper presented to the Inaugual Conference of the British Academy of Management, 13 – 15 September

ULLMAN, D. G., and DIETTERICH, T. A. (1987): 'Mechanical design methodology: Implications on future developments of computer-aided design and knowledge-based systems', *Engineering with Computers*, **2**, pp.21 – 29

VAN DER ULEIT (1985): 'Where Lucas sees the light', *Management Today*

VOLLMAN, T. E., BERRY, W. L., and WHYBARK, D. C. (1984): Manufacturing planning and control systems (Irwin, Homewood, Ill. USA)
VONDEREMBSE, M. A., and WOBSER, G. S. (1987): *Ind. Engineering*, **19**, (4)
VOSS, C. (1985): 'Success and failure in advanced manufacturing technology'. Working Paper, Warwick University
VOSS, C. A. (1987): 'Just in Time Manufacturing' (IFS Publications)

WALKER, W. (1990): 'The usage of computer-integrated manufacturing technology within manufacturing organisations located in the north east of England'. *Int. J. Operations and Production Management*, **9**, (7)
WANG, H. P., and WYSK, R. A. (1987): 'Intelligent reasoning for process planning', *Computers in Industry*, **8**, pp.293 – 309
WATKINS, A. J., and LEECH, D. J. (1988): 'Reliability engineering and system safety' (to be published)
WEILL, R., SPUR, G., and EVERSHEIM, W. (1982): 'Survey of computer-aided process planning', *Ann. CIRP*, **31**, (2), pp.539 – 551
WHEELWRIGHT, S. C. (1981): 'Japan, where operations really are strategic', *Harvard Business Review*, July/August
WIDROW, B. (1971): 'Adaptive filters' *in* KALMAN, R. E., and DECLARIS, N. (Eds.): 'Aspects of network and systems theory' (Holt, Rinehart, and Winston)
WIDROW, B. *et al.* (1976): 'Stationary and nonstationary learning characteristics of the LMS adaptive filter', *Proc. IEEE*, **64**, pp.1151 – 1162

WILLIS, R. G., and SULLIVAN, K. H. (1984): *Industrial Engineering*, February

WIRA REPORT (1976): 'Report on Kirby-Lester TA7 tension analyser (phase 1)'. Ref. 1/302, February

WIRA REPORT (1981): 'Static tests on needle thread tension system'. Report on Sewing Tension Variations, December

WOOD, J. (1986): 'Just in Time manufacturing'. Proceedings 1st International Conference, JIT Manufacturing

WOODGATE, H. S. (1986): 'Linked management graphics for production control'. Proc. Eurographics, Lisbon, pp.43 – 52

WYSK, R. A. (1985): 'Automated process planning systems – An overview of ten years of activities', 1st CIRP Working Seminar on Computer-Aided Process Planning, pp.13 – 18

Index